RECENT ADVANCES IN THE SUPERWORLD

INTERNATIONAL WORKSHOP ON

RECENT ADVANCES IN THE SUPERWORLD

Houston Advanced Research Center
The Woodlands, Texas, USA
14 – 16 April 1993

Editors

J L Lopez & D V Nanopoulos
Texas A & M University
College Station, Texas, USA
and
Houston Advanced Research Center
The Woodlands, Texas, USA

World Scientific
Singapore • New Jersey • London • Hong Kong

Published by

World Scientific Publishing Co. Pte. Ltd.

P O Box 128, Farrer Road, Singapore 9128

USA office: Suite 1B, 1060 Main Street, River Edge, NJ 07661

UK office: 73 Lynton Mead, Totteridge, London N20 8DH

International Workshop on
Recent Advances in the SuperWorld

ISBN 981-02-1730-7

Printed in Singapore by Utopia Press.

Table of Contents

---

* Conference speaker

Preface

The realization that precision LEP data lends support to the idea of supersymmetric grand unification and supersymmetry in general, thrusted us to organize the First HARC International Workshop on Recent Advances in the SuperWorld. Our idea was to have a workshop that discussed all aspects of the supersymmetric world — the SuperWorld as it should be properly known. These aspects are so diverse that they are usually studied by high-energy physicists with vastly different interests. The selection of speakers included the whole spectrum in "energy", from experimental physicists to specialists in low-energy phenomenology, supersymmetric grand unification, supergravity, superstrings, and supercosmology. In the end, our general impression — as corroborated by the many unsolicited positive comments by the participants — has been that the workshop accomplished its purpose completely, that is, a coherent discussion of the many issues relating to the SuperWorld.

The workshop took place on April 14–16, 1993 on the Mitchell Campus of the Houston Advanced Research Center in The Woodlands, Texas. During the three days of presentations and intensive dicussions, we heard the most up-to-date advances in low-energy supersymmetry (H. Haber and V. Barger), supersymmetric grand unification (S. Raby, A. Zichichi and P. Ramond), supergravity and model-building (R. Arnowitt and S. Kelley), superstrings and phenomenology (H. Tye, J. Lopez and V. Kaplunovsky), superstrings and cosmology (D. Nanopoulos), supercosmology (D. Schramm and K. Yuan), and experimental prospects (J. White and R. Webb). There were several shorter presentations by younger people working actively on these fields as well.

This workshop would not have been possible without the invaluable support of the World-Laboratory and Professor A. Zichichi. We also like to thank the members of the local organizing committee (D. Norton, Vice-president for Research–HARC, G. Park, D. Stewart and K. Yuan), the World-Laboratory Fellows who served as Scientific Secretaries (R. Gandhi, I. Giannakis, G. Park, K. Yuan and X. Wang), and last but not least, the HARC Office for Research (D. Stewart and her staff) and the HARC Public Affairs Office (B. Peyton, J. Schoger and R. Hoy).

Jorge Lopez and Dimitri Nanopoulos

Opening Lecture

A Liouville String Approach to Microscopic Time and Cosmology

John Ellis[a], N.E. Mavromatos[b], D.V. Nanopoulos[c]

[a] Theory Division, CERN, CH-1211, Geneva, Switzerland,
[b] Theoretical Physics Laboratory, ENSLAPP, Chemin de Bellevue, Annecy-le-Vieux, France, and
S.E.R.C. Advanced Fellow, University of Oxford, Dept. of Physics (Theoretical Physics), 1 Keble Road, Oxford OX1 3NP, United Kingdom,
[c] Center for Theoretical Physics, Dept. of Physics, Texas A & M University, College Station, TX 77843-4242, USA, and
Astroparticle Physics Group, Houston Advanced Research Center (HARC), The Mitchell Campus, The Woodlands, TX 77381, USA

Abstract

In the non-critical string framework that we have proposed recently, the time t is identified with a dynamical local renormalization group scale, the Liouville mode, and behaves as a statistical evolution parameter, flowing irreversibly from an infrared fixed point - which we conjecture to be a topological string phase - to an ultraviolet one - which corresponds to a static critical string vacuum. When applied to a toy two-dimensional model of space-time singularities, this formalism yields an apparent renormalization of the velocity of light, and a t-dependent form of the uncertainty relation for position and momentum of a test string. We speculate within this framework on a stringy alternative to conventional field-theoretical inflation, and the decay towards zero of the cosmological constant in a maximally-symmetric space.

1 Introduction

The subject of fundamental constants in field or string theory is a fascinating challenge for any physicist. Unfortunately, despite vigorous attempts, the topic remains controversial today. We would not exaggerate if we said that it is still at a very primitive stage of understanding. Following Weinberg [1], we can define as "fundamental constants" a list of constants whose "value we cannot calculate in terms of more fundamental constants ... because we do not know of anything more fundamental ..." . Thus, Weinberg concludes that "the membership of such a list reflects our present understanding of fundamental physics."

We shall not attempt in this talk to offer any solutions to this fundamental problem in field theory. Instead, we shall examine the consequences of a recently-proposed approach [2] to the origin of time in non-critical String Theory [3, 4] for some "constants" in string theory, namely the speed of light c, and the fundamental (minimal) length λ_s derived from the uncertainty principle [5].

According to Veneziano [6], string theory appears to have no fundamental constants *per se*, but these constants arise as a result of spontaneous breaking of symmetries in the selection of a ground state, thereby leading a (string) physicist to conclude that [6] "the constants of the vacuum are the constants of the world". To prove or disprove this statement is a formidable task. Presumably, to understand, in String Theory, the origin of the fundamental constants that characterize the vacuum would require first an answer to the important and still unresolved issue of what (if anything) lifts the huge degeneracy that appears at present to characterize the string ground states. This question is tied up with the question on the origin of space-time.

As a step in this direction, we have formulated [2, 7] time in string theory as a renormalization group scale on the world-sheet, which we identify with the Liouville mode. The connection of time with the Liouville field in non-critical string theory has been considered in [3, 4] in order to incorporate an expanding Universe in string theory. In [2, 7] we took one step further, and interpreted the Liouville mode as a local renormalization group scale, which flows irreversibly according to a generalization of Zamolodchikov C-theorem [8]. This scale is introduced in theories with target space-time singularities or, more generally, event or particle horizons, as a result of the breaking of conformal invariance by the truncation of the matter theory to observable local low-energy modes. In the toy example of a two-dimensional black hole that we used to develop these ideas [9, 7], these low-energy propagating modes are coupled to quasi-topological delocalized higher-level string modes, as a result of infinite level-mixing gauge stringy symmetries [10]. These discrete modes are believed to be remnants of a topological phase in string theory [11]. The information carried by these modes, which is hidden from a low-energy observer who 'measures' using local scattering methods, is the root of the flow of time and the non-equilibrium nature of the low-energy field theory derived from the string. As we have noted [2], the fact that information is lost to these hidden modes causes apparent decoherence of the observable modes, reflected in a collapse of their (part of the full) wave function that is rapid for more complex systems.

In this scenario, the resulting low-energy two-dimensional string-inspired theory appears not to have any 'fundamental' constants, however there are dimensionful parameters that relax to their *finite* critical-string values asymptotically as the Liouville time $t \to \infty$. These quantities cannot be calculated in terms of anything more fundamental in the model. According to Weinberg's definition, then, the asymptotic values of these quantities, which in this two-dimensional example include the speed of light and the minimal uncertainty in length, or $\hbar$, would characterize the vacuum state and hence would be the dimensionful fundamental constants of that ground state. In addition, there appears in the vacuum an arbitrary expectation value of the dilaton field, which could be considered as an arbitrary dimensionless constant, the analogue of the fine structure constant in four-dimensional models. In the non-equilibrium model the dilaton field is a time-dependent scale factor that relaxes to its asympotic (vacuum) value. We find it interesting, but also quite natural, that the non-equilibrium nature of the non-critical Liouville string induces the appearance of constant quantities only in the asymptotic vacuum state via a (quantum) relaxation process. After all, the best way of understanding an equilibrium state in statistical mechanics is the study of the approach to it by perturbing the relevant system away from equilibrium. This is precisely the way we studied time in string theory. Using the concept of 'measurement' through localized fields, we have perturbed the conformal string theory by operators that spoil conformal invariance and, thus, introduced the concept of time as an evolution parameter of the relaxation process[2, 7, 9].

As a byproduct of our approach we speculate on cosmological applications, with emphasis on the possibility for a string alternative to conventional field-theoretical inflation. We also deal with the issue of the target-space cosmological constant in generic D-dimensional maximally-symmetric spaces, in models with a dilaton such as certain no-scale supergravities [12], and find [9] that it relaxes to zero asymptotically.

2 String-Inspired Density Matrix Mechanics

In this section we review briefly the formalism of first-quantized non-critical strings, and its connection with the Density Matrix Mechanics that we proposed recently [2]. We emphasize the restoration of criticality by turning a $(1,1)$ deformation into an exactly marginal one by Liouville dressing, thereby leading, in our interpretation, to time-dependent backgrounds. Consider a two-dimensional critical conformal field theory model described by an action $S_0(r)$ on the world sheet, where the $\{r\}$ are matter fields spanning a D-dimensional target manifold of Euclidean signature, that we term "space". Consider, now, a deformation

$$S = S_0(r) + g\int d^2zV_g(r) \tag{1}$$

Here V_g is a (1,1) operator, i.e. its anomalous dimension vanishes, but it is not *exactly marginal* in the sense that the operator-product expansion coefficients C_{ggg} of V_g with itself are non-zero in any renormalization scheme, and hence 'universal' in the Wilsonian sense. The scaling dimension α_g of V_g in the deformed theory (1) is, to $O(g)$ [13],

$$\alpha_g = -gC_{ggg} + \ldots \tag{2}$$

Liouville theory [14] requires that scale invariance of the theory (1) be preserved. The non-zero scaling dimension (2) would jeopardize this, but scale invariance is restored if one dresses V_g gravitationally on the world-sheet as

$$\int d^2zgV_g(r) \to \int d^2zge^{\alpha_g\phi}V_g(r) = \int d^2zgV_g(r) - \int d^2zg^2C_{ggg}V_g(r)\phi + \ldots \tag{3}$$

where ϕ is the Liouville field. The latter acquires dynamics through integration over world-sheet covariant metrics, $\gamma_{\alpha\beta}$, after conformal gauge fixing $\gamma_{\alpha\beta} = e^{\phi}\hat{\gamma}_{\alpha\beta}$ in the way discussed in [14, 15]. Scale invariance is guaranteed through the definition of renormalized couplings g_R , given in terms of g through the relation

$$g_R \equiv g - C_{ggg}\phi g^2 + \dots \tag{4}$$

This equation leads to the correct β-functions for g_R

$$\beta_g = -C_{ggg} g_R^2 + \dots \tag{5}$$

implying a renormalization-group scale ϕ-dependence of g_R. The reader might have noticed that above we viewed the Liouville field as a local scale on the world-sheet [2, 7]. Local world-sheet scales have been considered in the past [16, 17], but the crucial difference of our Liouville approach is that this scale is made dynamical by being integrated out in the path-integral. In this way, in Liouville strings the local dynamical scale acquires the interpretation of an additional target coordinate. If the central charge of the matter theory is $c_m > 25$, the signature of the kinetic term of the Liouville coordinate is opposite to that of the matter fields r, and thus the Liouville field is interpreted as Minkowski target time [3, 4]. In this interpretation the renormalization group evolution of the matter system alone, ignoring Liouville dynamics, is interpreted as time evolution, in which the time appears as an 'external' parameter. As argued in ref. [18, 7], the actual time flow is *opposite* to the usual renormalization group flow, the latter being from ultraviolet to infrared fixed points. This has to do with metastability interpretation of the renormalization group flow in Liouville strings [18, 7], where the infrared fixed point is viewed as a 'bounce' point of the flow [19].

It is instructive to review the properties of such a flow.

(i) Due to the assumed unitarity of the matter system, the flow is *irreversible*, as can be seen using the Zamolodchikov C-theorem [8]. The effective central charge (Zamolodchikov C-function [8]) $C(g)$ varies with time as

$$\partial_t C(g) = \beta^i < V_i V_j > \beta^j \equiv \beta^i G_{ij} \beta^j \geq 0 \tag{6}$$

where G_{ij} plays the rôle of a metric in coupling constant space. This implies a time *arrow* in all such models.

(ii) Defining a density matrix for the system, in a coupling constant phase space g^i, p_i, one has the following modified Liouville equation [2, 20]

$$\partial_t \rho = \{H, \rho\} + \beta^i G_{ij} \frac{\partial \rho}{\partial p_j} \tag{7}$$

where $\{,\}$ denote the appropriate Poisson bracket and H is a Hamiltonian function. The latter is defined in terms of the effective action, which in string theories coincides with the Zamolodchikov C-function [21]

$$C(g) = \int dt p_i \beta^i - H \tag{8}$$

The total probability $P = \int dg^i dp_i tr\rho(g,p,t)$ is conserved, since its rate of change is a boundary term in phase space, and the latter is assumed to have no boundary.

(iii) Quantization of the coupling constants g^i is achieved as follows: the g^i actually represent target-space background fields in string theory, and as such they are quantum field operators in a string field theory concept, where summation over genera is assumed. Hence the situation is similar to a two-dimensional wormhole calculus. From a quantum point of view, the Poisson brackets $\{,\}$ are then replaced by appropriate commutators, and the couplings g^i are viewed as Heisenberg operators [2, 7].

(iv) Entropy S is defined in terms of the density matrix ρ as usual, by

$$S = -Tr\rho ln\rho \tag{9}$$

and it increases with time t

$$\partial_t S = \beta^i G_{ij} \beta^j S \geq 0 \tag{10}$$

thereby implying, if any $\beta^i \neq 0$, non-equilibrium for the non-critical matter theory. The latter approaches asymptotically its critical state as an ultraviolet fixed point of the flow.

(v) Energy H is conserved on the average, in this quantum version, as a result of the renormalizability of the σ-model [2]

$$\partial_t << H >> \equiv \partial_t Tr H\rho = Tr(\rho\partial_t H) + Tr(H\partial_t\rho) = \partial_t(p_i\beta^i) = 0 \tag{11}$$

where we used (7) and (8). The vanishing result is a consequence of the fact that neither $p_i = \frac{\delta}{\delta g^i}$ nor β^i have an exlicit scale dependence in renormalizable theories. However, fluctuations in the energy are time-dependent. For instance,

$$\partial_t << H^2 >> = << \frac{d\beta^i}{dt} G_{ij}\beta^j >> \neq 0 \tag{12}$$

in general, which affects energy uncertainty relations for the string.

(vi) The above picture leads naturally [2, 7] to collapse of part of the wave function describing observable modes, in a way similar to that proposed in ref. [22] within a version of the wormhole calculus, namely the asymptotic vanishing (as $t \to \infty$) of the off-diagonal elements in space of the matter density matrix ρ :

$$\rho_{out}(x - x') = \rho_{in}(x, x') e^{-Dt(x-x')^2 + \dots} \tag{13}$$

where x, x' are target spatial coordinates, and D is proportional to the sum of the squares of the anomalous dimensions of the relevant perturbations of the σ-model [7].

3 The Two-Dimensional String Black Hole Model

As an illustration of our approach to non-critical string theory, we now discuss the two-dimensional black hole model of ref. [23]. We regard it as a toy laboratory that gives us insight into the nature of time in string theory and contributes to the physical effects mentioned in the previous section.

The action of the model is

$$S_0 = \frac{k}{2\pi}\int d^2z[\partial r\overline{\partial}r - tanh^2 r\partial t\overline{\partial}t] + \frac{1}{8\pi}\int d^2zR^{(2)}\Phi(r) \tag{14}$$

where r is a space-like coordinate and t is time-like, $R^{(2)}$ is the scalar curvature, and Φ is the dilaton field. The customary interpretation of (14) is as a string model with $c =$ 1 matter, represented by the t field, interacting with a Liouville mode, represented by the r field, which has $c < 1$ and is correspondingly space-like [3, 4]. As an illustration of the approach outlined in the previous section, however, we re-interpret (14) as a fixed point of the renormalization group flow in the local scale variable t. In our interpretation, the "matter" sector is defined by the spatial coordinate r, and has central charge $c_m = 25$ when $k = 9/4$ [23].Thus the model (14) describes a critical string in a dilaton/graviton background. The fact that this is static, i.e. independent of t, reflects the fact that one is at a fixed point of the renormalization group flow [7, 9].

We now outline how one can use the machinery of the renormalization group in curved space,with t introduced as a local renormalization scale on the world sheet, to derive the model (14). A detailed technical description is given in [9, 7] There are two contributions to the kinetic term for t in our approach, one associated with the Jacobian of the path integration over the world-sheet metrics, and the other with fluctuations in the background metric.

To exhibit the former, we first choose the conformal gauge $\gamma_{\alpha\beta} = e^{\rho}\hat{\gamma}_{\alpha\beta}$ [14, 15], where ρ represents the Liouville mode. We will later identify ρ with an appropriate function of ϕ, thereby making the local scale ϕ a dynamical σ-model field. Ref.[15] contains an explicit computation of the Jacobian using heat-kernel regularization, which yields

$$-\frac{1}{48\pi}[\frac{1}{2}\partial_\alpha\rho\partial^\alpha\rho + R^{(2)}\rho + \frac{\mu}{\epsilon}e^{\rho} + S'_G] \tag{15}$$

where the counterterms S'_G are needed to remove the non-logarithmic divergences associated with the induced world-sheet cosmological constant term $\frac{\mu}{\epsilon}e^{\rho}$, and depend on the background fields. This procedure reproduces the critical string results of ref. [23] when one identifies the Liouville field ρ with $2\alpha'\phi$. Equation (15) contains a negative (time-like) contribution to the kinetic term for the Liouville (time) field, but this is not the only such contribution, as we now show.

We recall that the renormalization of composite operators in σ-models formulated on curved world sheets is achieved by allowing an arbitrary dependence of the couplings g^i on the world-sheet variables $z, \bar{z}$ [16, 17]. This induces counterterms of "tachyonic" form, which take the following form in dimensional regularization with $d = 2 - \epsilon$ [17]:

$$\int d^2 z \Lambda_0 \tag{16}$$

where

$$\Lambda_0 = \mu^{-\epsilon}(Z(g)\Lambda + Y(g)) \tag{17}$$

Here $Z(g)$ is a common wave function renormalization that maps target scalars into scalars, Λ is a residual renormalization factor, and the remaining counterterms $Y(g)$ can be expanded as power series in $1/\epsilon$, with the the one-loop result giving a simple pole. Simple power-counting yields the following form for $Y(g)$:

$$Y(g) = \partial_\alpha g^i \mathcal{G}_{ij} \partial^\alpha g^j \tag{18}$$

where $\mathcal{G}_{ij}$ is the analogue of the Zamolodchikov metric [8] in this formalism, which is positive for unitary theories. It is related to the divergent part of the two-point function $< V_i V_j >$ [17] that cannot be absorbed in the conventional renormalization of the operators V_i. We need to consider a σ-model propagating in a graviton background G_{MN}, in which case a standard one-loop computation [17] yields the following result for the simple ϵ-pole in Y:

$$Y^{(1)} = \frac{\lambda}{16\pi\epsilon} \partial_\alpha G_{MN} \partial^\alpha G^{MN} \tag{19}$$

where $\lambda \equiv 4\pi\alpha'$ is a loop-counting parameter. We note that the wave-function renormalization $Z(g)$ vanishes at one-loop. In ref. [17] G_{MN} was allowed to depend arbitrarily on the world-sheet variables, and all world-sheet derivatives of the couplings were set to zero at the end of the calculation. In our Liouville mode interpretation, we assume that such dependence occurs only through the local scale $\mu(z, \bar{z})$, so that

$$\partial_\alpha g^i = \hat{\beta}^i \partial_\alpha \phi(z, \bar{z}) \tag{20}$$

where $\hat{\beta}^i = \epsilon g^i + \beta^i(g)$ and $\phi = ln\mu(z, \bar{z})$. Taking the $\epsilon \to 0$ limit, and separating the finite and $O(\frac{1}{\epsilon})$ terms, we obtain for the former

$$O(1) - terms: \qquad Res Y^{(1)} = \alpha'^2 R \partial_\alpha \phi \partial^\alpha \phi \tag{21}$$

where R is the scalar curvature in target space, and we have used the fact that the one-loop graviton β-function is

$$\beta^G_{MN} = \frac{\lambda}{2\pi} R_{MN} \tag{22}$$

The non-logarithmic divergent terms

$$\frac{1}{\epsilon}\beta^i \mathcal{G}^{(1)}_{ij}\beta^j \tag{23}$$

do not contribute to the renormalization group, and can be removed explicitly by target-space metric counterterms

$$S_G = \frac{1}{\epsilon}G_{\phi\phi}\partial_\alpha\phi\partial^\alpha\phi + \delta S(\phi, r) \tag{24}$$

where the coefficients $G_{\phi\phi}$ are fixed by the requirement of cancelling the $\frac{1}{\epsilon}$ terms (23). The δS denotes arbitrary finite counterterms, which are invariant under the simultaneous conformal rescalings of the fiducial world-sheet metric, $\hat{\gamma} \to e^\sigma \hat{\gamma}$, and local shifts of the scale $\phi \to \phi - \sigma$. This last requirement arises as in the conventional approach to Liouville gravity [14, 15], where the local renormalization scale ϕ is identified with the Liouville mode ρ, after appropriate normalization. In our interpretation one is forced to treat the scale ϕ simultaneously as the target time coordinate.

In the case of the Minkowski black hole model of ref. [23], the Lorentzian curvature is

$$R = \frac{4}{cosh^2 r} = 4 - 4tanh^2 r, \tag{25}$$

which we substitute into equation (21) to obtain the form of the second contribution to the kinetic term for the Liouville field ϕ. Combining the world-sheet metric Jacobian term in (15) with the background fluctuation term (21,25), we finally obtain the following terms in the effective action

$$\frac{1}{4\pi\alpha'}\int d^2z[\partial_\alpha r\partial^\alpha r - tanh^2 r\partial_\alpha\phi\partial^\alpha\phi + dilaton - terms] \tag{26}$$

Thus we recover the critical string σ-model action (14) for the Minkowski black-hole. Dilaton counterterms are incorporated in a similar way, yielding the dilaton background of [23]. In addition, as standard in stringy σ-models, one also obtains the necessary counterterms that guarantee target-space diffeomorphism invariance of the Weyl-anomaly cefficients [16]. Details are given in ref. [9].

It should be noticed that the renormalization group yields automatically the Minkowski signature, due to the $c_m = 25$ value of the matter central charge [3, 4]. However, as we remarked in ref. [9, 7], one can also switch over to the Euclidean black hole model, and still maintain the identification of the compact time with some appropriate function of the Liouville scale ϕ that takes into account the compactness of t in that case. This Euclidean version is better studied from the point of view of discussing exactly marginal deformations that turn on matter in the model (14). In ref. [24] it was argued that the exactly marginal deformation that turned on a static tachyon background for the black hole of ref. [23] necessarily involved the higher-level topological string modes, which are non-propagating delocalized states. This is a consequence of the operator product expansion of the tachyon zero-mode operator $\mathcal{F}^c_{-\frac{1}{2},0}$ [24]:

$$\mathcal{F}^c_{-\frac{1}{2},0} \circ \mathcal{F}^c_{-\frac{1}{2},0} = \mathcal{F}^c_{-\frac{1}{2},0} + W^{hw}_{-1,0} + W^{lw}_{-1,0} + \dots \tag{27}$$

where we only exhibit the appropriate holomorphic part for reasons of economy of space. The W operators and the ... denote level-one and higher string states. The latter cannot be detected in local scattering experiments, due to their delocalized character. From a formal field-theoretic point of view such states cannot exist as asymptotic states to define scattering, and also cannot be integrated out in a local path-integral. They can only exist as marginal deformations in a string theory. An 'experimentalist' therefore sees necessarily a truncated matter theory, where the only deformation is the tachyon $\mathcal{F}^c_{-\frac{1}{2},0}$, which is a (1,1) operator in the black hole σ-model (14), but is not exactly marginal. This truncated theory is non-critical, and hence Liouville dressing in the sense of (3) is essential, thereby implying time-dependence of the matter background. Due to the fact that the appropriate exactly-marginal deformation associated with the tachyon in these models involves all higher-level string states, one can conclude that in this picture the ensuing non-equilibrium time-dependent backgrounds are a consequence of information carried off by the unobserved topological string modes. The rôle of the space-time singularity[1] was crucial for this argument. Indeed, in flat target-space matrix models [25] the tachyon zero-mode operator $\mathcal{F}$ is exactly marginal. As we shall argue later on, these flat models can be regarded as an asymptotic ultraviolet limit in time of the Wess-Zumino black hole. Hence, any time-dependence of the matter disappears in the vacuum, leading to equilibrium.

In ref. [7] an additional deformation was considered in parallel with the tachyon. This was the instanton vertex [26]

$$V_I = g^I \int d^2\rho \frac{1}{\rho^4} d^2 z c^{\rho\partial(sinhre^{-it})+hc} \tag{28}$$

where $g^I << 1$ in a dilute-gas approximation. The physics behind such a deformation lies in the world-sheet interpretation of the Euclidean (Minkowski) black hole as representing a world-sheet vortex (spike) [27]. The charge of such objects is proportional to the black-hole mass, and hence the instantons, that induce transitions between defects of different charge, can represent mass changes of black holes. Higher-genus (quantum) effects are known [28] to produce black hole decay, and it is sensible to consider instanton effects as being related to these. In addition, from a conformal field theory point of view, it is known that deformations consisting purely of higher-level string modes do induce global rescalings to the black-hole metric [24], and hence

[1]We would like to stress that the notion of 'singularity' is clearly a low-energy effective-theory concept. The existence of infinite-dimensional stringy symmetries associated with higher-level string states (W_∞-symmetries [10]) 'smooth out' the singularity, and render the full string theory finite.

shifts in its mass. Therefore we associate instanton deformations with local renormalization counterterms representing collectively higher-level and higher-genus, i.e. quantum string, effects.

Instantons in these models appear to have logarithmically-divergent world-sheet actions and topological charge, but their contribution to the path integral is finite. From a renormalization point of view they are by themselves irrelevant operators. As we mentioned before, their effect is similar to that of higher genera [29]. They induce [7] world-sheet ultraviolet-divergent (logarithmic) shifts to the dilaton field, and when considered in the model deformed by a tachyon, their effect is to lead to extra infinities that cannot be absorbed in conventional renormalization of the tachyon fields. The presence of matter is essential in this framework, given that the instantons alone cannot affect the renormalization group flow induced by relevant operators (tachyons). Instead, in the matter-instanton deformed theory one has a renormalization of the Wess-Zumino level parameter, which, thus becomes scale-dependent. It can be shown [7, 9] that close to the ultraviolet fixed point the renormalized k_R in the dilute-gas approximation is given by

$$k_R(t) \simeq e^{-4\pi\beta^I T_0 t} \tag{29}$$

where t is the target time, and T_0 denotes the coupling of the (relevant) tachyon perturbation. The instanton β-function itself depends on k, e.g. [26] $\beta^I = -\frac{k}{2}g^I$ in the case of large k, where the instantons form a dilute gas, giving an exponential increase of k_R with t [7, 9].

4 Time-Dependent "Fundamental Constants"

The consequences of the renormalization effect (29) of the effective level parameter $k_R(t)$ are dramatic for the physics of the string. First, we recall that the mass M of the black hole is given in units of the Planck mass by [23, 7]

$$M/M_{planck} = \sqrt{\frac{1}{k_R(t)-2}}e^{const} \tag{30}$$

implying asymptotically a vanishing-mass black-hole, i.e. flat target-space time [23, 10]. This makes the connection with the flat $c = 1$ matrix model at the ultraviolet fixed point of the flow.

Secondly, we note that the exact target-space background metric of the Wess-Zumino σ-model (for finite k) has the following asymptotic form for large $r \to \infty$ [31]

$$ds^2 = 2(k-2)(dr^2 - \frac{k}{k-2}dt^2) \tag{31}$$

In view of (29), the above relation implies an effective time dependence of the velocity of light:

$$c_q = c\sqrt{\frac{k(t)}{k(t)-2}} \tag{32}$$

where $c \equiv c(\infty)$.

The result (32), implying an increase of the velocity of light as one approaches the infrared fixed point, can be re-derived from a local-renormalization-group viewpoint. Consider the two-dimensional string model (14) perturbed by an irrelevant deformation g^i. As shown in ref. [7] and reviewed in section 3, following the local-renormalization-scale formalism of ref. [16, 17] in dimensional regularization $d = 2-\epsilon$, an induced metric counterterm for the time-component of the metric tensor assumes the form,

$$g^i \mathcal{G}^{(1)}_{ij} \beta^j \partial\phi\overline{\partial}\phi \tag{33}$$

where ϕ is the (time-like) Liouville field, and $\mathcal{G}^{(1)}_{ij}$ is the residue of the first pole in ϵ of the Zamolodchikov metric in coupling constant space. This term contributes to the wave function renormalization of ϕ. For irrelevant deformations, like instantons in the black-hole, the $\beta^i < 0$ and the counterterm (33) is negative. Taking into account the discussion in section 3 showing that in the case of two-dimensional black-hole backgrounds the Liouville mode induces a kinetic term for ϕ of the form $-tanh^2 r\partial\phi\overline{\partial}\phi$ [9], we observe that (33) corresponds to an additive shift in the time-component of the background metric. Hence, in the limit $r \to \infty$ this would imply an increase in the velocity of light. The presence of a relevant operator, on the other hand, leads to a decrease of the effective velocity of light. In the Wilsonian procedure of renormalization, one integrates out the irrelevant deformations, in the sense that one sets the respective β-functions to zero and solves the irrelevant couplings in terms of the relevant ones in order to obtain the known Gell-Mann-Low β-functions for the latter. In the σ-model approach, this would correspond to considering the tachyon perturbations in the presence of an instanton background.

We now remark that in the background defined by (31) one may consider (low-energy) particle excitations and derive the conventional Lorentz transformations for coordinates r and time t and particle velocities $u < c_q$. The arrow of time does not appear in these Lorentz transformations: to see the arrow of time we must consider transitions among theories characterized by different k_R, in which case one abandons flat space-times and equilibrium situations. Such transitions are induced by the interaction of the propagating light string modes with the *'ether'* of the topological string states. Then Lorentz transformations are only approximate, and should be replaced by general coordinate transformations appropriate for non-critical string theory. Such transformations are viewed as local redefinitions of σ-model fields, and undergo renormalization group flow. The limit $c_q \to \infty$ appears 'Galilean', but it it should be remembered that in this limit the space-time picture changes completely, since the back-reaction effects are enormous as one approaches the topological phase. The dominance of delocalized topological modes with discrete energies and momenta in such a case is consistent with the *instantaneous influence* implied by $c_q \to \infty$. However, the above perturbative framework, is certainly inapplicable near the topological phase. Thus, until we understand the theory in the infrared, the above remarks should be considered as speculative.

The result (32) is reminiscent of the well-known variation of the velocity of light in a medium with its density and temperature [32]. Indeed, there the effective velocity of light decreases as the temperature of the heat-bath increases. However, the mechanism here is different, since the 'medium' of non-local topological string modes does not yield a plasmon mass.

The time-dependence of the string as it approaches the ultraviolet fixed point is reflected in a computation of the string position-momentum uncertainty relation in the σ-model deformed by tachyons and instantons. The result for the position-momentum uncertainty, defined appropriately to incorporate curved gravitational backgrounds [9], can be expressed as

$$(\Delta X \Delta P)_{min} \equiv \hbar_{eff}(t) = \hbar(1 + O(\frac{1}{k(t)})) \tag{34}$$

where $\Delta A \equiv (< A^2 > - < A >^2)^{\frac{1}{2}}$, $< \ldots >$ denotes σ-model vacuum expectation values, and $\hbar$ is the critical-string Planck's constant. The string uncertainty relation introduces a minimum length λ_s [5], that in our case decreases with time too :

$$\lambda_s(t) \equiv (2\frac{\hbar(t)\alpha'(t)}{c_q(t)^2})^{\frac{1}{2}} = \lambda_s^0(1 + O(\frac{1}{k(t)})) \tag{35}$$

where the superscript 0 a denotes quantity evaluated at the ultraviolet fixed point in the critical string theory. We also mention the following important relation that can be proven [9] while deriving (35):

$$\frac{\alpha'(t)}{\alpha'_0} = \frac{c_q^2(t)}{c^2} \tag{36}$$

which stems from the relation between k and the Regge slope in the model (14) [7, 9]. The relation (36) is consistent with the topological phase transition that we conjecture occurs at the infrared fixed-point of the flow, where $\alpha'(t) \to \infty$ [11].

5 Applications to Inflation and the Cosmological "Constant" problem

We saw in the previous section that fundamental constants vary in the non-critical string Universe, because of the interaction of the low-energy world with the topological string modes. Fundamental "constants" attain fixed values asymptotically, in the particular string vacuum which is attained as the ultraviolet fixed point of the σ-model renormalization group flow. So what physical meaning is attached to the concept of the "running" velocity of light $c_q(t)$ (32) or of the Planck "constant" ((34),(35)) ?

The velocity of light can always be absorbed in a rescaling of the time coordinate, provided that it depends only on time and that this time dependence does not imply causality violations. For instance, in the case of the asymptotic black-hole metric (31), one can redefine time $t \to t'(t)$ in such a way that

$$dt' = \sqrt{\frac{k}{k-2}}dt \tag{37}$$

This rescaling implies that such a time dependence of the velocity of light will be unobservable locally, in the sense that in a co-moving frame the velocity of light will appear as constant that can always be set to unity. From a renormalization-group point of view, the redefinition (37) represents a local scheme change. As long as there is a finite maximum speed in the low-energy world, this will not have physical consequences locally. However, we conjecture that the velocity of light becomes ill-defined at the topological phase transition in the infrared limit, where the normal concepts of space and time break down.

However, the evolution of the velocity of light may have important cosmological consequences. If we consider a Friedmann-Robertson-Walker Universe, the horizon distance d in co-moving coordinates over which an observer can look back is

$$d = \int dt c_q(t) = \int dt \sqrt{\frac{k(t)}{k(t)-2}} \tag{38}$$

which is larger than the naive estimate $d = ct$, because the effective $k(t)$ was smaller at earlier times. Indeed, the horizon distance could even become infinite if $k(t) \to 2$ in a suitable way as $t \to 0$, but this conjecture takes us beyond the dilute-gas approximation where we can compute reliably. Moreover, the time-dependence (35) of the minimum string length indicates that the unit of length in the low-energy world appears to shrink with time. Both these features are reminiscent of inflationary cosmology, although we do not have an additional inflaton field.

The correspondence to inflation is reinforced by the possible appearance of Jeans-like instabilities [33], which may affect low-energy string modes at finite k, as we now argue. We start from the observation that the $SL(2,R)/U(1)$ quantum σ-model metric (31) contains an overall scale factor

$$a(t)^2 = 2(k(t)-2) \tag{39}$$

which grows with time t. This scale factor supplements any time-dependent dilaton scale factor (cf the first, simpler, linear example in ref. [3]) that could be absorbed in the metric background after appropriate redefinition. For large times, and hence large $k \to \infty$, the background (31) is identical after the time redefinition $t \to e^{\delta t}$, $\delta \equiv -4\pi\beta^I T_0$, to the linearly-expanding universe of ref. [33], for which there are no Jeans-like instabilities. However, this is no longer the case if $\beta^I T_0$ is not constant and $k_R(t)$ is no longer a simple exponential, as could occur away from the ultraviolet fixed point when instanton effects become strong. As is shown elsewhere [34], for general $k_R(t)$, the evolution equation for $a(t)$ in the cosmic time frame is

$$\frac{\ddot{a}}{a} = \frac{1}{2k_R(k_R-2)}\{\partial_t^2 k_R - \frac{k_R - 1}{k_R(k_R-2)}(\partial_t k_R)^2\} \tag{40}$$

where the dot denotes differentiation with respect to the cosmic time t_c, defined as

$$dt_c = a(t')dt' \tag{41}$$

The first factor in the square brackets in (40) might dominate the second near the infrared fixed point, in which case $\frac{\ddot{a}}{a} > 0$ and a Jeans-like instability occurs.

We re-emphasize, though, that a physical expansion of the Universe occurs whether or not Jeans instabilities appear. It is natural to enquire into their physical meaning, should they appear. To understand this, it is better to go to higher dimensions, where there are transverse modes. It can be shown [33] that the temporal evolution of these modes is similar to the longitudinal ones. The point is that due to the existence of these unstable modes, the proper size of the string does not grow in the same way as the expansion of the Universe. Hence strings stretch as one goes back in time [2]. We expect that at early times, close to the topological string phase [11], there will be string regeneration through the breaking of strings larger than the Hubble horizon size R_0 [35]. In our picture, these effects would be attributed to purely quantum effects of the string, given the connection of the instanton perturbations with higher-genus instabilities of the black hole background [28]. This quantum string regeneration scenario is similar to the string-driven inflation proposed by Turok [35]. Indeed, were it not for this regeneration phenomenon, the exponential expansion of the Universe in an inflationary scenario would imply a strong suppression of the string density, due to conformal stretching of the string size which grows like the scale factor of the Universe.

In this simplified scenario, one can have a qualitative picture of the entropy production rate in the Universe. In our framework, the rate of entropy increase with time is given by [2, 27]

$$\partial_t S = \beta^i G_{ij} \beta^j S \quad ; \quad G_{ij} = < V_i V_j > \tag{42}$$

[2] There is no stretching in two space-time dimensions. However, even in that case the existence of unstable modes in the longitudinal amplitude, makes things similar to the higher-dimensional case.

where the unitarity requirement of the world-sheet theory implies the positivity of the Zamolodchikov metric [8] $G_{ij} > 0$. Using the C-theorem [8], especially in its string formulation [36] on the fiducial-metric world-sheet, one may write

$$\beta^i G_{ij} \beta^j = -\frac{1}{12}\partial_t C(g) \quad : \quad C(g) = \int d^D y \sqrt{G} e^{-2\Phi} < TT > + \ldots \tag{43}$$

In this expression, the y denote target spatial coordinates, Φ is the dilaton field, and $T \equiv T_{zz}$ is a component of the world-sheet stress tensor. The ... denote the remaining two-point functions that appear in the Zamolodchikov C-function [8], which involve the trace Θ of the stress tensor, i.e. $< T\Theta >$ and $< \Theta\Theta >$. Taking into account the off-shell corollary of the C-theorem, $\frac{\delta C(g)}{\delta g^i} = G_{ij}\beta^j$, it can readily be shown [37] that such terms can always be removed by an appropriate renormalization-scheme choice, that is by appropriate redefinitions of the renormalized couplings g^i, and hence play no rôle in the physics. Thus, one can solve (42) for the entropy S in terms of the Zamolodchikov C-function

$$S(t) = S_0 e^{-\frac{1}{12}\int_0^t \int d^D y \sqrt{G} e^{-2\Phi} <TT> + \ldots} \tag{44}$$

where the minus sign in the exponent indicates the opposite flow of the time t with respect to the renormalization-group flow. Expression (44) reduces a complicated target-space computation of entropy production in an inflationary scenario to a conformal-field-theory computation of two-point functions involving components of the stress tensor of a first-quantized string. We observe from (42) that the rate of entropy increase is maximized on the maximum-β^i surface in coupling constant space. At late stages of the inflationary era, i.e. close to the ultraviolet fixed point, the rate of change of S is strongly suppressed, due to the smallness of the β^i.

The non-critical string scenario for the expanding Universe described in the preceeding paragraphs offers the prospect of solving the three basic problems of the standard-model cosmology in a manner reminiscent of conventional inflation [38]. The *horizon problem* could be solved by the enhanced look-back distance (38), and/or the breakdown of the normal concepts of space and time in a transition to a topological phase close to the infrared fixed point. The *flatness* problem could be solved by an epoch of exponential expansion, induced by a Jeans-like instability in (40). The *entropy problem* could be solved by the enhanced rate of entropy production (44) at early times. However, the crucial difference in our approach is that the fundamental scalar field, usually termed the *inflaton*, is replaced by a world-sheet field, the Liouville mode, in our approach. Fluctuations of this field create the renormalization group flow of the system that leads to the generation of propagating matter, in the way described above and in previous works[27, 7]. Of course, this mode is associated with the appearance of a target space scalar, the dilaton, but the latter is part of the metric background. This can be seen clearly in the two-dimensional Wess-Zumino string theory, which may be considered as a prototype for the description of a spherically-symmetric (s-wave) four-dimensional Universe [39]. In this model the dilaton belongs to the graviton level-one string multiplet, which is a non-propagating (discrete) string mode, and as such can only exist as a background, in contrast to a massless 'tachyon' mode, which propagates and scatters.

As a final comment, we discuss the cosmological "constant" problem in our framework. We adopt a toy-model approach, and concentrate on the one-loop β-functions for the graviton and dilaton fields only. Tachyonic effects are summarized by the running of the level parameter, and we shall not deal with the explicit form of the tachyon potential. As argued in [40], its particular form is renormalization-scheme dependent and, thus, irrelevant for our purposes. We also remark that, as will become clear from the analysis below, *our results can be extended to any number of dimensions for the target space*, provided one maintains the requirement of maximal space symmetry.

Consider the following one-loop results for the dilaton and graviton β-functions in bosonic σ-models [41, 17]:

$$\begin{aligned} \beta^{\Phi} \equiv \frac{d\Phi}{d\phi} &= -\frac{2}{\alpha'}\frac{\delta c}{3} + \nabla^2\Phi - (\nabla\Phi)^2 \\ \beta^{G}_{MN} \equiv \frac{dG_{MN}}{d\phi} &= -R_{MN} - 2\nabla_M\nabla_N\Phi \end{aligned} \tag{45}$$

where ϕ denotes our covariant Liouville cutoff (c.f. the relative minus sign compared with the notation of ref. [41, 17] where the cutoff is defined with the dimensions of mass), and $\delta c = c - 26$, the 26 coming from the space-time reparametrization ghosts. If the central charge of the theory is not 26, as is the case of non-critical bosonic strings, then a cosmological constant term appears in the target space effective action. The form of this target-space action, whose variations yield the β-functions (45), reads:

$$\mathcal{I} = \frac{2}{\alpha'}\int d^D y \sqrt{G} e^{-\Phi}\{\frac{1}{3}\delta c - \alpha'(R + 4(\nabla\Phi)^2 + \ldots)\} \tag{46}$$

where the ... denote other fields in the theory that we shall not use explicitly. We now notice that the effects of the tachyons in our two-dimensional target-space string model amount to a shift of the level parameter $k(\phi)$ with the renormalization group scale. This is the result of the combined effects of tachyon and instanton deformations, the latter representing higher-genus instabilities. The instantons alone, as irrelevant deformations, produce an initial instability by inducing an increase of the central charge, which then flows downhill towards 26 in the presence of relevant matter (tachyon) couplings. Hence there is a running central charge $c(\phi) > 26$, according to the C-theorem [8], that will, in general, imply a non-vanishing, time-varying (running), *positive* cosmological constant, $\Lambda(\phi)$, for the background of (31). Its precise form is determined by consistency with the equations (45).

For simplicity, we assume that the only effect of the dilaton is a constant contribution to the scale anomaly, which is certainly the case of interest. This allows one to decouple Φ in the field equations obtained from (46). Then the latter read

$$\begin{aligned} \frac{\delta \mathcal{I}}{\delta \Phi} &= \Lambda(\phi) - R \\ \frac{\delta \mathcal{I}}{\delta G_{MN}} &= -R_{MN} + \frac{1}{2} G_{MN} R \end{aligned} \tag{47}$$

In two dimensions the second equation is satisfied identically. Decoupling of the dilaton field also implies that the first equation yields

$$R = \Lambda(\phi) \tag{48}$$

We note at this stage that the fact that the background equation for graviton is satisfied automatically implies that the metric in coupling constant space of the full two-dimensional target-space string is singular. This seems to be in agreement with the time-like signature of the Liouville field, which implies non-unitary partial contributions to the σ-model, and should be compared with the situation of ref. [3], between compact spaces and four dimensional uncompactified Minkowski space. In the present case, the conformal field theory associated with the space coordinate r and the running coupling $G_{rr}(\phi)$ is unitary and satisfies the C-theorem [8]. The non-unitarity of the Liouville part is essential in assuring the Weyl invariance of the full theory.

The metric background (31) has a maximal symmetry in its space part. To make the analysis more general, we extend the background to $d = 2 + \epsilon$ dimensions, keeping the maximal symmetry in the spatial part of the metric [3]:

$$G_{ij} = a^2(\phi)\hat{G}_{ij}(x) \quad ; \quad R_{ij} = \frac{R}{d-1} G_{ij} \tag{49}$$

where the scale factor $a^2(\phi)$ is related to $k(\phi)$ as in (31). The curvature of G_{ij} is assumed to be that of a scaled sphere

$$R = \frac{b}{d-2} \frac{1}{a^2(\phi)} = \frac{b}{\epsilon} \frac{1}{a^2(\phi)} \tag{50}$$

where b is a constant. From (48) one finds the following relation between the cosmological constant and scale factor

$$a^2(\phi) = \frac{b/(d-2)}{\Lambda(\phi)} \tag{51}$$

[3]The G_{00}-component depends at most on the time ϕ and can be absorbed in a redefinition of the time variable. It will not be of interest to us here.

Substituting into the β-function equation (45) for the spatial component of G_{ij}, it is immediate to see that the d-dependent factors cancel out, leaving an effective equation for the running cosmological constant

$$\frac{1}{\Lambda^2}\frac{d\Lambda}{d\phi} = \frac{1}{d-1} \tag{52}$$

which should be applicable to higher dimensions. As we have explained above, the actual target-time flow is opposite to that of the renormalization group, and hence it is the negative real ϕ-axis that is relevant, given also that $\phi = ln(a^2/L^2)$, where a and L are ultraviolet and infrared length cut-offs respectively. This yields as a solution for $\Lambda(\phi)$[4]

$$\Lambda(t) = \frac{\Lambda(0)}{1+t\frac{\Lambda(0)}{d-1}} \quad ; \quad t \equiv -\phi > 0 \tag{53}$$

which for positive Λ implies an asymptotically-free cosmological constant β-function, thereby leading to a vanishing cosmological constant at the ultraviolet fixed point on the world-sheet.

The rate of the decrease of $\Lambda(\phi)$ is determined by its initial value at the infrared fixed point, where we conjectured that the theory makes a transition to a topological (twisted $N = 2$ supersymmetric) σ-model. It is of great interest to estimate this value in our two-dimensional model. This can be done by noticing that

$$\Lambda(0) = \frac{2}{\alpha'(0)}\frac{c(0)-26}{3} \tag{54}$$

where $c(0) - 26 = \frac{3k(0)}{k(0)-2} - 27 \simeq \frac{3k(0)}{k(0)-2}$, given that $k(\phi) \to 2$ as $\phi \to 0$. Thus, taking into account (36) one observes that $\Lambda(0)$ is determined by the critical string tension, as it should be, given that α'_0 is the only scale in the problem (or equivalently the minimal string length): the result is

$$\Lambda(0) = \frac{2}{\alpha'_0} \tag{55}$$

The latter result implies a really fast decay of the cosmological constant in this model. Notice that the finite initial value of $\Lambda(0)$ implies from (48) no curvature singularity in the Euclidean model at the origin of target space $r = 0$, as is indeed the case of the two-dimensional black-hole model of ref. [23], given that this point is a pure coordinate singularity. The above analysis for the cosmological constant, therefore, applies most likely to singularity-free inflationary universes. It is understood that

[4] We remark that a similar equation has also been considered in ref. [42], but the flow of time in that reference coincides with the renormalization group flow. In such a case, one gets sensible results only for negative initial values of the cosmological constant, contrary to our case where we have a vanishing cosmological constant asymptotically, starting from positive initial values.

until the precise behaviour of the running couplings near the infrared fixed point is found, there will always be uncertainties in the above estimates. String perturbation theory is not applicable near the topological phase transition, and the infinities we get in the various running couplings constitute an indication of this. In the complete theory, these infinities should be absent.

At this stage, we notice that such a scenario for an asymptotic (in time) vanishing of the cosmological constant was conjectured in ref. [3], but no explicit example was provided. In our case, it is the association of the Liouville mode with a (local) renormalization group scale, flowing irreversibly, that makes possible a realization of such a scenario via (53).

In order to asses the physical consequences of the equation (53) for the asymptotic vanishing of the cosmological "constant", we also need to explore further the relationship between the renormalization group time $t = -\phi$ and the physical time measured, for instance, by atomic clocks. As already mentioned, a time-dependent scale factor appears (31) already in the σ-model metric, supplementing any possible time-dependent dilaton scale factors which may appear [3] in the "physical" metric. The latter is defined as a co-moving frame, where such a time-dependent scale factor is absorbed in an appropriate redefinition (37,41) so as to yield a conventional Friedmann-Robertson-Walker form for the target-space metric:

$$ds^2 = 2(k(t'') - 2)dr^2 - (dt'')^2 \tag{56}$$

where t'' denotes the redefined time variable. In the previous two-dimensional example, where the Wess-Zumino model's dilaton field is space like outside the horizon [23], the redefined time t'' (56) is given simply by the scale factor (39)

$$dt'' = \sqrt{2k(t)}dt \tag{57}$$

To integrate (57), we need to know the full t-dependence of $k(t)$, which is lacking away from the ultraviolet limit.

It is premature to discuss quantitatively the implications of the vanishing mechanism (53), for this and other reasons. However, we finish this talk by noting a few relevant points. One is that the present age of the Universe in "physical" time is about 10^{60} in natural (Planck) units. Another is that one-loop calculations at the point-like field theory level in no-scale supergravity models [12, 43]) yield a negative contribution to the cosmological constant that is $O((m_w/M_{Plnack})^4) \simeq 10^{-60}$ in Planck units. Finally, we note that astrophysical and cosmological observations are compatible with a present-day value of the cosmological constant that is about 10^{-120} in Planck units. Moreover, a cosmological constant of this order of magnitude could even be a welcome adjunct to Cold Dark Matter models. Thus it may even be desirable that the cosmological "constant" has not yet completely relaxed.

Acknowledgements

The work of N.E.M. is supported by a EC Research Fellowship, Proposal Nr. ERB4001GT922259 . That of D.V.N. is partially supported by DOE grant DE-FG05-91-GR-40633.

References

[1] S. Weinberg, *The Constants of Physics*, Philos. Trans. Roy. Soc. London A310 (1983), 249.

[2] J. Ellis, N.E. Mavromatos and D.V. Nanopoulos, Phys. Lett. B293 (1992), 37.

[3] I. Antoniadis, C. Bachas, J. Ellis and D.V. Nanopoulos, Phys. Lett. B211 (1988), 393; Nucl. Phys. B328 (1989), 117; Phys. Lett. B257 (1991), 278.

[4] J. Polchinski, Nucl. Phys. B324 (1989), 123;

D.V. Nanopoulos, in *Proc. Int. School of Astroparticle Physics*, HARC-Houston (World Scientific, Singapore, 1991), p. 183.

[5] D. Amati, M. Ciafaloni and G. Veneziano, Phys. Lett. B197 (1987), 81; *ibid* B289 (1990), 87; Nucl. Phys. B347 (1990), 550;

G. Veneziano, CERN preprint CERN-TH.5366/89 (1989), proc. Workshop on *Superstrings '89* , Texas A & M Univ., College Station , March 13-18 1989, 86-103.

[6] G. Veneziano, *Fundamental Constants in Field and in String Theory*, Proc. Erice Workshop June 1992, eds. N Sanchez and A. Zichichi (World Scientific, Singapore, 1993), 552.

[7] J. Ellis, N.E. Mavromatos and D.V. Nanopoulos, CERN, ENS-LAPP and Texas A & M Univ. preprint, CERN-TH.6897/93, ENS-LAPP-A427-93, CTP-TAMU-30/93; ACT-10/93 (1993).

[8] A.B. Zamolodchikov, JETP Lett. 43 (1986), 730; Sov. J. Nucl. Phys. 46 (1987), 1090.

[9] J. Ellis, N.E. Mavromatos and D.V. Nanopoulos, to appear.

[10] J. Ellis, N. Mavromatos and D.V. Nanopoulos, Phys. Lett. B267 (1991), 465; *ibid* B272 (1991), 261.

[11] J. Ellis, N.E. Mavromatos and D.V. Nanopoulos, Phys. Lett. B288 (1992), 23.

[12] A. Lahanas and D.V. Nanopoulos, Phys. Rep. 145 (1987), 1.

[13] See, for instance, P. Ginsparg, Nucl. Phys. B295 (1988), 153.

[14] F. David, Mod. Phys. Lett. A3 (1988), 1651;
J. Distler and H. Kawai, Nucl. Phys. B321 (1989), 509.

[15] N.E. Mavromatos and J.L. Miramontes, Mod. Phys. Lett. A4 (1989), 1847.

[16] G. Shore, Nucl. Phys. B286 (1987), 349.

[17] H. Osborn, Nucl. Phys. B294 (1987), 595; *ibid* B308 (1988), 629; Phys. Lett. B222 (1989), 97.

[18] I. Kogan, Phys. Lett. B265, (1991), 269.

[19] S. Coleman, Phys. Rev. D15 (1977), 2929;
C. Callan and S. Coleman, Phys. Rev. D16 (1977), 1762.

[20] J. Ellis, J.S. Hagelin, D.V. Nanopoulos and M. Srednicki, Nucl. Phys. B241 (1984), 381.

[21] N. E. Mavromatos and J.L. Miramontes, Phys. Lett. B226 (1989), 291.

[22] J. Ellis, S. Mohanty and D.V. Nanopoulos, Phys. Lett. B221 (1989), 113; *ibid* B235 (1990), 305.

[23] E. Witten, Phys. Rev. D44 (1991), 314.

[24] S. Chaudhuri and J. Lykken, Nucl. Phys. B396 (1993), 270.

[25] E. Brézin and V.A. Kazakov, Phys. Lett. B236 (1990), 144;
M. Douglas and A. Shenker, Nucl. Phys. B335 (1990), 635;
D. Gross and A.A. Migdal, Phys. Rev. Lett. 64 (1990), 127;
For a recent review see, e.g., I. Klebanov, in *String Theory and Quantum Gravity*, Proc. Trieste Spring School 1991, ed. by J. Harvey et al. (World Scientific, Singapore, 1991), and references therein.

[26] A.V. Yung, SISSA preprint 97/92/EP (1992), Int. J. Mod. Phys. A, to appear.

[27] J. Ellis, N. Mavromatos and D.V. Nanopoulos, Phys. Lett. B289 (1992), 25; *ibid* B296 (1992), 40.

[28] J. Ellis, N.E. Mavromatos and D.V. Nanopoulos, Phys. Lett. B276 (1992), 56.

[29] E. Cohen, H. Kluberg-Stern, and R. Peschanski, Nucl. Phys. B328 (499) (1989);
E. Cohen, H. Kluberg-Stern, H. Navelet, and R. Peschanski, Nucl. Phys. B347 (802) (1990).

[30] H. Partovi, Phys. Rev. Lett. 50 (1983), 1885;

R. Blankenbecler and H. Partovi, Phys. Rev. Lett. 57 (1986), 2887), and references therein.

[31] R. Dijkgraaf, E. Verlinde and H. Verlinde, Nucl. Phys. B371 (1992), 269;

H.J. de Vega, J. Ramirez Mittelbrun, M. Ramón Medrano and N. Sánchez, LPTHE 93/14 preprint (1993).

[32] R. Tarrach, Phys. Lett. B133 (1983), 253.

[33] N. Sanchez and G. Veneziano, Nucl. Phys. B333 (1990), 253.

[34] J. Ellis, N.E. Mavromatos and D.V. Nanopoulos, to appear.

[35] N. Turok, Phys. Rev. Lett. 60 (1988), 549.

[36] N.E. Mavromatos and J.L. Miramontes, Phys. Lett. B212 (1988), 33;

N.E. Mavromatos, Phys. Rev. D39 (1989), 1659.

[37] N.E. Mavromatos, Mod. Phys. Lett. A3 (1988), 1079.

[38] See, for instance, A.H. Guth, *Phase Transitions in the Embryo Universe*, Phil. Trans. Roy. Soc. Lond. A307 (1982), 141.

[39] J. Ellis, N. Mavromatos and D.V. Nanopoulos, Phys. Lett. B278 (1992), 246.

[40] T. Banks, Nucl. Phys. B361 (1991), 166.

[41] A.A. Tseytlin, Phys. Lett. B178 (1986), 34.

[42] I. Kogan, Univ. of British Columbia preprint, UBCTP 91-13 (1991).

[43] I. Antoniadis, G.F.R. Ellis, J. Ellis, C. Kounnas and D.V. Nanopoulos, Phys. Lett. B191 (1987), 393;

J. Barrow and N. Deruelle, Nucl. Phys. B297 (1988), 733.

Low-Energy Supersymmetry

The Supersymmetric Top-Ten Lists

HOWARD E. HABER
Santa Cruz Institute for Particle Physics
University of California, Santa Cruz, CA 95064

Abstract

Ten reasons are given why supersymmetry is the leading candidate for physics beyond the Standard Model. Ultimately, the experimental discovery of supersymmetric particles at future colliders will determine whether supersymmetry is relevant for TeV scale physics. The grand hope of supersymmetry enthusiasts is to connect TeV scale supersymmetry with Planck scale physics. The ten most pressing theoretical problems standing in the way of this goal are briefly described.

1. Introduction

The organizers of this workshop asked me to give a talk entitled "Low-Energy Supersymmetry Basics". I assume that the organizers intended this to serve as an introduction to subsequent talks on supersymmetry in particle physics. In this regard, I thought that it would be useful to present a collection of the main reasons we all have attended this workshop. Namely, why is it that we are all partisans of supersymmetry, and what are the main theoretical problems that keep us interested.

Seven months prior to this workshop, I was attending another workshop, not all that dissimilar to this one, held in Erice. During a sumptuous Italian meal, Gordon Kane made the provocative statement that there are already at least seven strong experimental indications that low-energy supersymmetry is a correct description of nature. A number of us joined in on the discussion, sometimes challenging Gordy's assertions and sometimes trying to augment or modify his list. Since that evening in Erice, Gordy's list has increased to nine phenomenological indications for low-energy supersymmetry which now appear in ref. 1.

However, this talk differs from ref. 1 in a number of important respects. First, any list that describes the reasons for pursuing supersymmetry should contain both theoretical arguments as well as phenomenological hints. In particular, the phenomenological hints are often ambiguous or open to a variety of interpretations. That such hints lend their support to a supersymmetric interpretation is strengthened by the theoretical motivations for supersymmetry. Second, I believe that it is important to highlight the main theoretical challenges for supersymmetric theories. The supersymmetric framework is a very ambitious one: it attempts to connect physics at low energies (the TeV scale and below) with the ultimate energy scale of fundamental physics—the Planck scale. A list of the main unsolved problems of the supersymmetric approach will illustrate how far we are from achieving this ultimate goal.

The plan of this talk is as follows. I will provide two "top-ten" lists (hence the use of the plural in the title). In section 2, I will give ten reasons why I (and many supersymmetry enthusiasts) favor supersymmetry. Here, there is substantial overlap with Gordy's list; although I do not subscribe to all the phenomenological hints of ref. 1. In section 3, I provide ten challenges for theorists who attempt to use supersymmetry to connect low-energy physics and the Planck scale. In some sense, the second list is the more important one. The question of whether low-energy supersymmetry exists (or more specifically, whether supersymmetry is responsible for the scale of electroweak symmetry breaking) is ultimately an experimental one. Yet even if explicit evidence for low-energy supersymmetry is eventually revealed, the issues addressed in the second top-ten list will remain. Of course, one hopes that additional information such as a detailed supersymmetric particle spectrum might significantly improve our chances to successfully address some of the items on the list in section 3.

An addiction of "top-ten" lists may be indicative of a night person (which is true in this case). However, unlike the lists of David Letterman, I will not be overly dramatic and save the best bits for the end. The order of the points on each list is somewhat arbitrary. I invite the reader to re-order the items on the list, as well as to subtract and/or add various items to them. In the end, I hope that these lists serve as an overview to the status of supersymmetry in particle theories today, as well as providing some impetus as to where we must focus our attention in the future.

2. The First Top-Ten List: Why Believe in Supersymmetry?

Low-energy supersymmetry is the leading candidate for physics beyond the Standard Model. The simplest model of this type is the Minimal Supersymmetric extension of the Standard Model (MSSM), which is reviewed in ref. 2. Of course, there are other candidates as well, including technicolor,[3] composite models,[4] models based on effective four-fermi Lagrangians (*e.g.*, top-mode condensate models[5]), and perhaps the true model of nature based on physical principles not yet invented. But, an informal poll, taken with the help of SPIRES, indicates that supersymmetry has attracted the most attention of both theorists and experimentalists. Here are ten reasons why.

1. *Supersymmetry is elegant.*

Supersymmetry is a symmetry that associates fermionic and bosonic degrees of freedom. It allows one to evade the famous Coleman-Mandula theorem[6] which asserted the impossibility of putting together space-time symmetries and internal symmetries in a non-trivial way. The twentieth century has seen the triumph of gauge symmetries as the underlying structure of all theories of fundamental forces and particles. Supersymmetry is a beautiful generalization of the concept of continuous symmetries; it would be surprising if nature did not make use of it.

2. *Gravity exists.*

Supersymmetry may be the link between theories of elementary particles and a fundamental theory of gravity. Local supersymmetric theories necessarily contain gravity.[7] Moreover, the only consistent quantum theories that incorporate gravity are superstring theories which possess supersymmetry at some stage in the theory.[8] It is important to note that this argument by itself does not set the energy scale at which supersymmetry breaks. In particular, it is theoretically conceivable that supersymmetry is relevant only at the Planck scale ($M_P \simeq 10^{19}$ GeV) in which case it would never affect the physics we see at present or future colliders.

3. *The gauge hierarchy problem.*[9]

It is very puzzling how the ratio $m_W^2/M_P^2 \simeq 10^{-34}$ emerges from the fundamental Planck scale theory. Low-energy supersymmetric models have the potential to solve this problem. In such theories, the effective scale of supersymmetry breaking lies below 1 TeV and provides the connection to the scale of electroweak symmetry breaking. The most successful model of this type is the radiative symmetry breaking scenario of minimal supergravity models.[10] In this picture, the effective theory at the Planck scale is a globally supersymmetric model broken by soft-supersymmetry breaking mass terms of order 1 TeV. The low-energy consequences of such a theory is revealed by using renormalization group equations (RGEs) with Planck scale boundary conditions. One of the Higgs scalar squared-masses, which is positive at the Planck scale, is driven negative due to the effects of the large top-quark Yukawa coupling; this triggers electroweak symmetry breaking at the required scale. Note that because the renormalization group evolution is logarithmic in nature, an exponential hierarchy between the electroweak scale and the Planck scale can develop.[11]

4. *Naturalness.*[12]

Despite the simplicity of the Higgs mechanism in the Standard Model, the existence of fundamental scalars in field theory is problematical. If the electroweak model is embedded in a more fundamental structure characterized by a much larger energy scale (*e.g.*, the Planck scale), the Higgs boson would tend to acquire mass of order the large scale due to radiative corrections. Only by adjusting (*i.e.*, "fine-tuning") the parameters of the Higgs potential "unnaturally" can one arrange a large hierarchy between the Planck scale and the scale of electroweak symmetry breaking. This requires new physics beyond the Standard Model. The virtue of the supersymmetric solution to the hierarchy problem is that it is natural. That is, the no-renormalization theorems of supersymmetry[13] guarantee that if the hierarchy is established at tree-level, it is not upset when radiative corrections are included.

5. *Unification of gauge couplings.*

Consider the three SU(3)×SU(2)×U(1) gauge couplings (g_1, g_2, and g_3) evaluated at m_Z. Now, run these couplings up to the Planck scale. In a low-energy supersymmetric model, where the supersymmetry breaking scale is characterized by $M_{\rm SUSY}$, one uses the RGEs of the Standard Model for energies between m_Z and $M_{\rm SUSY}$ and the RGEs of the supersymmetric model for energies between $M_{\rm SUSY}$ and M_P. Do the three gauge coupling constants meet at a single point (call it M_X)? In the Standard Model, the answer is no. In the MSSM, the answer is yes! Unification occurs at $M_X \simeq 10^{16}$ GeV for $M_{\rm SUSY} \sim 1$ TeV. This result has been known for some time,[14,15] although a re-analysis by Amaldi and co-workers a few years ago based on LEP data caused a great stir in the particle physics community.[16] Is this result truly significant?

Before being carried away by all the hype, consider the reactions of the optimist, the pessimist, and the cynic. The optimist says: the unification of coupling constants is the first experimental verification of the low-energy supersymmetric scenario. The pessimist says: the unification of coupling constants only rules out the simplest GUT extensions of the Standard Model. It may imply new physics at any scale between the weak scale and the GUT scale and says nothing about TeV scale physics. The cynic says: in the GUT extension of low-energy supersymmetry, there are three unknown parameters: g_U (the unified coupling constant at the GUT scale), M_X (the GUT scale or unification point) and $M_{\rm SUSY}$. Thus the RGEs for g_1, g_2 and g_3 provide three equations and three unknowns. A unique solution is essentially guaranteed, so the unification of coupling constants is no surprise at all. The optimist clearly overstates the (experimental) case for supersymmetry. On the other hand, the pessimist admits that the unification of couplings implies that the desert hypothesis of no new physics between the electroweak scale and the GUT scale is incorrect. New physics must enter somewhere between m_Z and M_X. This is an exciting result! Clearly, low-energy supersymmetry is one possible model for such new physics. Although there is no guarantee that the new physics is associated with the TeV scale, the arguments based on the hierarchy and naturalness problems of the Standard Model strongly suggest that new TeV scale physics must exist. The simplest possible scenario would be one in which this TeV scale physics also accounts for the unification of couplings. Finally, the cynic's remarks that the unification of couplings is guaranteed is technically true (if we ignore the effects of supersymmetric thresholds). However, in solving the RGEs for $M_{\rm SUSY}$ and M_X, there was no guarantee that the coupling constant unification that emerges would be consistent with sensible values for these parameters. The fact that such values correspond precisely to the expected range of a successful grand unified extension of low-energy supersymmetry may be more than coincidental and should not be simply dismissed.

Of course, there are numerous complications to the conclusion that coupling constant unification is a hint for low-energy supersymmetry.[17] One must consider the effects of thresholds, both at the low-energy scale (*e.g.*, the various MSSM particle masses) and at the high-energy scale (*e.g.*, the superheavy grand unified particle masses). Non-renormalizable operators induced at the Planck scale can also affect the unification of couplings.[18] Nevertheless, I believe that the message is clear. The unification of couplings is a strong hint for grand unification at scales near M_P. The failure of coupling constant unification in the Standard Model means that there is no desert between m_Z and M_X—new physics at some intermediate scale must exist. The fact that coupling constant unification does occur in the MSSM presents an intriguing clue that the physics of the desert may have been identified!

6. *Proton decay has not yet been observed.*[19]

If one accepts the grand unification scenario just discussed, then one must consider carefully the predictions of proton decay. It is interesting to note that non-supersymmetric grand unified models tend to predict proton decay rates that are incompatible with current experimental bounds (primarily because M_X tends to lie below 10^{15} GeV). In contrast, in supersymmetric grand unified models, M_X turns out to be significantly larger, and the conventional proton decay modes are unobservable. One must still check that other decay modes that are induced by new (dimension-five) operators particular to supersymmetric models are consistent with present experimental bounds. This imposes interesting constraints in some cases but does not rule out supersymmetric grand unified models.[20]

7. *Relations between third generation quark and lepton masses.*

If one uses Standard Model RGEs and assumes no new physics between the electroweak scale and M_X, then the prediction of $m_b = m_\tau$ at the unification scale is not compatible with low-energy data.[21] However, the relation $m_b = m_\tau$ at M_X is still viable in supersymmetric grand unified models.[22] Relations among other quark and lepton masses require a more complicated structure at the grand unification scale. Supersymmetric models can accommodate such a structure, although it is not clear whether this constitutes a real hint for low-energy supersymmetry.[23]

The result $m_b = m_\tau$ at M_X may be a less compelling clue than the unification of gauge coupling constants. For example, in some string models, unification of gauge couplings can occur without grand unification, whereas the Yukawa couplings depend in part on the structure of the compactification manifold and do not necessarily satisfy standard unification relations.[24,25]

8. *The existence of cold dark matter.*

Most theoretical cosmologists believe that the ratio of the matter density in the universe to the critical density, $\Omega \equiv \rho/\rho_c = 1$. This result follows from theories of inflation, and there are some observational hints that also support this conclusion.[26] But, the baryonic matter density cannot contribute more than $0.2\rho_c$, which strongly suggests the existence of dark matter making up a significant portion of the total matter density of the universe. The precise nature of the dark matter has been much debated in the astrophysical community. Evidence based on theories of galaxy formation and the fluctuations of the microwave background radiation suggest that a substantial fraction of the dark matter is likely to be "cold". The lightest supersymmetric particle (LSP) is an ideal candidate for cold dark matter. Ranges of MSSM parameter space exist where the primordial abundance of the LSP provides exactly the right amount of "missing mass" to reach the critical closure density.[27] See ref. 28 for further details.

9. *Precision electroweak measurements at LEP show no deviation from the Standard Model.*

Suppose that the origin of the electroweak scale lies with new physics beyond the Standard Model. By the naturalness arguments of section 2.4, the energy scale at which this new physics enters must not lie much above 1 TeV. One must then check that the effects of virtual new heavy particle exchange is not in conflict with the precision electroweak measurements at LEP, which at present show no evidence of departures from the Standard Model. One might be tempted to conclude that the effects of new heavy physics should decouple from LEP observables. However, this is not always true, since violation of decoupling can occur in spontaneously broken gauge theories. In many cases, the effects of new heavy particles on electroweak radiative corrections can be neatly summarized by three different combinations of vector boson self-energies.[29,30] (These are called oblique radiative corrections.) T is proportional to the shift in the ρ-parameter and S counts the number of very massive degenerate chiral weak multiplets. A third parameter, U, also enters although it is typically smaller than S and T.

Recent LEP measurements show that the contributions to S, T and U from new physics beyond the Standard Model must be less than 1.[31] This is not a trivial constraint. For example, it has been shown that S can be reliably estimated in a class of technicolor models.[29,32] New heavy technifermion doublets do not decouple from S, so precision electroweak measurements can potentially rule out such models. In contrast, supersymmetric models have the property that their contributions to S, T and U precisely decouple in the limit of large supersymmetry breaking scale, M_{SUSY}. Still, one must check the coefficient of the leading terms to determine the numerical importance of the supersymmetric contributions.[33,34] I have computed the contributions of the various supersymmetric sectors to S, T and U as a function

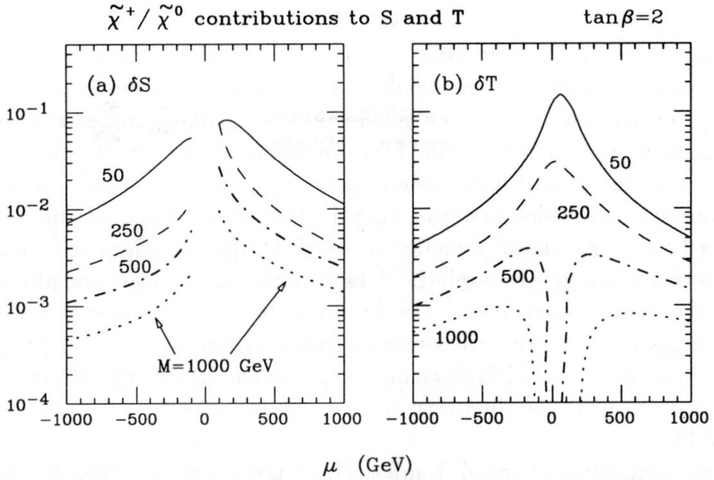

Fig. 1. The contribution to the S and T parameters from the neutralino and chargino sector of the MSSM as a function of μ for $\tan\beta = 2$. The four curves shown correspond to $M = 50$, 250, 500 and 1000 GeV [with $M_2 \equiv M$ and $M_1 = (5g'^2/3g^2)M$]. In (a), curves in the region of $|\mu| \leq 100$ GeV are not shown, since in this region of parameter space the light chargino mass is less than of order m_Z. In (b), T is related to the ρ parameter via $\delta\rho = \alpha\delta T$ which is an experimental observable over the entire mass parameter region. Taken from ref. 34.

of the MSSM parameters.[34] An example of these results is shown in fig. 1. I find that once supersymmetric particle masses all become larger than about 150 GeV, the effects of supersymmetry on the oblique radiative corrections become negligible. Thus, the non-observation of deviations from the Standard Model at LEP is easily compatible with supersymmetric extensions of the Standard Model, in contrast to other excursions beyond the Standard Model mentioned above.

To be complete, it is important to note that some non-oblique radiative corrections (*e.g.*, vertex corrections) can arise in supersymmetric models that are observable. Perhaps the most interesting example of this type is $b \to s\gamma$ which is induced in the Standard Model at one-loop.[35] In the MSSM, new contributions enter which could alter the Standard Model prediction. Recent bounds from CLEO[36] can already place interesting limits on the MSSM parameter space. Nevertheless, in the limit of large supersymmetric masses, these contributions vanish as well. Thus, it is likely that supersymmetric particles, if they exist, will be discovered by direct production at future colliders before their virtual effects are uncovered.

10. *LEP has not discovered the Higgs boson.*[37]

The tenth reason is admittedly given with a little tongue in cheek. (I am sure that technicolor enthusiasts would place this point prominently on their top-ten list.) Nevertheless, low-energy supersymmetry leads to important constraints on the Higgs sector. The experimental discovery or absence of the Higgs boson in future experiments will have a significant impact on the validity of low-energy supersymmetry. Consider first the MSSM. The Higgs sector of the MSSM is a constrained two-Higgs-doublet model, in which all quartic Higgs self-couplings are given in terms of the gauge couplings.[38] This means that at least one physical Higgs boson of the model cannot be arbitrarily heavy. It is easy to show that the lightest CP-even Higgs scalar satisfies the following tree-level bound: $0 \leq m_{h^0} \leq m_Z|\cos 2\beta|$, where $\tan\beta$ is the ratio of Higgs vacuum expectation values. In light of this result, it seems that the LEP non-discovery of the Higgs is somewhat of an embarrassment for the MSSM.

This perception changed dramatically a few years ago when it was realized that both the lower and upper bounds on the lightest Higgs mass in the MSSM are significantly affected by radiative corrections. This is mainly a result of the incomplete cancellation between top quark and top-squark loop corrections to the Higgs two-point function. For example, the most significant one-loop radiative correction to the neutral CP-even Higgs masses grows as $m_t^4 \ln(M_{\tilde{t}}^2/m_t^2)$. Since these effects were first uncovered independently by three groups,[39] there have been many papers in the literature examining the impact of the radiative corrections on the MSSM Higgs sector.[40–50] Here, I shall quote only two results. First, if $\tan\beta = 1$, then the tree-level prediction for the light CP-even Higgs mass is $m_{h^0} = 0$. In this case, the physical mass of the h^0 is entirely due to radiative corrections. Marco Diaz and I have performed an exact one-loop calculation of the neutral Higgs masses for values of $\tan\beta$ near 1.[49] We confirmed that for $\tan\beta$ near 1, large radiative corrections to the light Higgs mass can easily push m_{h^0} to values beyond the current LEP experimental lower bound of 60 GeV. Thus, the possibility that the light Higgs mass is due entirely to radiative corrections is not ruled out! Second, in fig. 2, I show the predicted value of m_{h^0} *vs.* $\tan\beta$ for various values of the CP-odd Higgs mass, m_{A^0}. Two graphs are shown corresponding to $m_t = 150$ GeV and 200 GeV, respectively. For values of $m_{A^0} > 150$ GeV (assuming characteristic supersymmetric particle masses of order 1 TeV), one sees that the predicted value for m_{h^0} lies above the current LEP Higgs mass bound.

The fact that a significant region of MSSM parameter space leads to a predicted value of $m_{h^0} > 60$ GeV is a consequence of enhanced radiative corrections driven by a large top-quark mass. To put it another way, given the large value of m_t suggested by LEP precision electroweak experiments,[51] the most probable MSSM parameters would put the light Higgs boson outside the reach of LEP-I.

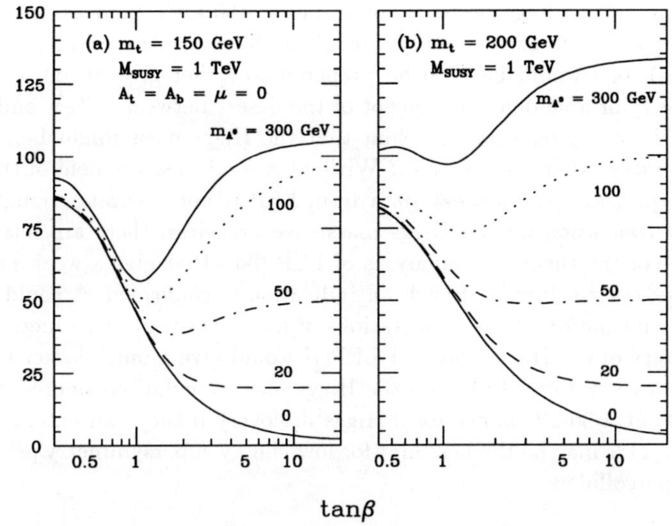

Fig. 2. RGE-improved Higgs mass m_{h^0} as a function of $\tan\beta$ for (a) $m_t = 150$ GeV and (b) $m_t = 200$ GeV. Various curves correspond to $m_{A^0} = 0$, 20, 50, 100 and 300 GeV as labeled in the figure. All A-parameters and μ are set equal to zero. The light CP-even Higgs mass varies very weakly with m_{A^0} for $m_{A^0} > 300$ GeV. Taken from ref. 50.

So far, the above discussion has focused on the MSSM. What about non-minimal models of low-energy supersymmetry? In such models, the main new feature is the possibility of Higgs-self couplings that are not related to the gauge couplings. Top-quark mass enhanced radiative corrections would still tend to raise the tree-level value of the lightest Higgs boson. But, the upper Higgs mass limit is seemingly unconstrained, since it depends on a new unknown parameter. In this case, it is tempting to make use of the observation of gauge coupling constant unification to conclude that *no new physics enters between the TeV scale and* M_X. If this is true, then it seems likely that all couplings of the model remain perturbative below M_X. In the case of the Standard Model, the Higgs squared-mass is proportional to the Higgs self-coupling. The renormalization group scaling of this coupling indicates that the Landau pole would be reached below the Planck scale (indicating that new physics must enter) if the Higgs self-coupling at the electroweak scale lies above a certain value. This result translates into the Higgs mass bound: $m_H \lesssim 175$ GeV.[52] In non-minimal low-energy supersymmetric models, new Higgs self-coupling parameters enter which must be bounded in the same way as in the Standard Model.[53,54] This yields a similar Higgs mass upper bound, which according to ref. 54 is around 150 GeV.

Thus, if no Higgs boson is found below 150 GeV (and assuming that such a Higgs scalar is not unexpectedly difficult to detect by the standard experimental techniques), one would have to be prepared to either give up on low-energy supersymmetry or abandon the concept of the desert between 1 TeV and the Planck scale. Is there any experimental hint that the Higgs mass might be light (*i.e.*, of order m_Z rather than, say, 1 TeV)? Without a good measurement of the top-quark mass, the precision electroweak data from LEP is not accurate enough for one to reach any conclusion on the Higgs mass. Nevertheless, there are intriguing hints from some of the theoretical analyses of LEP data that give a weak preference for light Higgs mass values.[55] If such an indication is confirmed, it could provide yet another confirmation of the expectations of low-energy supersymmetry. Of course, the discovery of the Higgs boson at LEP-II would give a much larger boost to low-energy supersymmetry. Based on the Higgs mass calculations quoted above, I give LEP-II about a 50-50 chance for a Higgs discovery if they can extend their search up to m_Z. This may be the best hint for low-energy supersymmetry prior to turning on the supercolliders.

3. The Second Top-Ten List: Challenges to a Supersymmetric Theory of Particle Physics.

The primary motivation for supersymmetry is that it has the potential for providing a consistent, natural embedding of the Standard Model of particle physics in a more fundamental theory whose natural scale is M_P. The unification of coupling constants discussed in section 2.5 provides the strongest hint that one can extrapolate from the TeV scale all the way up to energies near M_P. However, a fundamental supersymmetric theory of particles remains an elusive goal. Many theorists insist that the fundamental supersymmetric theory can be truly understood only in the context of superstring theory. This is a very ambitious point of view which proposes that superstring theory is the "theory of everything". For example, in such a framework, one could in principle derive the effective low-energy broken-supergravity model that emerges at the Planck scale. If this is your point of view, then perhaps you should replace the following list with the top-ten list of the outstanding problems in string theory and string model building. Such a list would contain questions such as: (i) what is the correct string vacuum? (ii) how does one compute the effective Planck-scale broken-supergravity model parameters from string theory? (iii) *etc.* These questions lie beyond the scope of this talk (and my expertise), and I refer you to the string talks of this workshop. Nevertheless, it is certainly worthwhile to contemplate the solutions to the questions posed in the list below in the context of string theory.

Here are ten theoretical problems that must be overcome on the way to constructing a successful supersymmetric theory of particle physics from the TeV scale to the Planck scale.

1. *The origin of supersymmetry breaking.*

The origin of supersymmetry breaking is one of the most pressing theoretical problem in fundamental theories of supersymmetry. Here, I shall only briefly outline the most common scenario for producing low-energy supersymmetry from a more fundamental broken supergravity model. This scenario has been called the *hidden sector* scenario.[56] In this scenario, one posits two sectors of fields. One sector (called the "visible" sector) contains all the fields of the Standard Model (and perhaps additional heavy fields in a grand unified model of the strong and electroweak forces). A second "hidden" sector contains fields which lead to the breaking of supersymmetry at some large scale $\Lambda_{\rm SUSY}$. One assumes that none of the fields in the hidden sector carry quantum numbers of the visible sector. Thus, the two sectors are nearly decoupled; they communicate only by weak gravitational interactions. Thus, the visible sector only finds out about supersymmetry breaking through its very weak gravitational couplings to the hidden sector. In the visible sector, the effective scale of supersymmetry breaking (denoted by $M_{\rm SUSY}$) is therefore much smaller than $\Lambda_{\rm SUSY}$. A typical result is

$$M_{\rm SUSY} \simeq \frac{\Lambda^n_{\rm SUSY}}{M_P^{n-1}}, \tag{3.1}$$

depending on the mechanism for supersymmetry breaking in the hidden sector. Two popular models for the breaking mechanism are the Polonyi model (where $n = 2$) based on F-type breaking in the hidden sector, and gaugino condensate models[57] (where $n = 3$). In both cases, $\Lambda_{\rm SUSY}$ can be quite large, above 10^{10} GeV, while still producing $M_{\rm SUSY}$ of order 1 TeV or less.

In these scenarios, supersymmetry has the potential for solving the hierarchy and naturalness problems described sections in 2.3 and 2.4. However, at this point, we have only scenarios rather than realistic models. The gaugino condensate model is indicative of the difficulty in constructing a realistic and viable fundamental model of supersymmetry breaking. It suggests that the origin of supersymmetry breaking is probably nonperturbative. As a result, reliable calculations are difficult. This situation is somewhat reminiscent of the status of technicolor approaches to electroweak symmetry breaking; *i.e.*, a number of scenarios have been advanced, but no standard model of technicolor exists. In this regard, I would like to make a plea to the advocates of technicolor and related strong interaction approaches to electroweak symmetry breaking. Lend us your skills of nonperturbative analysis and help us to unravel the secrets of the fundamental origin of supersymmetry breaking!

In string models, supersymmetry breaking should emerge as a consequence of the dynamics of the model. If one were able to successfully solve the string theory and determine the correct vacuum, one would in principle have the tools for determining the "low-energy" effective broken supergravity model at the Planck scale. The soft-supersymmetry breaking parameters would then be computable, and would serve as boundary conditions for renormalization group evolution down to the electroweak scale. Recently, there have been some attempts to explore model-independent features of the soft-supersymmetry breaking terms that emerge from string theory.[58] Kaplunovsky will summarize some of these results later in this workshop.[59]

2. *The cosmological constant problem.*

Even if one is successful in making use of supersymmetry to solve the gauge hierarchy problem with no fine-tuning, there is one unsolved fine-tuning problem which remains, called the cosmological constant problem.[60] Once we take gravity into account in particle theory, the vacuum energy density (Λ_0) is a physical quantity which in principle is calculable in a fundamental theory of gravity. Theoretically, the vacuum energy density is naively expected to be of order M_P^4, but this is not our universe. (A universe with such a large vacuum energy density would have a lifetime of order the Planck time, $\hbar/M_P c^2 \simeq 10^{-43}$ sec!) Thus, based on the fact that the universe has endured over 10 billion years (and looks very flat at large scales), $\Lambda_0/M_P^4 < 10^{-121}$. This is the mother of all fine-tuning and naturalness problems!

The extent of this fine-tuning problem is slightly alleviated in broken supersymmetric models. But at best, M_P^4 is replaced by $M_{\rm SUSY}^4$. Most theorists simply put this question aside, perhaps to be solved at an undetermined future time. However, it is not clear that this is justified. One could ask whether one should accept the theoretical motivation of low-energy supersymmetry to solve the gauge hierarchy and fine-tuning problems while ignoring the most severe fine-tuning problem of them all. Perhaps a solution of the cosmological constant problem will automatically solve all other fine-tuning problems by some presently unknown theoretical mechanism, thus rendering supersymmetry unnecessary. Although it is difficult to argue against such a proposition, the fact that supersymmetry seems to significantly reduce the severity of the cosmological constant problem (as indicated above) may be a hint that supersymmetry will play a key role in its eventual solution.

In the absence of a solution to the cosmological constant problem, one must simply be prepared to accept for now the required fine-tuning in models that incorporates both particle physics and gravity. In models of spontaneously-broken supergravity, there is some freedom that allows a fine-tuning of parameters to set the cosmological constant to zero. All model builders must do this in order to have a theoretically consistent framework. In some string models, the cosmological constant cannot be adjusted by hand, rather it is fixed by the theory. For example, in the models of ref. 61, the cosmological constant is of order $M_{\rm SUSY}^4$. While this is certainly an improvement over the natural value of M_P^4, it is difficult to ascertain whether one can make sense of such models by pretending that the cosmological constant is zero.

3. *The origin of the unification scale (M_X).*

In section 2.5, we saw that the unification of gauge coupling constants takes place at $M_X \simeq 10^{16}$ GeV. This implies that $M_X \neq M_P$. How is the scale M_X generated? Perhaps threshold effects of super-heavy particles are sufficient to push M_X toward M_P such that the two scales are not distinct. In some superstring theories, the unification of coupling constants is predicted to occur at M_P, so the fact that $M_X \neq M_P$ is somewhat problematical. One possible way around this problem is to add extra multiplets to the theory to delay the unification to M_P; some examples can be found in ref. 62.

4. *The gauge hierarchy and tree-level fine-tuning problem.*

One of the theoretical motivations of supersymmetry is to solve the gauge hierarchy and naturalness problems, as discussed in sections 2.3 and 2.4. In supersymmetric grand unified models, the ratio $m_W^2/M_X^2 \simeq 10^{-28}$ is stable under radiative corrections. This follows from the supersymmetric no-renormalization theorems,[13] which imply that the parameters of the superpotential (where the above ratio is set) are not renormalized. Still, one can ask: where does such a small number arise in the first place? For example, in supersymmetric SU(5), this small number must be inserted into the theory (*i.e.*, in the superpotential) by hand at tree-level. Specifically, one must fine-tune tree-level parameters of the theory to an accuracy of one part in 10^{28}, in order that the uncolored doublet Higgs fields remain light (of order m_W) while the color triplet Higgs fields are superheavy (with masses of order M_X). This is the famous doublet-triplet mass splitting problem which is shared by most grand unified theories. In ordinary SU(5), this hierarchy is unstable under radiative corrections. In supersymmetric SU(5), one can "set it and forget it", but this is not a desirable attribute of a fundamental theory. Possible solutions to the tree-level fine-tuning problem do exist. One such example is the missing-partner mechanism;[63] other solutions have also been proposed.[64] However, having eliminated tree-level fine-tuning does not guarantee that the gauge hierarchy problem has been solved. One must check that higher dimensional operators (which are suppressed by inverse powers of the Planck scale) do not re-introduce fine-tuning (which may be less severe than the original fine-tuning, but may still be too large for comfort) in order to maintain the required gauge hierarchy.[65] String theory also provides another mechanism for generating (approximately) massless Higgs doublets, in models where no formal grand unification occurs.[24]

5. *The μ-problem.*[66]

A low-energy supersymmetric model is specified by its superpotential and collection of soft-supersymmetry breaking terms. The latter are dimension two or three terms with coefficients with units of mass to the appropriate power. The scale of the soft-supersymmetry-breaking terms is the electroweak scale; the origin of this scale is tied to the mechanism of supersymmetry-breaking, discussed in section 3.1. But what about the terms in the superpotential with units of mass? For example, the superpotential of the MSSM contains the term $\mu\widehat{H}_1\widehat{H}_2$ (where the $\widehat{H}_i$ are the two Higgs superfields), The parameter μ has dimensions of mass, which must be no larger than about 1 TeV in order to preserve the naturalness of the electroweak theory. What is the origin of the scale μ? There is danger that in the fundamental theory at the Planck scale, μ could be generated with a value of order M_P. Perhaps the most natural solution is to demand that only dimensionless parameters in the superpotential can be nonzero. However, this is not acceptable in the case of the MSSM, since by setting $\mu = 0$, the theory would have a Peccei-Quinn symmetry, leading to a weak scale axion which is experimentally untenable.[⋆] Other solutions to the μ-problem have been proposed. One solution is to add a singlet superfield $\widehat{N}$ to the theory and eliminate the μ-term in the superpotential in favor of the term $\widehat{H}_1\widehat{H}_2\widehat{N}$. Then, μ would be generated when $\widehat{N}$ acquires a vacuum expectation value. However, singlet superfields are dangerous in that they can destroy the gauge hierarchy.[67] A more natural mechanism is one in which $\mu = 0$ initially, but a nonzero value of μ of order 1 TeV (or less) is generated when supersymmetry-breaking effects are taken into account.

6. *The gravitino problem.*

In low-energy supersymmetry, one typically expects the gravitino to possess a mass of order $M_{\rm SUSY}$. The gravitino may or may not be the lightest supersymmetric particle.[†] Since its interactions with ordinary matter are gravitational in strength, its lifetime would exceed the lifetime of the universe by many orders of magnitude. Thus, the gravitino is another candidate for dark matter (in addition to the LSP mentioned in section 2.8). This is problematical, since as shown in ref. 68, a gravitino whose mass is of order $M_{\rm SUSY} \simeq 100$ GeV—1 TeV would lead to a mass density of the universe significantly larger than the critical density. That is, the number of primordial gravitinos is predicted to be too large. One solution to this problem is to suppose that the universe reheats only up to about 10^{10} GeV after inflation. Inflation dilutes the primordial gravitinos sufficiently and a low reheating temperature would insure that gravitinos are not regenerated in significant numbers.[69] In the early days of supersymmetry model building, such a solution was disfavored, since a successful model of baryogenesis at the GUT scale implied that the baryons were generated after inflation, which required a reheating temperature substantially above

---

⋆ In the standard low-energy supergravity approach, if $\mu = 0$, then so is the soft-supersymmetry breaking term $m_{12}^2 \equiv B\mu$ which is the coefficient of H_1H_2 in the scalar potential.

† I shall stick to the notation in which the LSP is the lightest supersymmetric particle, excluding the gravitino.

10^{10} GeV. Recently, there has been much theoretical work which indicates that baryogenesis at the electroweak scale[70] is possible.[‡] In this case, inflation and a low reheating temperature can be a viable solution to the gravitino problem. (See also ref. 72 for an alternative suggestion.)

7. *Flavor changing neutral current (FCNC) problems.*

One of the great successes of the Standard Model is that FCNCs are very suppressed, as required by experimental bounds on FCNC processes. The suppression of FCNCs is a consequence of the GIM-mechanism. On the other hand, the origin of flavor in the Standard Model is a complete mystery. Extended technicolor models attempt to solve both the origin of electroweak symmetry breaking and the origin of flavor with new physics in the energy range between 1 TeV and 1000 TeV. Perhaps it is not surprising that such an ambitious program is generally plagued with FCNCs which violate the strict experimental bounds.

Supersymmetry is often touted as being superior in that there is no FCNC problem. This is only partially correct. To avoid FCNCs, it must be true that the squark mass matrices are approximately diagonal in the same basis that the corresponding quark mass matrices are diagonal. In addition, since the dominant contributions to the squark masses arise from soft-supersymmetry breaking terms, one finds that the squarks must be roughly degenerate in mass.[73 §] In supergravity model building, a standard assumption is that all soft-supersymmetry-breaking scalar masses at the Planck scale are universal. Of course, flavor information does enter the renormalization group evolution, so that the low-energy squark mass parameters will not be exactly flavor independent. Nevertheless, it is easy to show that the assumption of universal soft scalar masses at the Planck scale is sufficient to keep FCNCs below their experimental upper limits.

Are universal soft scalar mass terms at the Planck scale natural? The answer appears to be model-dependent. Such a result appears automatically in supergravity models with canonical kinetic energy terms, although there is no fundamental reason why a theory of supergravity should only possess the simplest kinetic energy terms. In superstring models (at string tree-level), universal scalar masses appear in models in which the supersymmetry-breaking arises solely from the dilaton F-term.[74] More general models of low-energy supergravity generate non-universal soft scalar masses, although the corrections to universality is calculable and in some cases may be sufficiently small.[75] Another possibility, where the required squark degeneracy is

---

‡ In low-energy supersymmetric models, electroweak baryogenesis is possible in the context of the MSSM.[71]

§ Phenomenological requirements impose strong constraints only on the first two generations of squarks. In the $\widetilde{q}_L$–$\widetilde{q}_R$ basis, significant off-diagonal mixing in the bottom and top-squarks sector (which splits the corresponding squark masses from the common diagonal soft-supersymmetry-breaking mass) cannot be ruled out at present.

obtained by exploiting flavor symmetries at the Planck scale, is explored in refs. 76 and 77.

There are other dangers lurking if one begins to allow for new physics at intermediate scales (between the TeV scale and the Planck scale). As shown in ref. 78, integrating out the effects of physics at an intermediate scale can produce effective non-universal scalar mass terms at that scale. Evolving the parameters of the effective Lagrangian down to the electroweak scale can generate FCNCs larger than the allowed bounds. This is an important constraint on models that attempt to attribute the origin of flavor to an intermediate scale.

8. *The flavor puzzle.*

As mentioned in the previous item, the Standard Model and the MSSM treat the fermion generations in the same way. Neither provide any insight into the origin of quark and lepton masses and mixing angles. The discussion of section 3.7 suggests that in a fundamental supersymmetric model, the origin of flavor probably lies at the Planck scale.* At least two different scenarios are possible. In the first scenario, the physics of flavor is imprinted on the fermion-Higgs Yukawa couplings that arise from the underlying superstring theory. Examples are known in which the Yukawa couplings are computable and depend on topological properties of the compactified space that defines the string vacuum.[24,25] In the second scenario, quark and lepton mass matrices are generated from the dynamics at the grand unified scale. Examples of this approach have recently appeared in refs. 23 and 80. In this approach, supersymmetry does not play a fundamental role in the generation of the quark and lepton matrices; rather it is required in order to have consistent unification of couplings. One particularly elegant scenario suggests that the *three* third-generation Yukawa couplings (h_t, h_b and h_τ) all unify at some large scale M_X.[80,81] Such models predict that $m_t \sim 180$ GeV and $\tan\beta \sim m_t/m_b$. Large $\tan\beta$ models have a number of interesting phenomenological implications including enhanced radiative corrections to the light Higgs mass (if $m_{A^0} > m_{h^0}$) and enhanced Higgs couplings to the b-quark (and τ-lepton).

9. *The CP-violation puzzle.*

In the Standard Model, CP-violation arises from a complex phase of the CKM-matrix. This complex phase is also a source of CP-violation in the MSSM; in both cases, there is no clue to the fundamental origin of CP-violation, or the relevant energy scale involved. The solution to this problem may be intimately connected to the flavor puzzle discussed above, since the CKM-phase arises after

---

⋆ Flavor could arise from intermediate scale physics.[79] But realistic models of this kind (that satisfy, *e.g.*, the FCNC constraints) are difficult to construct. In addition, the presence of intermediate scales could seriously disrupt the successful unification of gauge coupling constants discussed in section 2.5.

diagonalizing the quark mass matrix. However, supersymmetric theories introduce new complex phases. For example, in the MSSM complex phases can appear in the gaugino Majorana mass terms and the A-parameters.[82] If these phases were $\mathcal{O}(1)$, one would compute an electric dipole moment for the neutron which is 2 to 3 orders of magnitude larger than the present experimental bounds. Thus, one must conclude that the new supersymmetric phases are no larger that 10^{-3}–10^{-2}. Can such a result arise in a natural way?

In the MSSM, one typically sets the unwanted phases to zero. But in a more fundamental supersymmetric theory, the question of these phases must be addressed. If these phases are zero at the Planck scale, then their magnitudes at the electroweak scale (driven by renormalization group evolution) are certainly small enough to avoid potential phenomenological problems. But, this then shifts the question to the Planck scale. What sets the Planck scale phases to zero?

String theory may provide a hint at a solution to this problem. It turns out that in string theory, the CP-transformation is in fact a gauge transformation in higher-dimensional space-time.[83] Since gauge symmetries cannot be explicitly broken, it follows that at a fundamental level, CP-violation must arise from a spontaneous breaking of CP at some higher energy scale (perhaps the Planck scale). This could provide a theoretical motivation for setting Planck scale phases to zero. Remarkably, in a model of spontaneous CP-breaking at a very high scale, upon integrating out the physics at the high scale, the effective low-energy CP-violation has precisely the form of a single phase in the CKM-matrix.[84–86] This scenario requires new intermediate scale (or Planck scale) physics associated with the scale of CP-violation.[85–87] It appears impossible to construct a viable model of spontaneous CP-violation solely in the context of the MSSM.[88]

Although the above scenario sounds compelling, there is a potential problem associated with the strong CP phase. Based on the present limits on the electric dipole moment of the neutron, it is known that $\bar{\theta} < 10^{-9}$. But in models of spontaneously broken CP, $\bar{\theta}$ is calculable. One can think of the mechanism as follows: set $\bar{\theta} = 0$ at the high scale where CP is a good symmetry. Below the scale of CP-breaking, a non-zero (finite) value of $\bar{\theta}$ is generated. If only Standard Model particles remained after integrating out the physics above the CP-breaking scale, one would find an incredibly small value for $\bar{\theta}$ well below the experimental limits.[89] This result follows because one needs to go to a high order in perturbation theory before the first nontrivial correction to $\bar{\theta}$ arises. But, in models with low-energy supersymmetry, contributions to $\bar{\theta}$ may arise at one-loop and yield a value for $\bar{\theta}$ larger than the experimental bound.[90,86] Thus, it is a challenge to model-builders to construct a viable supersymmetric model of spontaneously broken CP-violation which does not generate a value of $\bar{\theta}$ that is incompatible with the bound on the neutron electric dipole moment.

10. *The origin of low-energy discrete symmetries.*

One of the great triumphs of the Standard Model is that SU(3)×SU(2)×U(1) gauge invariance is sufficient to eliminate the possibility of baryon number (B) and lepton number (L) violating operators of dimension four or less.[91] This provides a natural explanation why the Standard Model conserves B and L to such great accuracy.[★] Unfortunately, this elegant result of the Standard Model is lost in the MSSM.[92] To see why, recall that the supersymmetric interactions are fixed once one specifies the superpotential. The most general gauge-invariant superpotential of the MSSM has the following form:

$$W = W_R + W_{NR} \,. \tag{3.2}$$

First, W_R is given by

$$W_R = \epsilon_{ij}\left[h_\tau \widehat{H}_1^i \widehat{L}^j \widehat{E} + h_b \widehat{H}_1^i \widehat{Q}^j \widehat{D} - h_t \widehat{H}_2^i \widehat{Q}^j \widehat{U} - \mu \widehat{H}_1^i \widehat{H}_2^j\right] \,. \tag{3.3}$$

In eq. (3.3), ϵ_{ij} is used to combine two SU(2) doublets [where $\epsilon_{ij} = -\epsilon_{ji}$ with $\epsilon_{12} = 1$]. The parameters introduced above are the Yukawa coupling matrices h_τ, h_b and h_t (generation labels are suppressed) and the Higgs superfield mass parameter, μ. Second, W_{NR} is given by

$$W_{NR} = \epsilon_{ij}\left[\lambda_L \widehat{L}^i \widehat{L}^j \widehat{E} + \lambda_L' \widehat{L}^i \widehat{Q}^j \widehat{D} - \mu' \widehat{L}^i \widehat{H}_2^j\right] + \lambda_B \widehat{U}\widehat{D}\widehat{D} \,, \tag{3.4}$$

where generation labels are again suppressed. One quickly observes that the terms in W_{NR} violate either baryon number (B) or lepton number (L). Specifically,

$$\begin{array}{cc} \widehat{L}\widehat{L}\widehat{E},\ \widehat{L}\widehat{Q}\widehat{D},\ \widehat{L}\widehat{H} & \Delta L \neq 0 \,, \\ \widehat{U}\widehat{D}\widehat{D} & \Delta B \neq 0 \,. \end{array} \tag{3.5}$$

In the MSSM, one sets $W_{NR} = 0$ in order to recover B and L symmetry. This can be implemented by introducing a discrete symmetry. There are two equivalent descriptions:

(*i*) Matter parity[93]

The MSSM does not distinguish between Higgs and quark/lepton superfields. One can define a discrete matter parity under which all quark/lepton superfields are odd while the Higgs superfields are even.

(*ii*) R-parity[94]

In the supersymmetric limit, one can show that the theory possesses a continuous $U(1)_R$ symmetry if $R = 2$ for all terms in the superpotential W.[†] It follows that in

★ In fact, B and L are not exact in the Standard Model but are violated due to the electroweak anomaly. But the size of such violations is exponentially suppressed and not relevant to the discussion here.

† My normalization of the R-quantum number differs by a factor of two from that of ref. 98.

order to set $W_{NR} = 0$, one may choose

$$\begin{aligned} R &= 1 \qquad &\text{for } \widehat{H}_1, \widehat{H}_2\,, \\ R &= \tfrac{1}{2} \qquad &\text{for } \widehat{L}, \widehat{E}, \widehat{Q}, \widehat{U}, \widehat{D}\,. \end{aligned} \tag{3.6}$$

The full continuous $\mathrm{U}(1)_R$ symmetry does not survive when the soft-supersymmetry-breaking terms are included; the $\mathrm{U}(1)_R$ symmetry breaks down to a discrete Z_2 symmetry called R-parity. It is easy to check that the R-parity quantum number is given by

$$R = (-1)^{3(B-L)+2S} \tag{3.7}$$

for particles of spin S.‡

If low-energy supersymmetry is correct, will it be R-parity invariant? From a purely phenomenological point of view, one cannot rule out the possibility that some of the operators listed in eq. (3.5) are present. One can set bounds on $\lambda_L, \lambda'_L, \mu'$ and λ_B [see eq. (3.4)], based on B and L violation limits in a variety of Standard Model processes.[99,100] There is one weak argument in favor of a conserved R-parity. If R-parity is violated, then the LSP is no longer stable and therefore cannot be the dark matter. If one regards the existence of dark matter as one of the selling points for low-energy supersymmetry (see section 2.8), then one must demand that R-parity is a good symmetry.

Whether one believes in R-parity or not, it is clear that phenomenological requirements prevent the simultaneous appearance of all the operators listed in eq. (3.5). Thus, it seems inevitable that some sort of discrete symmetry will be required to remove the unwanted operators. Any fundamental theory of supersymmetry must address the origin of such discrete symmetries. Examples of superstring models are known in which the discrete symmetries necessary for R-parity invariance emerge naturally.[101] This may provide a clue as to the origin of discrete symmetries in low-energy supersymmetry.

4. Conclusions

With the discovery of the W and Z gauge bosons in the early 1980s, particle physics entered the era of electroweak symmetry breaking. Nevertheless, the fundamental origin of electroweak symmetry breaking is still unknown. Theoretical

‡ Interesting alternative supersymmetric models exist in which the R-parity symmetry described above is modified. Among such models are R-parity-violating models, and models that promote the Z_2 R-parity of the MSSM to a larger discrete symmetry group[95] or even to the full continuous $\mathrm{U}(1)_R$ symmetry.[96] In the latter case, one must introduce new color octet fermions to mix with the gluinos,[97] in which case $\mathrm{U}(1)_R$-symmetric massive color-octet Majorana fermions are permitted. Such models represent interesting alternatives to the MSSM.

arguments have made a convincing case that the dynamics underlying electroweak symmetry breaking must be associated with physics at the TeV scale. Low-energy supersymmetry is a leading candidate for this dynamics, for the reasons presented in section 2. However, these ten reasons can never carry the weight of one reason—the discovery of supersymmetric particles at some future collider. It is in this regard that the next generation of supercolliders—the LHC and SSC—are indispensable. These machines are designed to have the capability of determining whether low-energy supersymmetry or some other dynamics is responsible for the masses of the gauge bosons. Without the supercolliders, progress at the forefront of particle physics will stop cold.

In the meantime, the top-ten list of section 3 provides a useful menu of theoretical problems that supersymmetric theorists must address. Perhaps some progress will occur as we await the deliberations of the politicians. But, I suspect that major theoretical breakthroughs will elude us until we have more experimental input. It will be instructive to see how the top-ten lists presented here are viewed ten and twenty years from now. If supersymmetry survives, its main promise will be that it provides a window to the Planck scale. May we be so fortunate!

Acknowledgements

The invitation and hospitality of Jorge Lopez and his colleagues is much appreciated. I would like to thank Gordy Kane whose advocacy of the phenomenological hints for supersymmetry inspired this talk. In addition, I gratefully acknowledge conversations with Michael Dine, Robert Leigh and Pierre Ramond. Finally, I thank the hospitality of the Aspen Center for Physics which provided me with another opportunity to give this talk, and where the written version of this work was completed. This work was supported in part by the Department of Energy and in part by the Texas National Research Laboratory Commission grant #RGFY93-330.

References

1. G.L. Kane, UM-TH-93-10 (1993), presented at the Coral Gables Conference on "Unified Symmetry in the Small and in the Large" (January, 1993), and at the XVIII Rencontre de Moriond, Les Arcs, France (March, 1993).
2. H.E. Haber, SCIPP 92/33 (1993), to appear in The Proceedings of the 1992 Theoretical Advanced Study Institute, Boulder, CO, June 1992.
3. E. Farhi and L. Susskind, *Phys. Rep.* **74** (1981) 277; R.K. Kaul, *Rev. Mod. Phys.* **55** (1983) 449.
4. I.A. D'Souza and C.S. Kalman, *Preons* (World Scientific, Singapore, 1992).
5. For a recent review, see M. Lindner, *Int. J. Mod. Phys.* **A8** (1993) 2167.
6. See, *e.g.*, P. West, *Introduction to Supersymmetry and Supergravity* (World Scientific, Singapore, 1990).

7. P. van Nieuwenhuizen, *Phys. Rep.* **68** (1981) 189.
8. M.B. Green, J.H. Schwarz and E. Witten, *Superstring Theory* (Cambridge University Press, Cambridge, England, 1987).
9. E. Gildener, *Phys. Rev.* **D14** (1976) 1667; S. Weinberg, *Phys. Lett.* **82B** (1979) 387; L. Susskind, *Phys. Rep.* **104** (1984) 181.
10. H.P. Nilles, *Phys. Rep.* **110** (1984) 1.
11. For a recent review, see L.E. Ibáñez and G.G. Ross, in *Perspectives on Higgs Physics*, edited by G.L. Kane (World Scientific, Singapore, 1993) p. 229.
12. S. Weinberg, *Phys. Rev.* **D13** (1976) 974; **D19** (1979) 1277; L. Susskind, *Phys. Rev.* **D20** (1979) 2619; G. 't Hooft, in *Recent Developments in Gauge Theories,* Proceedings of the NATO Advanced Summer Institute, Cargese, 1979, edited by G. 't Hooft *et. al.* (Plenum, New York, 1980) p. 135.
13. M.T. Grisaru, W. Siegel and M. Roček, *Nucl. Phys.* **B159** (1979) 429; I. Jack, D.R.T. Jones and P. West, *Phys. Lett.* **B258** (1991) 382.
14. M.B. Einhorn and D.R.T. Jones, *Nucl. Phys.* **B196** (1982) 475; W.J. Marciano and G. Senjanovic, *Phys. Rev.* **D25** (1982) 3092.
15. U. Amaldi *et. al.*, *Phys. Rev.* **D36** (1987) 1385.
16. U. Amaldi, W. de Boer and H. Furstenau, *Phys. Lett.* **B260** (1991) 447; U. Amaldi *et. al.*, *Phys. Lett.* **B281** (1992) 374.
17. R. Barbieri and L.J. Hall, *Phys. Rev. Lett.* **68** (1992) 752; P. Langacker and N. Polonsky, *Phys. Rev.* **D47** (1993) 4028; A.E. Faraggi, B. Grinstein, and S. Meshkov, *Phys. Rev.* **D47** (1993) 5018.
18. L.J. Hall and U. Sarid, *Phys. Rev. Lett.* **70** (1993) 2673.
19. P. Langacker, UPR-0539-T (1992), invited talk given at The Benjamin Franklin Symposium in Celebration of the Discovery of the Neutrino, Philadelphia, PA, April 29—May 1, 1992.
20. R. Arnowitt and P. Nath, *Phys. Rev. Lett.* **69** (1992) 725; *Phys. Lett.* **B287** (1992) 89; NUB-TH-3056/92 (1992); J. Hisano, H. Murayama and T. Yanagida, *Nucl. Phys.* **B402** (1993) 46.
21. H. Arason *et. al.*, *Phys. Rev.* **D46** (1992) 3945.
22. V. Barger, M.S. Berger and P. Ohmann, *Phys. Rev.* **D47** (1993) 1093.
23. S. Dimopoulos, L.J. Hall and S. Raby, *Phys. Rev. Lett.* **68** (1992) 1984; *Phys. Rev.* **D45** (1992) 4192; **D46** (1992) 4793.
24. E. Witten, *Nucl. Phys.* **B258** (1985) 75.
25. See, *e.g.*, G.G. Ross, in *The Santa Fe TASI-87*, Proceedings of the 1987 Theoretical Advanced Study Institute, Santa Fe, NM, edited by R. Slansky and G. West (World Scientific, Singapore, 1988) p. 628; in *Particles and Fields—3*, Proceedings of the 3rd Banff Summer Institute on Particles and Fields, Banff, Alberta, August 14—27, 1988, edited by A.N. Kamal and F.C. Khanna (World Scientific, Singapore, 1989) p. 223.
26. E.W. Kolb and M.S. Turner, *The Early Universe* (Addison-Wesley Publishing

Company, Reading, MA, 1990); T. Padmanabhan, *Structure Formation in the Universe* (Cambridge University Press, Cambridge, England, 1993).
27. S. Kelley, J.L. Lopez, D.V. Nanopoulos, H. Pois and K. Yuan, *Phys. Rev.* **D47** (1993) 2461; R. Arnowitt and P. Nath, *Phys. Lett.* **B299** (1993) 58 [E: **B309** (1993) 403]; R.G. Roberts and L. Roszkowski, *Phys. Lett.* **B309** (1993) 337.
28. K. Yuan, contribution to these Proceedings.
29. M.E. Peskin and T. Takeuchi, *Phys. Rev. Lett.* **65** (1990) 964; *Phys. Rev.* **D46** (1992) 381.
30. G. Altarelli and R. Barbieri, *Phys. Lett.* **B253** (1990) 161.
31. P. Langacker, in *Electroweak Physics Beyond the Standard Model*, Proceedings of the International Workshop on Electroweak Interactions Beyond the Standard Model, Valencia, Spain, October 2—5, 1991, edited by J.F.W. Valle and J. Velasco (World Scientific, Singapore, 1992) p. 75.
32. B. Holdom and J. Terning, *Phys. Lett.* **B247** (1990) 88; M. Golden and L. Randall, *Nucl. Phys.* **B361** (1991) 3.
33. R. Barbieri, M. Frigeni, F. Giuliani, and H.E. Haber, *Nucl. Phys.* **B341** (1990) 309.
34. H.E. Haber, SCIPP 93/06 (1993), to appear in the Proceedings of the 23rd Workshop of the INFN Eloisatron Project, "Properties of Supersymmetric Particles", Erice, Italy, September 28—October 4, 1992.
35. A Buras, P. Krawczyk, M.E. Lautenbacher and C. Salazar, *Nucl. Phys.* **B337** (1990) 284; S. Bertolini, F. Borzumati, A. Masiero and G. Ridolfi, *Nucl. Phys.* **B353** (1991) 591; J.L. Hewett, *Phys. Rev. Lett.* **70** (1993) 1045; V. Barger, M.S. Berger, and R.J.N. Phillips, *Phys. Rev. Lett.* **70** (1993) 1368; R. Barbieri and G.F. Giudice, *Phys. Lett.* **B309** (1993) 86.
36. R. Ammar *et. al.* [CLEO Collaboration], *Phys. Rev. Lett.* **71** (1993) 674.
37. See, *e.g.*, D. Decamp *et. al.* [ALEPH Collaboration], *Phys. Rep.* **216** (1992) 253.
38. See chapter 4 of J.F. Gunion, H.E. Haber, G.L. Kane and S. Dawson, *The Higgs Hunter's Guide* (Addison-Wesley Publishing Company, Reading, MA, 1990).
39. H.E. Haber and R. Hempfling, *Phys. Rev. Lett.* **66** (1991) 1815; Y. Okada, M. Yamaguchi and T. Yanagida, *Prog. Theor. Phys.* **85** (1991) 1; J. Ellis, G. Ridolfi and F. Zwirner, *Phys. Lett.* **B257** (1991) 83; **B262** (1991) 477.
40. Y. Okada, M. Yamaguchi and T. Yanagida, *Phys. Lett.* **B262** (1991) 54; R. Barbieri, M. Frigeni, and F. Caravaglios, *Phys. Lett.* **B258** (1991) 167; J.R. Espinosa and M. Quiros, *Phys. Lett.* **B266** (1991) 389.
41. R. Barbieri and M. Frigeni, *Phys. Lett.* **B258** (1991) 395; A. Yamada, *Phys. Lett.* **B263** (1991) 233.

42. A. Brignole, J. Ellis, G. Ridolfi and F. Zwirner, *Phys. Lett.* **B271** (1991) 123 [E: **B273** (1991) 550].
43. P.H. Chankowski, S. Pokorski and J. Rosiek, *Phys. Lett.* **B274** (1992) 191; **B281** (1992) 100; MPI-Ph/92-117 and DFPD 93/TH/13 (1993).
44. A. Brignole, *Phys. Lett.* **B277** (1992) 313.
45. M.A. Diaz and H.E. Haber, *Phys. Rev.* **D45** (1992) 4246.
46. D.M. Pierce, A. Papadopoulos, and S. Johnson, *Phys. Rev. Lett.* **68** (1992) 3678.
47. K. Sasaki, M. Carena and C.E.M. Wagner, *Nucl. Phys.* **B381** (1992) 66.
48. A. Brignole, *Phys. Lett.* **B281** (1992) 284.
49. M.A. Diaz and H.E. Haber, *Phys. Rev.* **D46** (1992) 3086.
50. H.E. Haber and R. Hempfling, SCIPP 91/33 (1992), *Phys. Rev.* **D48** (1993) in press.
51. D. Buskulic *et. al.* [ALEPH Collaboration], CERN-PPE/93-40 (1993).
52. N. Cabibbo, L. Maiani, G. Parisi, R. Petronzio, *Nucl. Phys.* **B158** (1979) 295.
53. H.E. Haber and M. Sher, *Phys. Rev.* **D35** (1987) 2206; M. Drees, *Phys. Rev.* **D35** (1987) 2910; *Int. J. Mod. Phys.* **A4** (1989) 3635; J. Ellis, J.F. Gunion, H.E. Haber, L. Roszkowski, and F. Zwirner, *Phys. Rev.* **D39** (1989) 844; J.R. Espinosa and M. Quiros, *Phys. Lett.* **B279** (1992) 92; **B302** (1993) 51.
54. G.L. Kane, C. Kolda and J.D. Wells, *Phys. Rev. Lett.* **70** (1993) 268.
55. J. Ellis, G.L. Fogli and E. Lisi, *Phys. Lett.* **B279** (1992) 169; *Phys. Lett.* **B285** (1992) 238; *Phys. Lett.* **B286** (1992) 85; *Nucl. Phys.* **B393** (1993) 3; F. del Aguila, M. Martinez and M. Quiros *Nucl. Phys.* **B381** (1992) 451.
56. See, *e.g.*, L. Hall, J. Lykken and S. Weinberg, *Phys. Rev.* **D27** (1983) 2359; S.K. Soni and H.A. Weldon, *Phys. Lett.* **126B** (1983) 215.
57. H.-P. Nilles, *Int. J. Mod. Phys.* **A5** (1990) 4199.
58. L.E. Ibáñez and D. Lust, *Nucl. Phys.* **B382** (1992) 305; V.S. Kaplunovsky and J. Louis, *Phys. Lett.* **B306** (1993) 269.
59. V. Kaplunovsky, contribution to these Proceedings.
60. S. Weinberg, *Rev. Mod. Phys.* **61** (1989) 1.
61. B. de Carlos, J.A. Casas and C. Muñoz, *Nucl. Phys.* **B399** (1993) 623.
62. I. Antoniadis, J. Ellis, R. Lacaze and D.V. Nanopoulos, *Phys. Lett.* **B268** (1991) 188; I. Antoniadis, J. Ellis, S. Kelley and D.V. Nanopoulos, *Phys. Lett.* **B272** (1991) 31; S. Kelley, J.L. Lopez and D.V. Nanopoulos, *Phys. Lett.* **B278** (1992) 140.
63. H. Georgi, *Phys. Lett.* **B108** (1982) 283; A. Masiero, D.V. Nanopoulos, K. Tamvakis, and T. Yanagida, *Phys. Lett.* **115B** (1982) 380; B. Grinstein, *Nucl. Phys.* **B206** (1982) 387; S.-C. Chao, *Nucl. Phys.* **B256** (1985) 705.
64. A.A. Anselm and A.A. Johansen, *Phys. Lett.* **B200** (1988) 331; G.R. Dvali, *Phys. Lett.* **B287** (1992) 101.
65. I thank Michael Dine for stressing this point to me.

66. For a recent discussion, see J.A. Casas and C. Muñoz, *Phys. Lett.* **B306** (1993) 288.
67. This point has recently been re-examined in J. Bagger and E. Poppitz, Johns Hopkins preprint JHU-TIPAC-93018 (1993).
68. S. Weinberg, *Phys. Rev. Lett.* **48** (1982) 1303.
69. J. Ellis, A.D. Linde and D.V. Nanopoulos, *Phys. Lett.* **118B** (1982) 59; S. Dimopoulos and S. Raby, *Nucl. Phys.* **B219** (1983) 479; L.M. Krauss, *Nucl. Phys.* **B227** (1983) 556; J. Ellis, J.E. Kim and D.V. Nanopoulos, *Phys. Lett.* **145B** (1984) 181; B.A. Ovrut, *Phys. Lett.* **147B** (1984) 263.
70. For a recent review, see A.G. Cohen, D.B. Kaplan and A.E. Nelson, UCSD-PTH-93-02 (1993).
71. A.G. Cohen and A.E. Nelson, *Phys. Lett.* **B297** (1992) 111.
72. J. Cline and S. Raby, *Phys. Rev.* **D43** (1991) 1781.
73. See, *e.g.*, F. Gabbiani and A. Masiero, *Nucl. Phys.* **B322** (1989) 235.
74. V. Kaplunovsky and J. Louis, ref. 58; R. Barbieri, J. Louis and M. Moretti, CERN-TH.6856/93 (1993).
75. B. de Carlos, J.A. Casas and C. Muñoz, *Phys. Lett.* **B299** (1993) 234.
76. For a recent discussion, see Y. Nir and N. Seiberg, *Phys. Lett.* **B309** (1993) 337.
77. M. Dine, R.G. Leigh and A. Kagan, SLAC-PUB-6147 (1993) and SCIPP 93/04 (1993).
78. L.J. Hall, V.A. Kostelecky and S. Raby, *Nucl. Phys.* **B267** (1986) 415.
79. M. Dine and A.E. Nelson, *Phys. Rev.* **D48** (1993) 1277.
80. G. Anderson, S. Raby, S. Dimopoulos, L.J. Hall and G. Starkman, LBL-33531 (1993).
81. B. Ananthanarayan, G. Lazarides and Q. Shafi, *Phys. Rev.* **D44** (1991) 1613; *Phys. Lett.* **B300** (1993) 245; L.J. Hall, R. Rattazzi and U. Sarid, LBL-33997 (1993).
82. For example, see M. Dugan, B. Grinstein, and L.J. Hall, *Nucl. Phys.* **B255** (1985) 413.
83. M. Dine, R.G. Leigh and D.A. MacIntire, *Phys. Rev. Lett.* **69** (1992) 2030; K. Choi, D.B. Kaplan and A.E. Nelson, *Nucl. Phys.* **B391** (1993) 515.
84. H.E. Haber and Y. Nir, *Nucl. Phys.* **B335** (1990) 363; L. Bento and G.C. Branco, *Phys. Lett.* **B245** (1990) 599.
85. A. Nelson, *Phys. Lett.* **136B** (1984) 387; S.M. Barr, *Phys. Rev. Lett.* **53** (1984) 329.
86. S.M. Barr and A. Masiero, *Phys. Rev.* **D38** (1988) 366; S.M. Barr and G. Segre, *Phys. Rev.* **D48** (1993) 302; M. Dine, R.G. Leigh and A. Kagan, SLAC-PUB-6090 and SCIPP 93/05 (1993), *Phys. Rev.* **D48** (1993) in press.
87. A. Dannenberg, L. Hall and L. Randall, *Nucl. Phys.* **B271** (1986) 574.
88. A. Pomarol, *Phys. Lett.* **B287** (1992) 331; *Phys. Rev.* **D47** (1992) 273.

89. J. Ellis and M.K. Gaillard, *Nucl. Phys.* **B150** (1979) 141.
90. R. Akhouri, I. Bigi and H.E. Haber, *Phys. Lett.* **135B** (1984) 113; R. Akhouri and I. Bigi, *Nucl. Phys.* **B234** (1984) 459.
91. S. Weinberg, *Phys. Rev. Lett.* **43** (1979) 1566; F. Wilczek, *Phys. Rev. Lett.* **43** (1979) 1571.
92. S. Weinberg, *Phys. Rev.* **D26** (1982) 287.
93. S. Dimopoulos, S. Raby and F. Wilczek, *Phys. Lett.* **112B** (1982) 133.
94. P. Fayet, *Nucl. Phys.* **B90** (1975) 104; *Phys. Lett.* **69B** (1977) 489.
95. L.J. Hall, *Mod. Phys. Lett.* **A5** (1990) 467.
96. L.J. Hall and L. Randall, *Nucl. Phys.* **B352** (1991) 289.
97. P. Fayet, *Phys. Lett.* **78B** (1978) 417.
98. J. Wess and J. Bagger, *Supersymmetry and Supergravity* (Princeton University Press, Princeton, NJ, 1992).
99. S. Dimopoulos and L.J. Hall, *Phys. Lett.* **B207** (1987) 210; S. Dimopoulos, R. Esmailzadeh, L.J. Hall, J.-P. Merlo and G.D. Starkman, *Phys. Rev.* **D41** (1990) 2099.
100. H. Dreiner and G.G. Ross, *Nucl. Phys.* **B365** (1991) 597.
101. B.R. Green, K.H. Kirklin, P.J. Miron and G.G. Ross, *Nucl. Phys.* **B278** (1986) 667; **B292** (1987) 606; R. Arnowitt and P. Nath, *Phys. Rev. Lett.* **62** (1989) 2225; *Phys. Rev.* **D40** (1989) 191; **D42** (1990) 2948.

SUPERSYMMETRY PHENOMENOLOGY*

V. Barger[a] and R.J.N. Phillips[b]

[a]*Physics Department, University of Wisconsin, Madison, WI 53706, USA*
[b]*Rutherford Appleton Laboratory, Chilton, Didcot, Oxon OX11 0QX, UK*

Abstract

We review some recent results and future prospects in the phenomenology of Supersymmmetry. We discuss the searches for superpartner states, the searches for Higgs bosons in the minimal SUSY model, and additional parameter constraints arising in SUSY-GUT models.

1 Introduction

Supersymmetry (SUSY) is a vital element in present thinking about fundamental particle physics. At this Conference it certainly needs no formal introduction nor motivation; we simply emphasise here two important aspects — it has perturbatively calculable consequences and can be experimentally tested. This talk will be about the associated phenomenology.

Experimental evidence for SUSY could come in various forms, such as

(a) discovery of one or more superpartners,

(b) discovery of a neutral Higgs boson with non-standard properties and/or a charged Higgs boson,

(c) discovery of $p \to K\nu$ decay,

(d) discovery that dark matter consists of heavy neutral particles.

This talk will address areas (a) and (b) above. We shall discuss the present status and future prospects of searches for superpartner particles, searches for Higgs bosons in the minimal SUSY model (MSSM) and additional parameter constraints that come from SUSY-GUT studies. We leave other phenomenological topics like proton decay, dark matter, R-parity breaking, etc., to other speakers.

*Talk presented by V. Barger

2 Experimental SUSY Signatures

The MSSM conserves a multiplicative R-parity: $R_p = (-1)^{2S+L+3B}$, where S, L, B are spin, lepton number, baryon number. Standard Model (SM) particles have $R = 1$; their superpartners have $R = -1$. Hence heavy sparticles must decay to lighter sparticles, while the lightest sparticle (LSP) is stable. If the LSP has no strong or electromagnetic interactions, as seems likely since it has defied detection for so long, it will carry away "missing" energy and momentum — a vital SUSY signature. The LSP will also be a candidate for cosmological dark matter; this focuses extra interest on the R_p-conserving case.

Grand Unified Theories (GUTs) are essential for SUSY phenomenology to economize parameters. A minimal set of GUT parameters with soft SUSY breaking consists of the gauge and Yukawa couplings g_i and λ_i, the Higgs mixing mass μ, the common gaugino mass at the GUT scale $m_{1/2}$, the common scalar mass at the GUT scale m_0, and two parameters A, B that give trilinear and bilinear scalar couplings. At the weak scale, the gauge couplings are experimentally determined. The Higgs potential depends upon m_0, μ, B (at tree level) and $m_{1/2}, A, \lambda_t, \lambda_b$ (at one loop). After minimizing the Higgs potential and putting in the measured Z and fermion masses, there remain 5 independent parameters, e.g. $m_t, \tan\beta, m_0, m_{1/2}, A$.

Beside the SM gauge bosons and fermions, the MSSM spectrum contains Higgs bosons $(h, H, A, H^{\pm})$, gluinos $(\tilde{g})$, squarks $(\tilde{q})$, sleptons $(\tilde{\ell}^{\pm})$, charginos ($\tilde{W}_i^{\pm}, i = 1, 2$; mixtures of winos and charged higgsinos), neutralinos ($\tilde{Z}_j, j = 1, 2, 3, 4$; mixtures of zinos, photinos and neutral higgsinos). An alternate notation is $\tilde{\chi}_i^{\pm}$ for $\tilde{W}_i^{\pm}$ and $\tilde{\chi}_j^0$ for $\tilde{Z}_j$. Figure 1 illustrates the evolution of the SUSY mass spectrum from the GUT scale.[1,2] The running masses are plotted versus μ and the physical value occurs approximately where the running mass $m = m(\mu)$ intersects the curve $m = \mu$. For the Higgs scalar H_2, the mass-square becomes negative at low μ due to coupling to top; in this region we have actually plotted $-|m(\mu)|$. Negative mass-square parameter is essential for spontaneous symmetry-breaking; here it is achieved by radiative effects. The running masses for the gauginos $\tilde{g}, \tilde{W}, \tilde{B}$ are given by

$$M_i(\mu) = m_{1/2}\,\frac{\alpha_i(\mu)}{\alpha_i(M_G)}\,, \tag{1}$$

where i labels the gauge symmetry; this applies before mixing with higgsinos to obtain the chargino and neutralino mass eigenstates. In Fig. 1 the squarks are heavier than the gluinos, but the opposite ordering $m_{\tilde{q}} < m_{\tilde{g}}$ is possible in other scenarios. Sleptons, neutralinos and charginos are lighter than squarks and gluinos in general. The usual soft SUSY-breaking mechanisms preserve the gauge coupling relations (unification) at M_G.

In order that SUSY cancellations shall take effect at low mass scales as required, the SUSY mass parameters are expected to be bounded by

$$m_{\tilde{g}},\ m_{\tilde{q}},\ |\mu|,\ m_A \lesssim 1\text{–}2\ \text{TeV}\,. \tag{2}$$

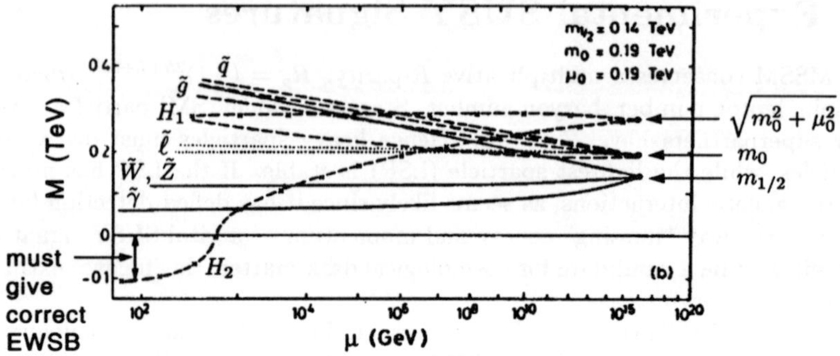

Fig. 1. Typical RGE results for spartner masses.[1]

The other parameter $\tan\beta$ is effectively bounded by

$$1 \lesssim \tan\beta \lesssim 65\,, \tag{3}$$

where the lower bound arises from consistency in GUT models and the upper bound is a perturbative limit (following Ref. 3). Proton decay gives $\tan\beta < 85$.[4,5]

At LEP I, sufficiently light SUSY particles would be produced through their gauge couplings to the Z. Direct searches for SUSY particles at LEP give

$$m_{\tilde{q}},\, m_{\tilde{\ell}},\, m_{\tilde{\nu}},\, m_{\tilde{W}_1} \gtrsim 40\text{–}45 \text{ GeV}\,. \tag{4}$$

There are also indirect limits on Z decays to SUSY particles,[6]

$$\begin{aligned} \Delta\Gamma_Z(\text{SUSY}) &< 25 \text{ MeV}\,, & (5)\\ \Gamma(Z \to \tilde{Z}_1\tilde{Z}_1 : \text{invisible}) &< 17 \text{ MeV}\,, & (6)\\ B(Z \to \tilde{Z}_1\tilde{Z}_1) &< \text{few} \times 10^{-5}\,. & (7) \end{aligned}$$

It is convenient to express the excluded regions in terms of the parameters that determine the gaugino masses, i.e. in the $(\mu, m_{1/2})$ or $(\mu, m_{\tilde{g}})$ plane: see Fig. 2.

Hadron colliders can explore much higher energy ranges, producing squarks and gluinos strongly. At the Tevatron, with luminosity 25 pb^{-1} expected in 1993, about 100 events would be expected for each of the channels $\tilde{g}\tilde{q}$ and $\tilde{q}\tilde{q}$ for $m_{\tilde{q}} \sim m_{\tilde{g}} \sim$ 200 GeV, reaching well beyond the LEP range.[6–8]

SUSY particle signatures depend on their decays. The most distinctive is the missing energy and momentum of the undetected LSP (usually assumed to be the lightest neutralino $\tilde{Z}_1$) in all SUSY decay chains with R-parity conservation. At hadron colliders one can only do book-keeping on the missing transverse momentum denoted $\not{p}_T$, where the missing momenta of both LSPs are added vectorially. The LSP momenta and hence the magnitude of $\not{p}_T$ depend on the decay chains.

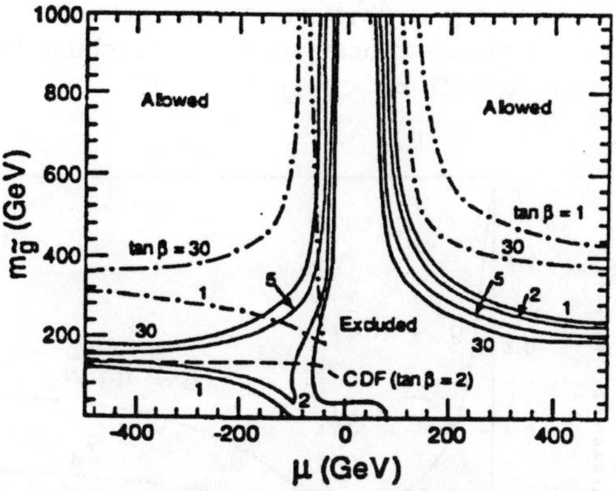

Fig. 2: MSSM parameter regions accessible to LEP, from Ref. 7. Solid curves surround regions excluded by LEP I data, dot-dashed curves show the potential reach of LEP II, and the dashed curve is the Tevatron CDF limit for $\tan\beta = 2$.

If squarks and gluinos are rather light ($m_{\tilde g}, m_{\tilde q} \lesssim 50$ GeV), their dominant decay mechanisms are strong decays to each other or decays to the LSP:

$$\left.\begin{array}{l}\tilde q \to q\tilde g \\ \tilde g \to q\bar q \tilde Z_1\end{array}\right\} \quad \text{if } m_{\tilde g} < m_{\tilde q}\,, \tag{8}$$

$$\left.\begin{array}{l}\tilde g \to q\tilde q \\ \tilde q \to q \tilde Z_1\end{array}\right\} \quad \text{if } m_{\tilde q} < m_{\tilde g}\,. \tag{9}$$

In these cases the LSPs carry a substantial fraction of the available energy and $\not p_T$ is correspondingly large. Assuming such decays and small LSP mass, the present 90% CL experimental bounds from UA1 and UA2 at the CERN p-$\bar p$ collider ($\sqrt{s} =$ 640 GeV) and from CDF at the Tevatron ($\sqrt{s} = 1.8$ TeV) are[9]

	$m_{\tilde g}$	$m_{\tilde q}$
UA1 (1987)	> 53 GeV	> 45 GeV
UA2 (1990)	> 79	> 74
CDF (1992)	> 141	> 126

The limits rise if the squark and gluino masses are assumed to be comparable.

For heavier gluinos and squarks, many new decay channels are open, such as decays into the heavier gauginos:

$$\tilde g \;\to\; q\bar q\tilde Z_i \; (i = 1,2,3,4),\; q\bar q'\tilde W_j \; (j = 1,2),\; g\tilde Z_1\,, \tag{10}$$

$$\tilde q_L \;\to\; q\tilde Z_i \; (i = 1,2,3,4),\; q'\tilde W j \; (j = 1,2)\,, \tag{11}$$

$$\tilde q_R \;\to\; q\tilde Z_i \; (i = 1,2,3,4)\,. \tag{12}$$

Some decays go via loops (e.g. $\tilde{g} \to g\tilde{Z}_1$); we have not attempted an exhaustive listing here. Figure 3 shows how gluino-to-heavy-gaugino branching fractions increase with $m_{\tilde{g}}$ in a particular example (with $m_{\tilde{g}} < m_{\tilde{q}}$).[10]

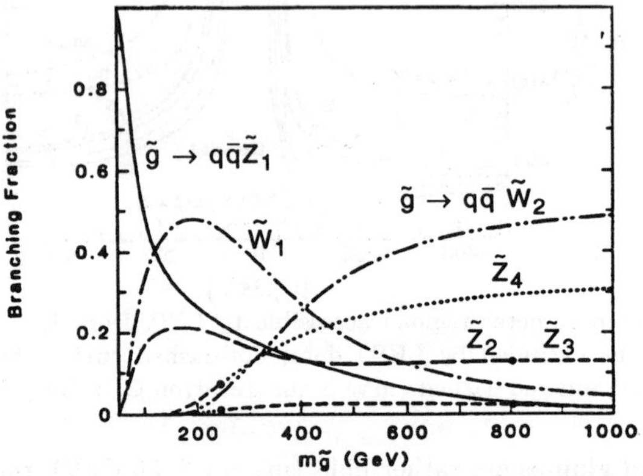

Fig. 3. Example of gluino decay branchings versus mass.[10]

The heavier gauginos then decay too:

$$\tilde{W}_j \;\to\; Z\tilde{W}_k,\; W\tilde{Z}_i,\; H_i^0\tilde{W}_k,\; H^{\pm}\tilde{Z}_i,\; f\tilde{f}\;, \tag{13}$$

$$\tilde{Z}_i \;\to\; Z\tilde{Z}_k,\; W\tilde{W}_j,\; H_i^0\tilde{Z}_k,\; H^{\pm}\tilde{W}_k,\; f\tilde{f}'\;. \tag{14}$$

Here it is understood that final W or Z may be off-shell and materialize as fermion-antifermion pairs; also Z may be replaced by γ. In practice, chargino decays are usually dominated by W-exchange transitions; neutralino decays are often dominated by sfermion exchanges because the $\tilde{Z}_2\tilde{Z}_1Z$ coupling is small. To combine the complicated production and cascade possibilities systematically, all these channels have been incorporated in the ISAJET 7.0 Monte Carlo package called ISASUSY.[11]

These multibranch cascade decays lead to higher-multiplicity final states in which the LSPs $\tilde{Z}_1$ carry a much smaller share of the available energy, so $\not{p}_T$ is smaller and less distinctive, making detection via $\not{p}_T$ more difficult. (Recall that leptonic W or Z decays, τ decays, plus semileptonic b and c decays, all give background events with genuine $\not{p}_T$; measurement uncertainties also contribute fake $\not{p}_T$ backgrounds.) Experimental bounds therefore become weaker when we take account of cascade decays. Figure 4 shows typical CDF 90% CL limits in the $(m_{\tilde{g}}, m_{\tilde{q}})$ plane; the dashed curves are limits assuming only direct decays (8)–(9), while solid curves are less restrictive limits including cascade decays (10)–(14).

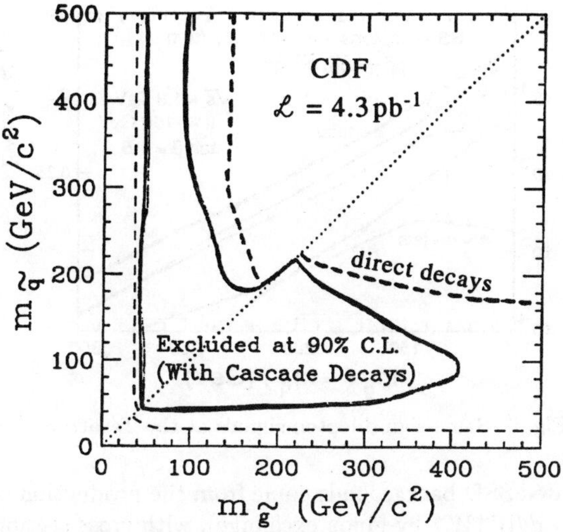

Fig. 4. 1992 CDF limits in the $(m_{\tilde{g}}, m_{\tilde{q}})$ plane, with or without cascade decays, for a typical choice of parameters.[8,9]

Cascade decays also present new opportunities for SUSY detection. Same-sign dileptons (SSD) are a very interesting signal,[12] which arises naturally from $\tilde{g}\tilde{g}$ and $\tilde{g}\tilde{q}$ decays because of the Majorana character of gluinos, with very little background. Figure 5 gives an example. Eqs. (10)–(14) show how a heavy gluino or squark can decay to a chargino $\tilde{W}_j$ and hence, via a real or virtual W, to an isolated charged lepton. If a gluino is present it can decay equally into either sign of chargino and lepton because it is a Majorana fermion. Hence $\tilde{g}\tilde{g}$ or $\tilde{g}\tilde{q}$ systems can decay to isolated SSD plus jets plus $\not{p}_T$. Cascade decays of $\tilde{q}\tilde{q}$ via the heavier neutralinos $\tilde{Z}_i$ offer similar possibilites for SSD, since the $\tilde{Z}_i$ are also Majorana fermions. Cross sections for the Tevatron are illustrated in Fig. 6.

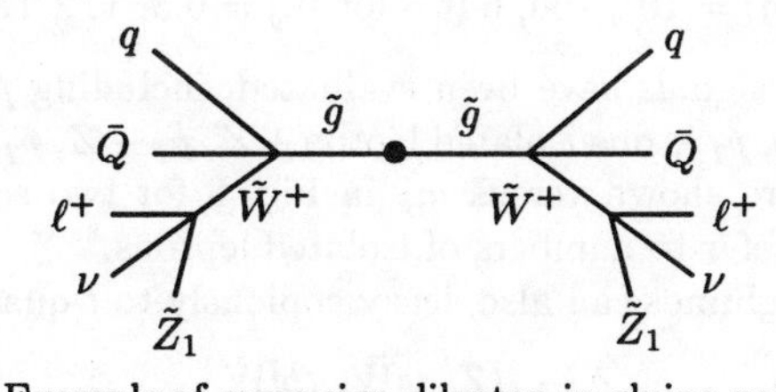

Fig. 5. Example of same-sign dilepton in gluino-pair decay.

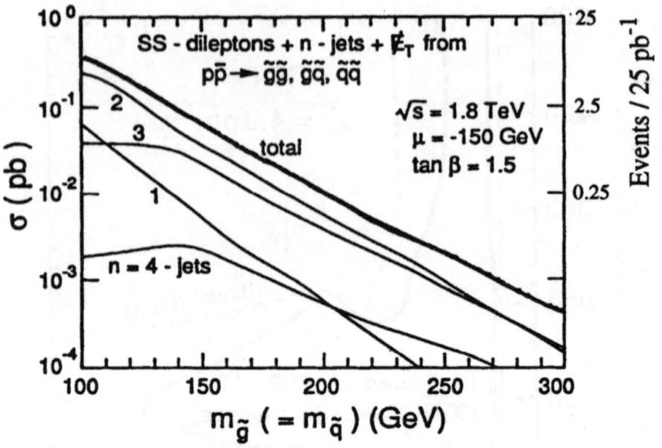

Fig. 6. Same-sign dilepton signals at the Tevatron.[13]

Genuinely isolated SSD backgrounds come from the production of WZ or $Wt\bar{t}$ or W^+W^+ (e.g. $uu \to ddW^+W^+$ by gluon exchange), with cross sections of order α_2^2 or $\alpha_2\alpha_3^2$ or $\alpha_2^2\alpha_3^2$ compared to α_3^2 for gluino pair production, so we expect to control them with suitable cuts. Very large $b\bar{b}$ production gives SSD via semileptonic b-decays plus B-$\bar{B}$ mixing, and also via combined $b \to c \to s\,\ell^+\nu$ and $\bar{b} \to \bar{c}\,\ell^+\nu$ decays, but both leptons are produced in jets and can be suppressed by stringent isolation criteria. Also $t\bar{t}$ gives SSD via $t \to b\,\ell^+\nu$ and $\bar{t} \to \bar{b} \to \bar{c}\,\ell^+\nu$, but the latter lepton is non-isolated. So SSD provide a promising SUSY signature.

The Tevatron can also search for trileptons,[14,15] arising for example from $W^* \to \tilde{W}\tilde{Z}_2 \to (\ell\nu\tilde{Z}_1)(\ell\bar{\ell}\tilde{Z}_1)$. Figure 7 shows predicted trilepton rates at the Tevatron for both the minimal SU(5) and no-scale flipped SU(5) supergravity models;[15] the flipped SU(5) model tends to give the bigger values.

Gluino production rates at SSC/LHC are much higher than at the Tevatron. At $\sqrt{s} = 40$ TeV, the cross section is

$$\sigma(\tilde{g}\tilde{g}) = 10^4,\ 700,\ 6\ \text{fb} \quad \text{for } m_{\tilde{g}} = 0.3,\ 1,\ 2\ \text{TeV}\ . \tag{15}$$

Many different SUSY signals have been evaluated, including $\not{p}_T + n\,$jets, $\not{p}_T +$ SSD, $\not{p}_T + n\,$isolated leptons, $\not{p}_T +$ one isolated lepton $+ Z$, $\not{p}_T + Z$, $\not{p}_T + Z + Z$. SSC signals from $\tilde{g}\tilde{g}$ production are shown versus $m_{\tilde{g}}$ in Fig. 8 for two scenarios, after various cuts; the labels 3,4,5 refer to numbers of isolated leptons.[8]

Sufficiently heavy gluinos can also decay copiously to t-quarks[8,16]:

$$\tilde{g} \to t\bar{t}\tilde{Z}_i,\ t\bar{b}\tilde{W}^-,\ b\bar{t}\tilde{W}^+\ . \tag{16}$$

Then $t \to bW$ decay leads to multiple W production. For a gluino of mass 1.5 TeV, the $\tilde{g} \to W$, WW, WWZ, $WWWW$ branching fractions are typically of order 30%, 30%,

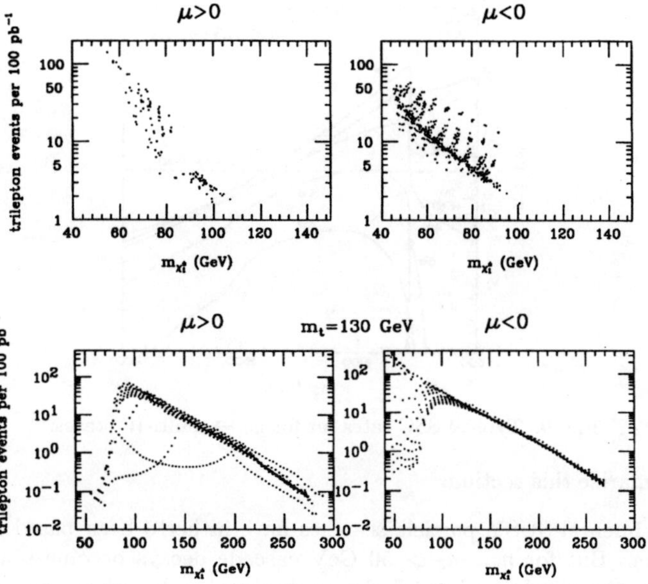

Fig. 7. Predicted Tevatron trilepton events per 100 pb^{-1} for minimal SU(5) and no-scale flipped SU(5) supergravity solutions (each represented by a point).[15]

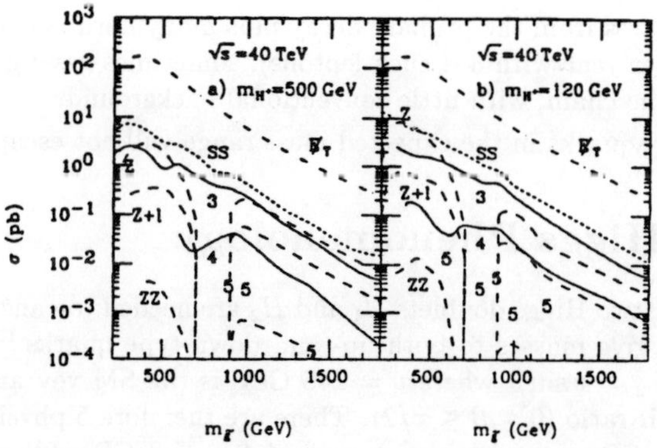

Fig. 8. SSC cross sections for various SUSY signals, after cuts.[8]

6%, 6%, respectively. Figure 9 illustrates SSC cross sections for multi-W production via gluino pair decays (assuming $m_{\tilde{g}} < m_{\tilde{q}}$). We see that for $m_{\tilde{g}} \sim 1$ TeV the SUSY rate for $4W$ production can greatly exceed the dominant SM $4t \to 4W$ mode, offering yet another signal for SUSY.[16]

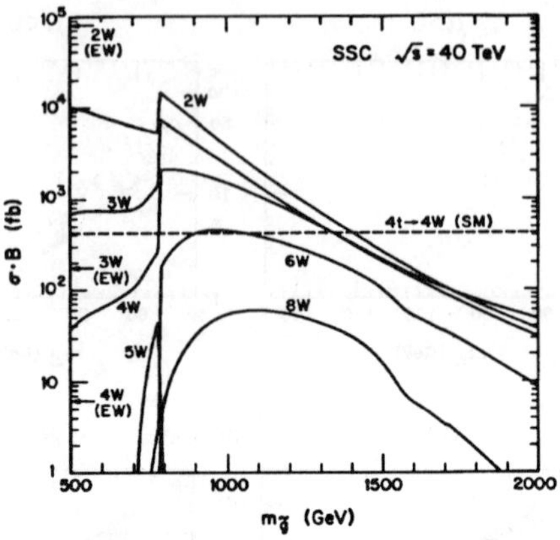

Fig. 9. Typical SSC rates for for $\tilde{g}\tilde{g} \to$ multi-W states.[16]

To summarize this section:

(a) Experimental SUSY particle searches have hitherto been based largely on $\not{p}_T$ signals. But for $m_{\tilde{g}}$, $m_{\tilde{q}} > 50$ GeV cascade decays become important; they weaken the simple $\not{p}_T$ signals but provide new signals such as same-sign dileptons and trileptons, which will be pursued at the Tevatron.

(b) For even heavier squarks and gluinos, the cascade decays dominate completely and provide further exotic (multi-W, Z and multi-lepton) signatures, which will be pursued at the SSC and LHC. They would find spectacular events, containing several hard jets from the primary decay plus many hard isolated leptons (10^3–10^4 events per year with 3–4 such leptons), sometimes having multiple t and b hadrons in the chain, with little conventional background.

(c) Gluinos and squarks in the expected mass range will not escape detection.

3 SUSY Higgs Phenomenology

In minimal SUSY, two Higgs doublets H_1 and H_2 are needed to cancel anomalies and at the same time give masses to both up- and down-type quarks.[17] Their vevs are $v_1 = v\cos\beta$ and $v_2 = v\sin\beta$ where $v = 246$ GeV is the SM vev and $\tan\beta = v_2/v_1$ parameterizes their ratio ($0 \leq \beta \leq \pi/2$). There are therefore 5 physical scalar states: h and H (neutral CP-even with $m_h < m_H$), A (neutral CP-odd) and $H^\pm$. At tree level the scalar masses and couplings and an h-H mixing angle α are all determined by two parameters, conveniently chosen to be m_A and $\tan\beta$. At tree level the masses obey $m_h \leq M_Z, m_A; m_H \geq M_Z, m_A; m_{H^\pm} \geq M_W, m_A$.

Radiative corrections can be big, however.[18] The most important new parameters entering here are the t and $\tilde{t}$ masses; we neglect for simplicity some other parameters related to squark mixing. One-loop corrections give h and H mass shifts of order $\delta m^2 \sim G_F m_t^4 \ln(m_{\tilde{t}}/m_t)$, arising from incomplete cancellation of t and $\tilde{t}$ loops. The h and H mass bounds get shifted up and for the typical case $m_t = 150$ GeV, $m_{\tilde{t}} = 1$ TeV (which we usually assume) we get

$$m_h \lesssim 115\,\mathrm{GeV} \lesssim m_H\,. \tag{17}$$

There are also corrections to cubic hAA, HAA, Hhh couplings, to h-H mixing, and smaller corrections to the $H^{\pm}$ mass. Figure 10 illustrates the dependence of m_h and m_H on m_A and $\tan\beta$.

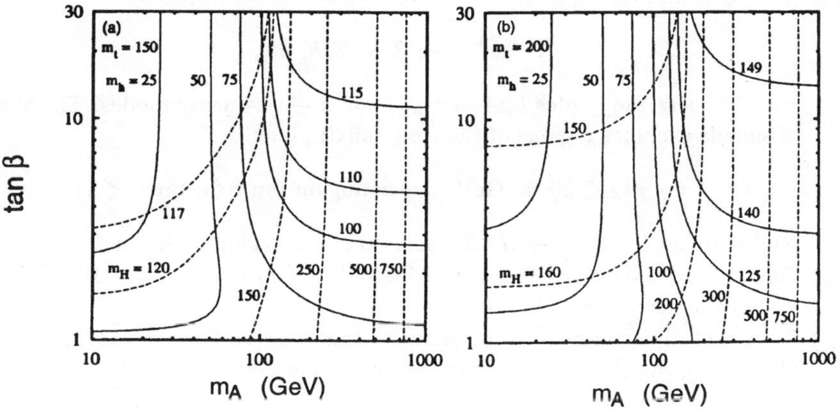

Fig. 10. h and H mass contours in the $(m_A, \tan\beta)$ plane for (a) $m_t = 150$ GeV and (b) $m_t = 200$ GeV, with $m_{\tilde{t}} = 1$ TeV.

In the next-to-minimal SUSY model (NMSSM), one Higgs singlet is added[17] giving 7 bosons $h^0, H_1^0, H_2^0, A_1^0, A_2^0, H^{\pm}$, with 6 parameters at tree level (because there are now more terms in the general scalar potential). This leads to a richer mass spectrum with looser limits. It is instructive to see what happens to the upper bound on the lightest scalar mass m_h, as constraints on the Higgs sector, the gauge sector and fermion sector are progressively relaxed while requiring perturbativity up to scale $\Lambda = 10^{16}$ GeV and a SUSY scale of order $M_{\mathrm{SUSY}} \sim 1$ TeV[19]:

within MSSM:	$m_h \leq M_Z$	[tree level]
	$m_h \lesssim 115\,\mathrm{GeV}$	[rad.corr., $m_t = 150$ GeV]
	$m_h \lesssim 130\,\mathrm{GeV}$	[rad.corr., any m_t]
beyond MSSM:	$m_h \lesssim 145\,\mathrm{GeV}$	[arbitrary gauge singlets]
	$m_h \lesssim 155\,\mathrm{GeV}$	[arbitrary Higgs sector]
	$m_h \lesssim 180\,\mathrm{GeV}$	[extra matter multiplets]

What are the constraints on the principal parameters m_A and $\tan\beta$?

(a) Perturbativity. This condition is a bit subjective. Requiring the $tbH^\pm$ couplings to be less than $g_3(M_Z) \simeq 1.2$ gives $m_t/(500\,\text{GeV}) \lesssim \tan\beta \lesssim (500\,\text{GeV})/m_b$.[20] Requiring instead (two-loop)/(one-loop) $< 1/4$ in the renormalization group equations (RGE) gives $\tan\beta \lesssim 65$.[3]

(b) Proton decay. Some models require $\tan\beta < 3$–15,[4] but a more conservative limit is $\tan\beta < 85$.[5]

(c) SUSY-GUT models[1–3,15,21] find solutions in the range $0.6 \lesssim \tan\beta \lesssim 65$ (with the upper bound from perturbativity).

(d) At LEP I, all four experiments[22] have searched for the processes

$$e^+e^- \to Z \to Z^*h, Ah\,, \tag{18}$$

with $Z^* \to \ell\ell, \nu\nu, jj$ plus $h, A \to \tau\tau, jj$ and $h \to AA$ decay modes. The absence of signals excludes regions of the $(m_A, \tan\beta)$ plane

$$m_h, m_A \gtrsim 20\text{–}45\ \text{GeV (depending on } \tan\beta \text{ for } \tan\beta \gtrsim 1)\,. \tag{19}$$

Null searches for $e^+e^- \to H^+H^-$ also exclude a region with $\tan\beta < 1$.[23]

(e) Global electroweak analysis with MSSM[24] prefers $m_A <$ a few 100 GeV, but without much statistical weight.

(f) To cancel divergences and to remain within the GUT scale, $m_A \lesssim 1$–2 GeV. Combining (a)–(f), there is a general consensus that

$$0.6 \lesssim \tan\beta \lesssim 65\,, \tag{20}$$

$$20\,\text{GeV} \lesssim m_A \lesssim 1\text{–}2\ \text{TeV}\,. \tag{21}$$

There is also indirect evidence from various quarters:

(g) The measured $B \to \tau\nu X$ branching fraction[25] is not far from the SM prediction and constrains possible virtual $H^\pm$ contributions, giving[26] $\tan\beta < 0.54 m_{H^\pm}/(1\ \text{GeV})$; this is weaker than Eqs. (20)–(21) however.

(h) The experimental bound[27] $B(b \to s\gamma) < 5.4 \times 10^{-4}$ is close to the SM prediction, based on $W + t$ loop calculations, setting stringent limits on non-SM contributions. If just $H^\pm + t$ loops are added, an important part of the $(m_A, \tan\beta)$ plane is excluded.[28] However, if spartners are relatively light, additional $\tilde{t} + \tilde{W}_i$ loops may contribute significantly with either sign.[29] We need more information about the SUSY spectrum to exploit this bound.

Future direct searches for MSSM Higgs bosons rely on future colliders (apart from marginal improvements at LEP I as higher luminosity accumulates). Several groups[30–33] have studied the prospects for discovering the different bosons at LEP II and SSC/LHC; their results broadly agree.

LEP II will have higher energy $\sqrt{s} \lesssim 200$ GeV, and should be able to discover h through the range $m_h \lesssim 90$ GeV, but will not be able to cover the full range of possible m_h. A might be light enough to discover, but H and $H^{\pm}$ are likely to be too heavy. The parameter sector where both $m_A \gtrsim 90$ GeV and $\tan\beta \gtrsim 5$ will not be accessible — i.e. will give no LEP II Higgs signals for $m_t \sim 150$ GeV.

Searches for neutral scalars at SSC and LHC will primarily be analogous to SM Higgs searches (see Figs. 11–13):

(i) untagged $\gamma\gamma$ signals from $pp \to (h, H, A) \to \gamma\gamma$ via top quark loops;
(ii) lepton-tagged $\gamma\gamma$ signals from $pp \to (h, H, A) \to \gamma\gamma$ plus associated $t\bar{t}$ or W, with leptons from $t \to W \to \ell\nu$ or $W \to \ell\nu$ decays;
(iii) "gold-plated" four-lepton signals from $pp \to (h, H) \to ZZ$ or $Z^*Z \to \ell^+\ell^+\ell^-\ell^-$.

Though qualitatively similar to SM signals, these will generally be smaller due to the different couplings that depend on β and α.

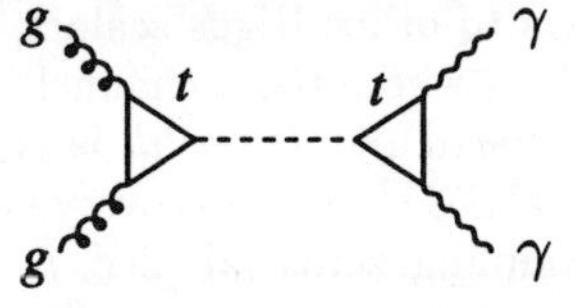

Fig. 11. Typical diagram for untagged Higgs → $\gamma\gamma$ signals.

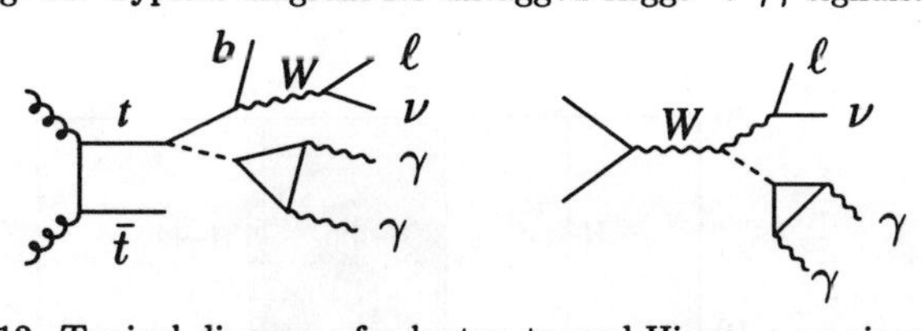

Fig. 12. Typical diagrams for lepton-tagged Higgs → $\gamma\gamma$ signals.

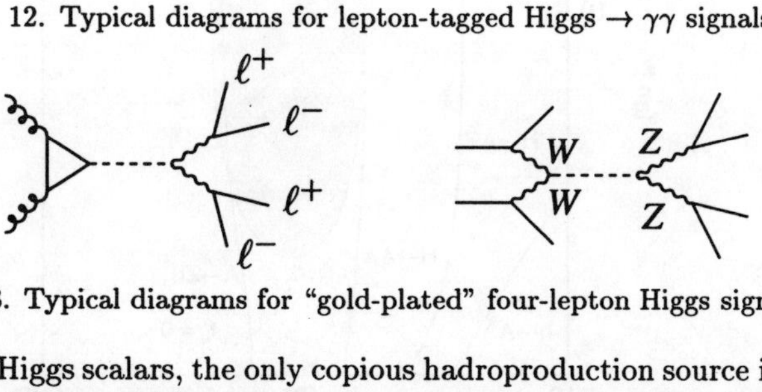

Fig. 13. Typical diagrams for "gold-plated" four-lepton Higgs signals.

For charged Higgs scalars, the only copious hadroproduction source is top production with $t \to bH^+$ decay (that requires $m_{H^{\pm}} < m_t - m_b$). The subsequent $H^+ \to c\bar{s}, \nu\bar{\tau}$ decays are most readily detected in the $\tau\nu$ channel (favored for $\tan\beta > 1$), with $\tau \to \pi\nu$ decay (Fig. 14).

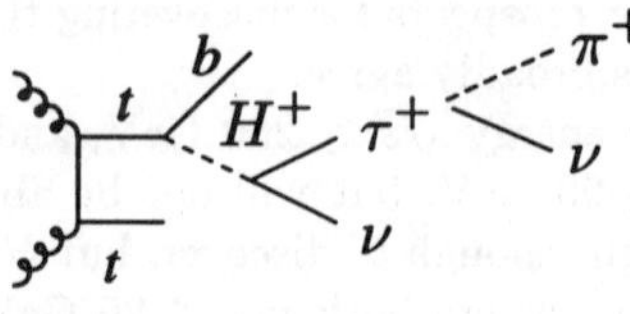

Fig. 14. Typical diagram for τ signals from top decay via charged-Higgs modes.

SM t-decays give equal probabilities for e, ν, τ leptons via $t \to bW \to b(e, \mu, \tau)\nu$, but the non-standard $t \to bH^+ \to b\tau\nu$ leads to characteristic excess of τ. The strategy is to tag one top quark via standard $t \to bW \to b\ell\nu$ decay and to study the τ/ℓ ratio in the associated top quark decay ($\ell = e$ or μ).

New non-SM complications must now be taken into account, however, especially for the neutral MSSM Higgs boson signals. New decay channels open in addition to the main SM modes $h, H \to WW, ZZ$ and $h, H, A \to t\bar{t}, b\bar{b}, c\bar{c}, \tau\bar{\tau}$ (and also $h, H, A \to \gamma\gamma$ via loops).

(a) Firstly there are decays to other Higgs scalars, $h \to AA$, $A \to Zh$, $H \to hh$ or AA or AZ; the regions where these channels are open are shown in Fig. 15, for $m_t = 150$ GeV. In particular $H \to hh$ is allowed and generally dominates (suppressing all the usual SM signatures) everywhere except in the shaded region or near the line of coupling zeros labelled $f_h = 0$. We already mentioned $h \to AA$ in passing; it changes the h signals in a small region of the LEP I searches.

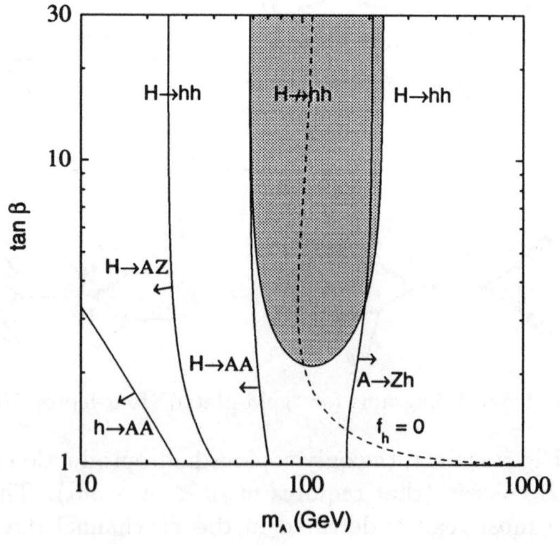

Fig. 15. Allowed regions for decays to other Higgses.[33]

(b) **Secondly, there are decays to SUSY particles, especially the invisible mode $h \to \tilde{Z}_1\tilde{Z}_1$; these have usually been ignored but could be significant.[34] They could give new signals and they could suppress old signals. Figure 16 shows contours of $B(h \to \text{SUSY}$, mostly $\tilde{Z}_1\tilde{Z}_1)$ versus μ and $\tan\beta$, the region excluded by LEP I SUSY searches being shaded (see Section 2); there are small regions where SUSY modes could rise to 70%–90%, severely suppressing standard h signals, but they are within the reach of LEP II SUSY searches (dashed lines) so we shall know. Here we shall mostly ignore SUSY modes, arguing that they may dilute other Higgs signals but will probably not efface them.**

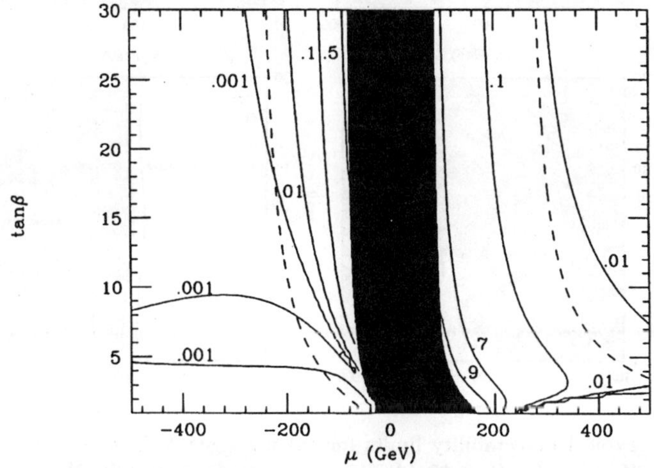

Fig. 16. Contours of $B(H \to \text{SUSY modes})$ versus μ and $\tan\beta$.[34]

Returning to MSSM Higgs searches at SSC/LHC,[30–33] Figs. 17(a) and 17(b) show typical limits of detectability for untagged and lepton-tagged $\gamma\gamma$ signals at SSC, assuming luminosities $\mathcal{L} = 20\,\text{fb}^{-1}$ (two years of running) and $m_t = 150$ GeV. Figure 17(c) shows a similar limit for the $H \to 4\ell$ search (no $h \to 4\ell$ signal is detectable). Figure 17(d) shows typical limits for detecting the $t \to H^+ \to$ excess τ signal; here the value of m_t is critical, since only the range $m_{H^+} < m_t - m_b$ can contribute at all. Putting all these discovery regions together with the LEP I and LEP II regions, we see that very considerable coverage of the $(m_A, \tan\beta)$ plane can be expected — but there still remains a small inaccessible region; see Fig. 18. For $m_t = 120$ GeV the inaccessible region is larger, for $m_t = 200$ GeV it is smaller.

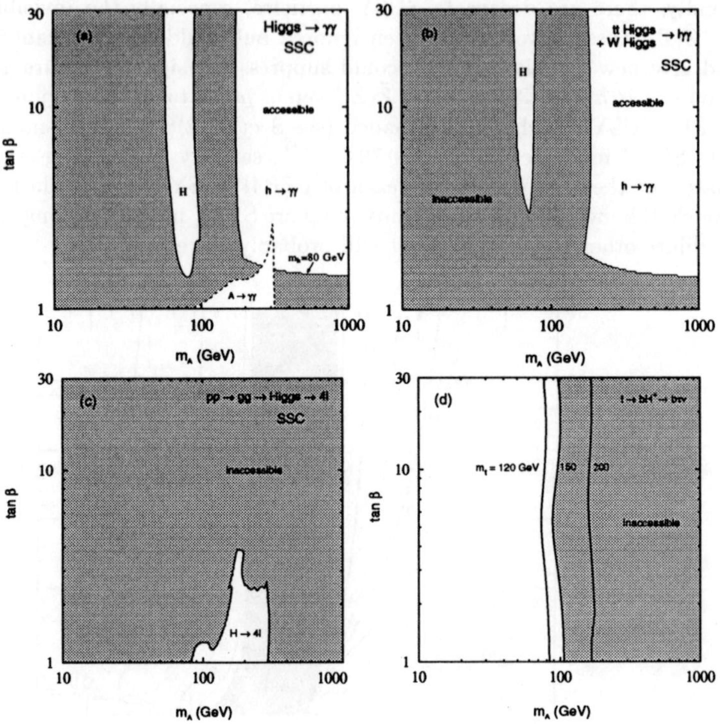

Fig. 17. Typical detectability limits for (a) untagged $h, H, A \to \gamma\gamma$, (b) lepton-tagged $h, H \to \gamma\gamma$, (c) $H \to 4\ell$, (d) $H^{\pm} \to \tau\nu$ signals at the SSC.[33]

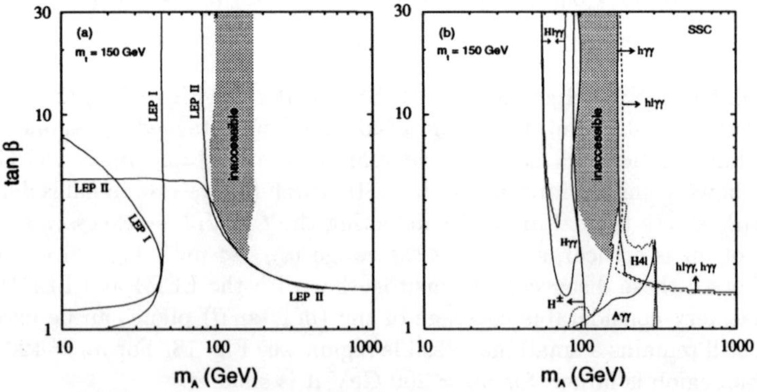

Fig. 18. Combined LEP and SSC discovery regions for $m_t = 150$ GeV from Ref. 33; similar results are obtained by other groups.[30–32]

Emphasizing first the positive side, Fig. 19 shows how many of the MSSM scalars $h, H, A, H^{\pm}$ would be detectable, in various regions of the $(m_A, \tan\beta)$ plane. In many regions two or more different scalars could be discovered, but for large m_A only h would be discoverable; in the latter region, the h couplings all reduce to SM couplings, the other scalars become very heavy and approximately degenerate, and the MSSM essentially behaves like the SM.

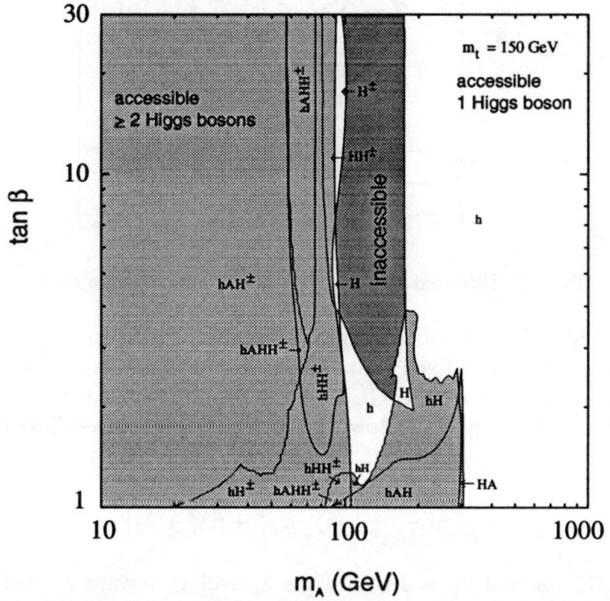

Fig. 19 How many MSSM Higgs bosons may be discovered (from Ref. 33).

Turning now to the negative side of Fig. 18, the inaccessible region, Fig. 20 shows that different groups broadly agree on the boundary. This region corresponds to intermediate masses for all Higgses: $m_h \sim 80$–115 GeV, $m_A \sim 100$–160 GeV, $m_H \sim$ 120–160 GeV, $m_{H^{\pm}} \sim 120$–160 GeV. Why are none detectable? What went wrong? It seems we are just unlucky here: the htt and hWW couplings that control $h \to \gamma\gamma$ become weaker; the Att coupling that controls both A production and two-photon decay gets weak; $H \to hh$ competition suppresses the H signals; $H^{\pm}$ is too heavy and $t \to bH$ is suppressed. Next, is there any indirect way to exclude this inaccessible region? One possibility is to understand and exploit the $B(b \to s\gamma)$ bound better: see above. We note also that if $m_t = 200$ GeV (instead of 150 GeV as in Fig. 18), the region is much smaller. Another possibility is to derive stronger parameter constraints from SUSY-GUT models, as we now discuss.

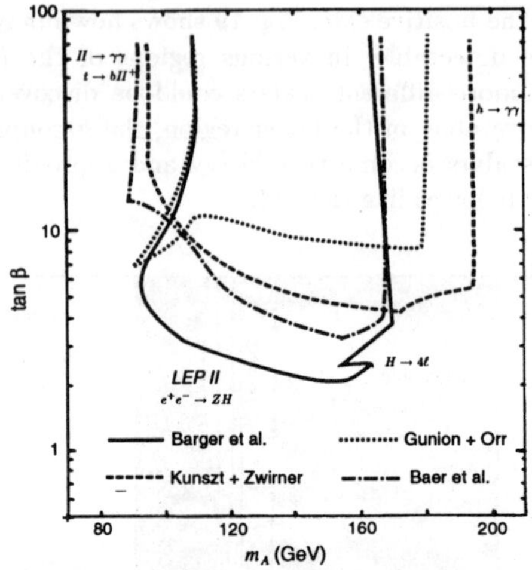

Fig. 20. Different groups agree about inaccessible region.

In minimal SUSY-GUT models with $M_{\rm SUSY} \lesssim 1$ TeV, the RGE have a solution dominated by the infrared fixed point of the top Yukawa coupling λ_t.[35] For large m_t, λ_t is plausibly large at the GUT scale M_G, in which case it evolves rapidly toward an infrared fixed point at low mass scales,[3,21,35–38] according to the one-loop renormalization group equation

$$\frac{d\lambda_t}{dt} = \frac{\lambda_t}{16\pi^2}\left[-\sum c_i g_i^2 + 6\lambda_t^2 + \lambda_b^2\right] , \tag{22}$$

with $c_1 = 13/15$, $c_2 = 3$, $c_3 = 16/3$. Thus λ_t evolves toward a fixed point close to where the quantity in square brackets in Eq. (22) vanishes. Then the known gauge couplings determine the running mass $m_t(m_t) = \lambda_t(m_t) v \sin\beta/\sqrt{2}$ and hence the pole mass $m_t(\text{pole}) = m_t(m_t)\left[1 + \frac{4}{3\pi}\alpha_s(m_t)\right]$; two-loop evaluations[3] give

$$m_t(\text{pole}) \simeq (200\,\text{GeV}) \sin\beta\,, \tag{23}$$

If $\lambda_t(M_G)$ is below the fixed point, its convergence to the fixed point is more gradual and Eq. (23) does not necessarily apply. But in practice large $\lambda_t(M_G)$ is favored in many SUSY-GUT solutions; large $\lambda_t(M_G)$ facilitates $\lambda_b(M_G) = \lambda_\tau(M_G)$ Yukawa unification[39] and allows intricate relationships between fermion masses and mixings. It is therefore interesting to pursue the phenomenological implication of Eq. (23).[40]

Figure 21 shows how $\lambda_t(M_G)$ and $\lambda_b(M_G)$ values relate to m_t(pole) and $\tan\beta$ in SUSY-GUT solutions with Yukawa unification; the lower (upper) shaded branches contains the m_t (m_b) fixed-point solutions. There is a small region at the upper right where both fixed point solutions are simultaneously satisfied. Figure 22 shows that the m_t fixed-point behavior is insensitive to GUT threshold corrections $\lesssim 10\%$ in the λ_b/λ_τ ratio. The sensitivity of the fixed point to threshold corrections is decreased for larger values of $\alpha_s(M_Z)$ where the solutions tend to have a stronger fixed point character, as indicated by Eq. (22). The perturbative limits of the Yukawa couplings near their Landau poles are shown in Fig. 22(a) as the dashed lines $\lambda_t^G = 3.3$ and $\lambda_b^G = 3.1$.

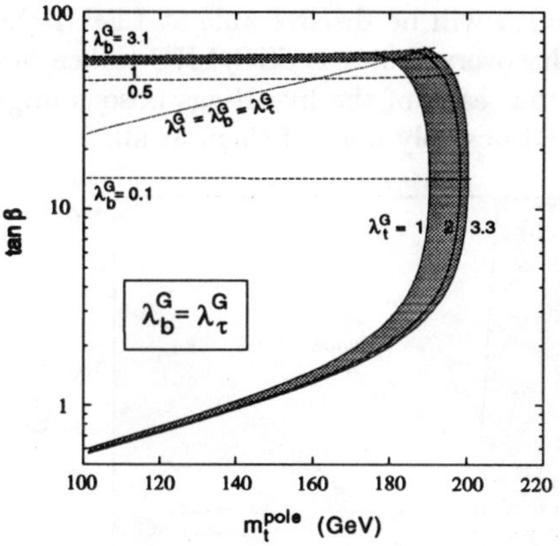

Fig. 21: Contours of constant Yukawa couplings $\lambda_i^G = \lambda_i(M_G)$ at the GUT scale in the $(m_t^{\rm pole}, \tan\beta)$ plane, obtained from solutions to the RGE with $\lambda_\tau^G = \lambda_b^G$ unification imposed. The GUT scale Yukawa coupling contours are close together for large λ^G. The fixed points describe the values of the Yukawa couplings at the electroweak scale for $\lambda_t^G \gtrsim 1$ and $\lambda_b^G \gtrsim 1$.

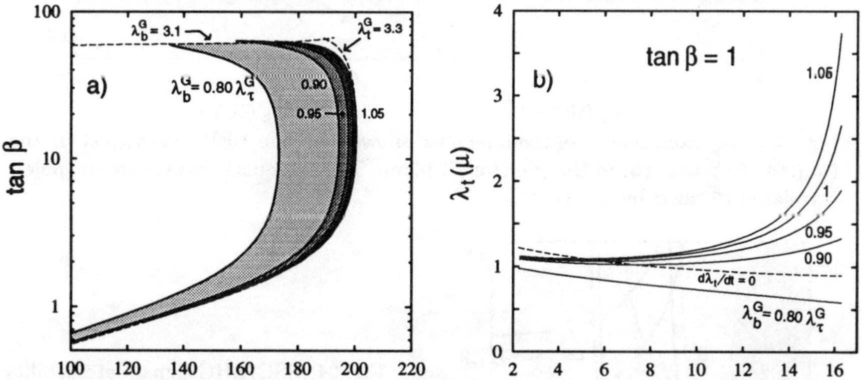

Fig. 22: RGE results for $\alpha_s(M_Z) = 0.118$ with the boundary condition $m_b(m_b) = 4.25$ GeV. (a) GUT threshold corrections to Yukawa coupling unification. The solutions strongly exhibit a fixed point nature, for threshold corrections $\lesssim 10\%$. Taking a larger supersymmetric threshold M_{SUSY} or increasing $\alpha_s(M_Z)$ moves the curves to the right, so that the fixed point condition becomes stronger. (b) Evolution of the top quark Yukawa coupling for $\tan\beta = 1$. The dashed line indicates $\frac{d\lambda_t}{dt} = 0$ which gives an approximation to the electroweak scale value of m_t with accuracy of order 10%.

There are immediate implications for phenomenology.[40] If $m_t \lesssim 160$ GeV, Eq. (23) constrains $\tan\beta$ to values near 1, where h is relatively light (recall the tree-level relation $m_h < M_Z|\cos 2\beta|$) and the couplings of h are close to those of a Standard Model Higgs boson. LEP Higgs searches[22,41] exclude a region of the $(m_A, \tan\beta)$ plane shown in Fig. 23(a): this translates to forbidden regions in $(m_h, \tan\beta)$ in Fig. 23(b). We see that the fixed-point condition predicts $m_t \gtrsim 130$ GeV, $m_h \gtrsim 60$ GeV, $m_A \gtrsim 70$ GeV; correspondingly $m_{H^\pm} \gtrsim 105$ GeV, $m_H \gtrsim 140$ GeV. If in fact $m_t \lesssim 160$ GeV, then $m_h \lesssim 85$ GeV as shown in Fig. 23, and h will be discoverable at LEP II (but none of the other Higgs bosons will). The discovery limits at SSC/LHC (taken here from Ref. 33) are shown in Fig. 24; we see that each of the five Higgs bosons might be discoverable there, but not all at once, and possibly none of them at all.

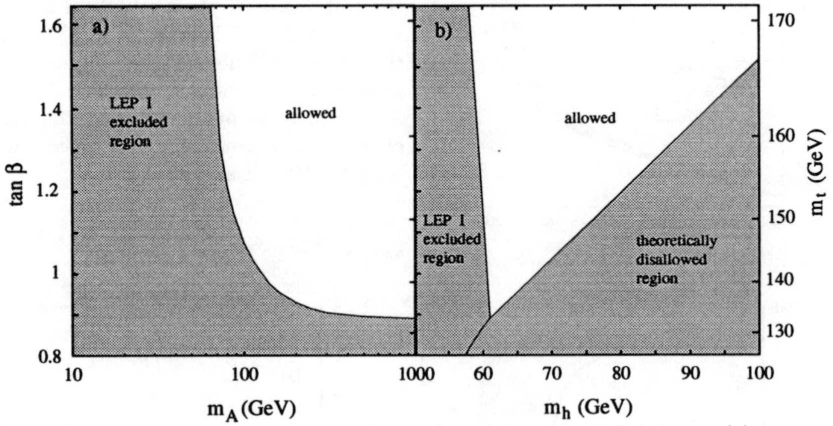

Fig. 23: m_t fixed-point solution regions allowed by the LEP I data: (a) in the $(m_A, \tan\beta)$ plane, (b) in the $(m_h, \tan\beta)$ plane. The top quark masses are m_t(pole), correlated to $\tan\beta$ by Eq. (23).

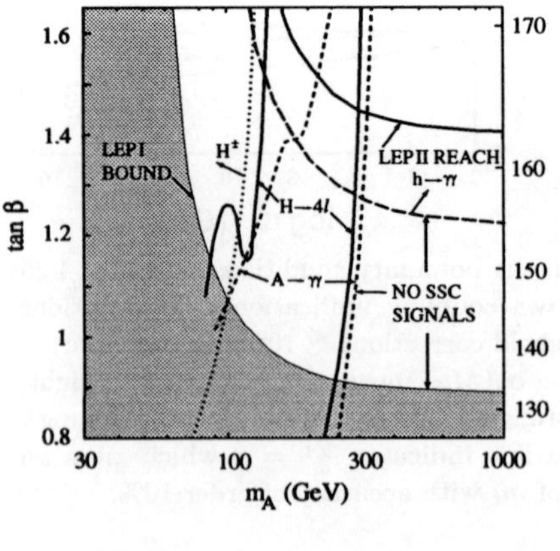

Fig. 24: SSC/LHC signal detectability regions, compared with the LEP I allowed region of m_t fixed-point solutions from Fig. 3(a) and the probable reach of LEP II. The top quark masses are m_t(pole).

Finally we may ask what a future e^+e^- collider could do. We have seen that part of the MSSM parameter space is inaccessible to e^+e^- collisions at $\sqrt{s} = 200$ GeV, $\mathcal{L} = 500\,\mathrm{pb}^{-1}$, for $m_t = 150$ GeV and $m_{\tilde{t}} = 1$ TeV. But a possible future linear collider with higher energy and luminosity could in principle cover the full parameter space. In is interesting to know what are the minimum s and $\mathcal{L}$ requirements for complete coverage, for given m_t. This question was answered in Ref. 42, based on the conservative assumption that only the channels $e^+e^- \to (Zh, Ah, ZH, AH) \to \tau\tau jj$ would be searched, with no special tagging. The results are shown in Fig. 25. We have estimated that including all $Z \to \ell\ell, \nu\nu, jj$ and $h, H, A \to bb, \tau\tau$ decay channels plus efficient b-tagging could increase the net signal S by a factor 6 and the net background B by a factor 4, approximately; this would increase the statistical significance $S/\sqrt{B}$ by a factor 3 and hence reduce the luminosity requirement by a factor 9 or so. In this optimistic scenario, the luminosity scale in Fig. 25 would be reduced by an order of magnitude.

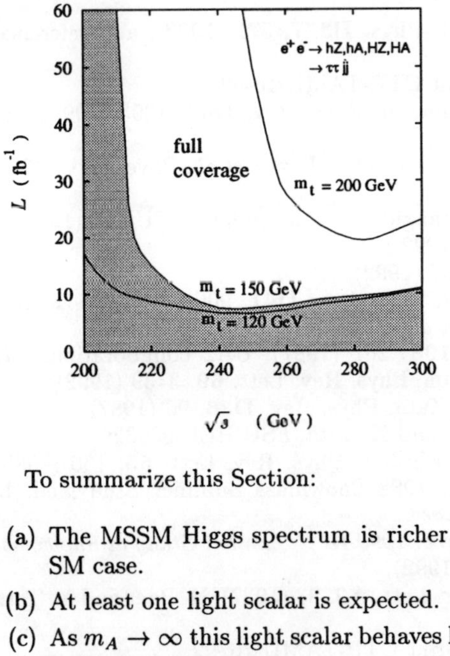

Fig. 25. Minimal requirements for a "no-lose" MSSM Higgs search at a future e^+e^- collider based on $\tau\tau jj$ signals alone. Curves of minimal $(\sqrt{s}, \mathcal{L})$ pairings are shown for $m_t = 120$, 150, 200 GeV; the no-lose region for $m_t = 150$ GeV is unshaded.[42]

To summarize this Section:

(a) The MSSM Higgs spectrum is richer but in some ways more elusive than the SM case.

(b) At least one light scalar is expected.

(c) As $m_A \to \infty$ this light scalar behaves like the SM scalar while the others become heavy and degenerate.

(d) LEP I, LEP II and SSC/LHC will give extensive but not quite complete coverage of the MSSM parameter space.

(e) For some parameter regions, several different scalars are detectable, but usually one or more remain undetectable.

(f) The $b \to s\gamma$ bound has the potential to exclude large areas of parameter space (possibly including the inaccessible region) but is presently subject to some uncertainty.

(g) A higher-energy e^+e^- collider could cover the whole MSSM parameter space, discovering at least the lightest scalar h.

(h) m_t fixed-point solutions in SUSY-GUT models are theoretically attractive and also strongly constrain the phenomenology; they predict $m_t \gtrsim 130$ GeV; if $m_t \lesssim 160$ GeV, they exclude the inaccessible region and predict that h will be discovered at LEP.

Acknowledgments: This work was supported in part by the University of Wisconsin Research Committee with funds granted by the Wisconsin Alumni Research Foundation, in part by the U.S. Department of Energy under contract no. DE-AC02-76ER00881, and in part by the Texas National Laboratory Research Commission under grant no. RGFY93-221.

References

1. G.G. Ross and R.G. Roberts, Nucl. Phys. **B377**, 571 (1992), and references therein.
2. S. Kelley *et al.*, Texas A&M preprint CTP-TAMU-16-92.
3. V. Barger, M.S. Berger, and P. Ohmann, Phys. Rev. **D47**, 1093 (1993), and unpublished calculations.
4. P. Nath and R. Arnowitt, NUB-TH-3056/92; J. Lopez *et al.*, Phys. Lett. **B299**, 262 (1993).
5. J. Hisaño, H. Murayama, and T. Yanagida, Tohoku preprint TU-400 (1992).
6. H. Baer and X. Tata, FSU-HEP-921222.
7. H. Baer et al., Phys. Rev. **D46**, 1067 (1992).
8. H. Baer, X. Tata, and J. Woodside, Phys. Rev. **D41**, 906 (1990); **D42**, 1568 (1991); **D45**, 142 (1992); H. Baer *et al.*, Phys. Rev. **D46**, 303 (1992).
9. UA1 collaboration, Phys. Lett. **B198**, 261 (1987); UA2 collaboration, *ibid.* **B235**, 363 (1990); CDF collaboration, Phys. Rev. Lett. **69**, 3439 (1992).
10. H. Baer, V. Barger, D. Karatas, X. Tata, Phys. Rev. **D36**, 96 (1987).
11. H. Baer, F. Paige, S. Protopopescu, and X. Tata, FSU-HEP-930329.
12. V. Barger, W.-Y. Keung, R. J. N. Phillips, Phys. Rev. Lett. **55**, 166 (1985); R. Barnett, J. Gunion, H. Haber, 1988 Snowmass Summer Study, ed. by S. Jensen, World Scientific (1989).
13. H. Baer *et al.*, FSU-HEP-901110, *Proc. 1990 DPF Summer Study at Snowmass*, ed. by E. Berger, World Scientific (1992).
14. P. Nath and R. Arnowitt, Mod. Phys. Lett. **A2**, 33 (1987); H. Baer and X. Tata, Phys. Rev. **D47**, 2739 (1993).
15. J. L. Lopez et al., Texas A&M preprint CTP-TAMU-76/92.
16. V. Barger, R.J.N. Phillips and A.L. Stange, Phys. Rev. **D45**, 1484 (1992).
17. J. F. Gunion, H. E. Haber, G. L. Kane and S. Dawson, *The Higgs Hunter's Guide*, Addison-Wesley, Reading, MA (1990).
18. S. P. Li and M. Sher, Phys. Lett. **B140**, 339 (1984); J. F. Gunion and A. Turski, Phys. Rev. **D39**, 2701, **D40**, 2325, 2333 (1989); M. S. Berger, Phys. Rev. **D41**, 225 (1990); Y. Okada *et al.*, Prog. Theor. Phys. **85**, 1 (1991); Phys. Lett. **B262**,

54 (1991); H. Haber and R. Hempfling, Phys. Rev. Lett. **66**, 1815 (1991); J. Ellis *et al.*, Phys. Lett. **B257**, 83 (1991); R. Barbieri *et al.*, Phys. Lett. **B258**, 167 (1991); J. Lopez and D. V. Nanopoulos, Phys. Lett. **B266**, 397 (1991); A. Yamada, Phys. Lett. **B263**, 233 (1991); M. Drees and M. N. Nojiri, Phys. Rev. **D45**, 2482 (1002); R. Hempfling, SCIPP-91/39; M. A. Diaz and H. E. Haber, Phys. Rev. **D45**, 4246 (1992); D. M. Pierce, A. Papadopoulos and S. Johnson, Phys. Rev. Lett. **68**, 3678 (1992).

19. T. Moroi and Y. Okada, Phys. Lett. **B295**, 73 (1992); G. L. Kane, C. Kolda and J. D. Wells, Phys. Rev. Lett. **70**, 2686 (1993); W. T. A. ter Veldhuis, Purdue preprint PURD-TH-92-11; J. R. Espinosa and M. Quiros, Phys. Lett. **B302**, 51 (1992); T. Elliott, S. F. King and P. L. White, ibid. **B305**,71 (1993); U. Ellwanger, Phys. Lett **B303**, 271 (1993).
20. V. Barger, J. L. Hewett and R. J. N. Phillips, Phys. Rev. **D41**, 3421 (1990).
21. S. Dimopoulos, L. Hall and S. Raby, Phys. Rev. **D45** 4192; **D46** 4793 (1992).
22. ALEPH Collaboration, D. Decamp *et al.*, Phys. Lett. **B246**, 306 (1990); **B265**, 475 (1991); DELPHI Collaboration: P. Abreu *et al.*, Phys. Lett. **B245**, 276 (1990); Nucl. Phys. **B373**, 3 (1992); L3 Collaboration: B. Adeva *et al.*, Phys. Lett. **B251**, 311 (1990); **B283**, 454 (1992); OPAL Collaboration: M. Z. Akrawy *et al.*, Z. Phys. **C49**, 1 (1991).
23. M. A. Diaz and H. E. Haber, Phys. Rev. **D45**, 4246 (1992).
24. J. Ellis, J. L. Fogli and E. Lisi, CERN-TH.6643/92.
25. ALEPH collaboration, Phys. Lett. **B298** 479 (1993).
26. G.J Isidori, Phys. Lett. **B298** 409 (1993); J. L. Hewett, ANL-HEP-CP-92-125.
27. CLEO collaboration, report to Washington APS meeting April 1993.
28. J. L. Hewett, Phys. Rev. Lett. **70** 1045 (1993); V. Barger, M. S. Berger and R. J. N. Phillips, ibid. **70** 1368 (1993).
29. S. Bertolini et al., Nucl. Phys. **B353** 591 (1991); R. Barbieri and G. F. Giudice, CERN-TH.6830/1993.
30. H. Baer et al., Phys. Rev. **D46**, 1067 (1992).
31. J. Gunion et al., Phys. Rev. **D46**, 2040, 2052 (1992); **D47**, 1030 (1993).
32. Z. Kunszt and F. Zwirner, Nucl. Phys. **B46**, 4914 (1992).
33. V. Barger, K. Cheung, R. J. N. Phillips, and A. L. Stange, Phys. Rev. **D46**, 4914 (1992).
34. H. Baer et al., Phys. Rev. **D47** 1062 (1993).
35. B. Pendleton and G. G. Ross, Phys. Lett. **98B**, 291 (1981); C. T. Hill, Phys. Rev. **D24**, 691 (1981).
36. H. Arason, et al., Phys. Rev. Lett. **67**, 2933 (1991); Phys. Rev. **D47**, 232 (1993).
37. C. D. Froggatt, I. G. Knowles and R. G. Moorhouse, Phys. Lett. **B249**, 273 (1990); Phys. Lett. **B298**, 356 (1993).
38. M. Carena, S. Pokorski, and C. E. M. Wagner, Munich preprint MPI-Ph/93-10.
39. M. Chanowitz, J. Ellis and M. Gaillard, Nucl. Phys. **B128**, 506 (1977).
40. V. Barger et al., University of Wisconsin-Madison preprint MAD/PH/755.
41. T. Mori, report to Dallas Conference 1992; E. Gross and P. Yepes, Int. J. Mod. Phys. **A8**, 407 (1993).
42. V. Barger, K. Cheung, R. J. N. Phillips, and A. L. Stange, Phys. Rev. **D47** 3041 (1993).

SUPERSYMMETRIC HIGGS BOSONS AT THE LIMIT

STEPHEN F. KING
Physics Department, University of Southampton,
Southampton, SO9 5NH, U.K.

ABSTRACT

Using a combination of renormalisation group and effective potential methods, we discuss the bound on the lightest CP-even Higgs boson mass m_h in the next-to-minimal supersymmetric standard model. We find $m_h \leq 146, 139, 149\ GeV$ for $m_t = 90, 140, 190\ GeV$.

1. Introduction

1.1. *Triviality*

The question of the origin of electroweak mass is one of the most urgent questions of present day particle physics. The discovery of a particle which resembles the Higgs boson of the minimal standard model, and the measurement of its mass, will provide clues as to the nature of new physics beyond the standard model. In this report we shall be concerned with the question of how heavy the lightest neutral CP-even supersymmetric Higgs boson, h^0, can be within the framework of supersymmetric grand unified theories (SUSY GUTs) [1]. In SUSY GUTs all the Yukawa couplings are constrained to remain perturbative in the region $M_{SUSY} \sim 1$ TeV to $M_{GUT} \sim 10^{16}$ GeV. This constraint provides a maximum value at low energies for those Yukawa couplings which are not asymptotically-free, and is obtained from the renormalisation group (RG) equations together with the boundary conditions that the couplings become non-perturbative at M_{GUT} – the so-called "triviality limit". In the minimal supersymmetric standard model (MSSM) [2], the triviality limits provide a useful bound on the top quark mass m_t. The upper bound on the h^0 mass, m_h, in the MSSM, including radiative corrections, has recently been the subject of much discussion[3, 4, 5, 6, 7, 8, 9]. However the MSSM is not the most general low energy manifestation of SUSY GUTs. It is possible that SUSY GUTs give rise to a low energy theory which contains an additional gauge singlet field, the so called next-to-minimal supersymmetric standard model (NMSSM) [10, 11, 12]. Here we shall concentrate on the question of the upper-bound on m_h in the NMSSM which is obtained from triviality limits of Yukawa couplings.

1.2. *The NMSSM*

The NMSSM provides an elegant extension of MSSM which eliminates the μ–problem, the appearance of a mass–scale in the superpotential. The NMSSM differs from the MSSM by the presence of a gauge singlet field whose vacuum expectation value (vev) plays the role of the mass parameter μ in the MSSM, and is defined by the superpotential

$$W = h_t Q H_2 t^c + \lambda N H_1 H_2 - \frac{1}{3} k N^3 + \dots, \tag{1}$$

where the superfield $Q^T = (t_L, b_L)$ contains the left–handed top and bottom quarks, and t^c contains the charge conjugate of the right–handed top quark; H_1 and H_2 are the usual Higgs doublet superfields and N is the Higgs gauge singlet superfield; the ellipsis represents terms whose relatively small couplings play no role in our analysis (in particular, we assume that the bottom quark Yukawa coupling, h_b, satisfies $h_b \ll h_t$). The fields $H_1^T = (H_1^0, H_1^-)$, $H_2^T = (H_2^+, H_2^0)$ and N develop vevs which may be assumed to be of the form [7]

$$< H_1 >= \begin{pmatrix} \nu_1 \\ 0 \end{pmatrix}, \quad < H_2 >= \begin{pmatrix} 0 \\ \nu_2 \end{pmatrix}, \quad < N >= x, \tag{2}$$

where ν_1, ν_2 and x are real, and $\sqrt{\nu_1^2 + \nu_2^2} = \nu = 174$ GeV. The low energy physical spectrum of the Higgs scalars consists of 3 CP-even neutral states, 2 CP-odd neutral states, and 2 charged scalars. A third CP-odd state is a Goldstone mode which becomes the longitudinal component of the Z^0, while a further two charged scalars become those of the $W^{\pm}$'s.

An upper bound on the lightest neutral CP-even scalar h^0 in the NMSSM may be obtained from the real symmetric 3×3 neutral scalar mass squared matrix, by using the fact that the smallest eigenvalue of such a matrix must be smaller than the smallest eigenvalue of its upper 2×2 block. The resulting bound at tree-level is [12]

$$m_h{}^2 \leq M_Z{}^2 + (\lambda^2 v^2 - M_Z{}^2) \sin^2 2\beta. \tag{3}$$

where $\tan\beta \equiv \frac{v_2}{v_1}$, and λ is regarded as a running parameter evaluated at M_{SUSY}. The upper bound on m_h is determined by the maximum value of $\lambda(M_{SUSY})$, henceforth denoted λ_{max}. The value of λ_{max} is obtained by solving the SUSY RG equations for the Yukawa couplings h_t, λ and k in the region $M_{SUSY} = 1$ TeV to $M_{GUT} = 10^{16}$ GeV [13, 14]. If the Yukawa couplings h_t, λ and k are all initially large at M_{GUT} then they approach low energy fixed point ratios [13]. However, if the boundary condition at M_{GUT} is $\lambda \gg k$, then larger values of $\lambda(M_{SUSY})$ can be achieved. We have repeated the calculation of ref.[14] and found that for $h_t(M_{SUSY}) = 0.5 - 1.0$, $\lambda_{max} = 0.87 - 0.70$ and for $h_t(M_{SUSY}) \to 1.06$,* $\lambda_{max} \to 0$ (with $k = 0$ always). Radiative corrections to the tree-level bound in Eq.(2) have been considered in refs.[15, 16, 17]. In ref.[16] these were estimated from a low energy RG analysis of the Higgs sector of the model between M_{SUSY} and a lower scale μ, assuming that only one Higgs boson has a mass below M_{SUSY}.

In Section 2 [18] we shall discuss a more general RG analysis in which both Higgs doublets and the Higgs singlet may be lighter than M_{SUSY}, making a simple approximation of hard decoupling below $M_{SUSY} = 1 TeV$ of the superpartners. In Section 3 [19] we shall go on to consider the more general case in which the squarks

*$h_t(M_{SUSY}) \leq 1.06$ is the triviality bound, which, together with $m_t = h_t(m_t) v \sin\beta$, where $h_t(m_t) \leq 1.12$, implies the bound $m_t \leq 195$ GeV.

have a more general mass spectrum, using the effective potential approach. Our numerical results for the bound are also presented in this section. Section 4 contains our concluding remarks.

2. Renormalisation Group Approach [18]

The basic assumption of this approach is that the squarks are degenerate at $M_{SUSY} = 1$ TeV while the Higgs bosons and top quarks have masses $\mu \approx 150$ GeV. The effective theory below M_{SUSY} is just the standard model with two light Higgs doublets and a light Higgs singlet. Thus the Higgs potential at some low energy scale $\mu < M_{SUSY}$ is given by the general expression

$$\begin{aligned} V_0(\mu) &= \frac{1}{2}\lambda_1({H_1}^\dagger H_1)^2 + \frac{1}{2}\lambda_2({H_2}^\dagger H_2)^2 + (\lambda_3+\lambda_4)({H_1}^\dagger H_1)({H_2}^\dagger H_2) \\ &- \lambda_4|{H_2}^\dagger H_1|^2 + \lambda_5|N|^2|H_1|^2 + \lambda_6|N|^2|H_2|^2 \\ &+ \lambda_7(N^{*2}H_1H_2 + H.c.) + \lambda_8|N|^4 \\ &+ {m_1}^2|H_1|^2 + {m_2}^2|H_2|^2 + {m_3}^2|N|^2 \\ &- m_4(H_1H_2N + H.c.) - \frac{1}{3}m_5(N^3 + H.c.). \end{aligned} \quad (4)$$

The running quartic couplings $\lambda_i(\mu)$ and the mass parameters $m_i(\mu)$ must satisfy the following boundary conditions at M_{SUSY}

$$\begin{aligned} \lambda_1 &= \lambda_2 = \frac{1}{4}({g_2}^2 + {g_1}^2), \quad \lambda_3 = \frac{1}{4}({g_2}^2 - {g_1}^2) \\ \lambda_4 &= \lambda^2 - \frac{1}{2}{g_2}^2, \quad \lambda_5 = \lambda_6 = \lambda^2, \quad \lambda_7 = -\lambda k, \quad \lambda_8 = k^2, \\ m_1 &= m_{H_1}, \quad m_2 = m_{H_2}, \quad m_3 = m_N, \\ m_4 &= \lambda A_\lambda, \quad m_5 = kA_k, \end{aligned} \quad (5)$$

where A_λ and A_k are soft parameters associated with the trilinear Higgs couplings in Eq.(1), and we have deferred a discussion of squark effects to Section 3. At energy scales μ below M_{SUSY}, the values of the quartic couplings may be obtained by solving the RG equations given in ref.[18]. The minimisation conditions implied by $\frac{\partial V_{Higgs}}{\partial v_i} = 0$ and $\frac{\partial V_{Higgs}}{\partial x} = 0$ allow us to eliminate the low energy parameters m_1, m_2, m_3. The remaining masses m_4 and m_5 are related to the parameters A_λ and A_k at M_{SUSY} by Eq.(5). Below this scale we shall regard m_4 and m_5 as free parameters.

The neutral CP-even (scalar) mass squared symmetric matrix in the basis $1,2,3 = H_1, H_2, N$ is

$$\begin{aligned} M^2 &= \begin{pmatrix} 2\lambda_1\nu_1^2 & 2(\lambda_3+\lambda_4)\nu_1\nu_2 & 2\lambda_5 x\nu_1 \\ 2(\lambda_3+\lambda_4)\nu_1\nu_2 & 2\lambda_2\nu_2^2 & 2\lambda_6 x\nu_2 \\ 2\lambda_5 x\nu_1 & 2\lambda_6 x\nu_2 & 4\lambda_8 x^2 - m_5 x \end{pmatrix} \\ &+ \begin{pmatrix} \tan\beta[m_4x - \lambda_7x^2] & -[m_4x - \lambda_7x^2] & -\frac{\nu_2}{x}[m_4x - 2\lambda_7x^2] \\ -[m_4x - \lambda_7x^2] & \cot\beta[m_4x - \lambda_7x^2] & -\frac{\nu_1}{x}[m_4x - 2\lambda_7x^2] \\ -\frac{\nu_2}{x}[m_4x - 2\lambda_7x^2] & -\frac{\nu_1}{x}[m_4x - 2\lambda_7x^2] & \frac{\nu_1\nu_2}{x^2}[m_4x] \end{pmatrix} \end{aligned} \quad (6)$$

The neutral CP-odd (pseudoscalar) mass squared symmetric matrix is

$$\tilde{M}^2 = \begin{pmatrix} \tan\beta[m_4 x - \lambda_7 x^2] & [m_4 x - \lambda_7 x^2] & \frac{\nu_2}{x}[m_4 x + 2\lambda_7 x^2] \\ [m_4 x - \lambda_7 x^2] & \cot\beta[m_4 x - \lambda_7 x^2] & \frac{\nu_1}{x}[m_4 x + 2\lambda_7 x^2] \\ \frac{\nu_2}{x}[m_4 x + 2\lambda_7 x^2] & \frac{\nu_1}{x}[m_4 x + 2\lambda_7 x^2] & 3m_5 x + \frac{\nu_1 \nu_2}{x^2}[m_4 x - 4\lambda_7 x^2] \end{pmatrix}. \quad (7)$$

In ref.[18] we obtained an upper bound on the mass of the lightest neutral scalar Higgs boson h^0, by using the fact that m_h^2 must not exceed the lower eigenvalue of the upper 2×2 block matrix. The resulting upper bound is a complicated function of ${m_c}^2$. It is easy to show that the bound reaches a maximum asymptotically for $m_c \to \infty$ [18]. It is easy to show that the bound is maximised for $\lambda > \lambda_{max}$ and $k = 0$, and so we shall use λ_{max} and $k = 0$, as in the case of the tree level bound discussed previously. With this information in hand, a bound may be calculated from the values of $\delta\lambda_i = \lambda_i(\mu) - \lambda_i(M_{SUSY})$ obtained from numerically integrating the RG equations, decoupling the top quark below its mass and choosing $\mu = 150$ GeV. For each value of m_t, we have determined the value of $h_t(M_{SUSY})$ and the corresponding value of λ_{max} which maximises the bound from a numerical analysis of the triviality condition as discussed previously. We defer a discussion of this bound until after squark effects have been considered.

3. Squark Contributions [19]

In order the calculate the shifts in the mass–squared matrices due to squarks, we employ the full one–loop effective potential. Much of our analysis of the squark spectrum is similar to that of the MSSM in Ref. [10], whose notation we follow closely. A similar approach has also been followed by Ellwanger [15]. The one–loop effective potential is given by

$$\begin{aligned} V_1(Q) &= V_0(Q) + \Delta V_1(Q), \\ \Delta V_1(Q) &= \frac{1}{64\pi^2}\mathrm{Str}\mathcal{M}^4\left(\ln\frac{\mathcal{M}^2}{Q^2} - \frac{3}{2}\right), \end{aligned} \quad (8)$$

where $V_0(Q)$ is the tree–level potential at the arbitrary $\overline{MS}$ scale Q, Str denotes the usual supertrace, and $\mathcal{M}^2$ is the field–dependent mass–squared matrix. We shall take $Q = M_{SUSY} = 1$ TeV. † V_0 is restricted to the pure Higgs part of the tree–level potential. Since the RG approach of Section 2 [18] has already included the logarithmic effects of top quark and Higgs boson loops we exclude these particles from the supertrace. The logarithmic contributions to $\Delta V_1(Q)$ from the top quark and Higgs bosons may be absorbed into $V_0(Q)$ where $Q = 1$ TeV, to yield $V_0(\mu)$ where $\mu = 150$ GeV, as in Eq.(4). It only remains to calculate $\Delta V_1(Q)$, where $Q = 1$ TeV, involving squark contributions.

To calculate the shifts in the Higgs mass matrices as a result of squark effects we require the field–dependent mass–squared matrices of the squarks. Taking the contribution to the soft SUSY breaking potential involving squarks to be

$$\Delta V_{soft} = h_t A_t(\tilde{Q}H_2\tilde{t^c} + h.c.) + m_Q^2|\tilde{Q}|^2 + m_T^2|\tilde{t^c}|^2 + m_B^2|\tilde{b^c}|^2, \quad (9)$$

†This value of Q is of course totally arbitrary. The value of 1 TeV is chosen to be consistent with our previous analysis. The results are independent of Q to one loop order.

where b^c contains the charge conjugate of the right–handed bottom quark, and tildes denote the scalar components of the superfields, together with the superpotential in Eq.(1), leads to the field–dependent squark mass–squared matrix (ignoring contributions proportional to gauge couplings and h_b)

$$\mathcal{M}^2 = \begin{pmatrix} m_Q^2 + h_t^2|H_2^0|^2 & \lambda h_t N H_1^0 + h_t A_t \bar{H}_2^0 & -h_t^2 \bar{H}_2^0 H_2^+ & 0 \\ \lambda h_t \bar{N} \bar{H}_1^0 + h_t A_t H_2^0 & m_T^2 + h_t^2(|H_2^0|^2 + H_2^+ H_2^-) & \lambda h_t \bar{N} H_1^+ - h_t A_t H_2^+ & 0 \\ -h_t^2 H_2^0 H_2^- & \lambda h_t N H_1^- - h_t A_t H_2^- & m_Q^2 + h_t^2 H_2^+ H_2^- & 0 \\ 0 & 0 & 0 & m_B^2 \end{pmatrix}, \tag{10}$$

in the basis $\{\tilde{t_L}, \tilde{\bar{t}}_R^c, \tilde{b_L}, \tilde{\bar{b}}_R^c\}$, where a bar denotes complex conjugation. Manifestly, one eigenvalue is field–independent and may be discarded. Only the upper 2×2 submatrix contributes to the CP-even and CP-odd mass–squared matrices, since the charged fields have zero vevs in order not to break QED, whereas the upper 3×3 submatrix contributes to the charged mass–squared matrix. By differentiating ΔV_1 once with respect to the fields we may obtain the shifts in the minimisation conditions induced by squark loops, and twice will yield the shifts in the mass matrices. ‡

In the basis $\{H_1, H_2, N\}$ we have the following mass-squared matrices, after ensuring that the full one–loop potential is correctly minimised. The couplings λ_i are those obtained from the potential in Eq.(4) renormalised at the scale $\mu = 150$ GeV. For notational simplicity we drop the tildes. Moreover, we work in the basis of mass eigenstates $\{t_1, t_2, b_1, b_2\}$ (though, since h_b, the bottom quark Yukawa coupling, is taken to be zero, b_2 never contributes to the mass–squared matrices), where the mass of the t_1 squark is m_{t_1}, and so on for the other squarks. The CP-even (scalar) mass–squared matrix is

$$M_s^2 = M^2 + \delta M^2, \tag{11}$$

where M^2 is given in Eq.(6), and

$$\delta M^2 = \begin{pmatrix} \Delta_{11}^2 & \Delta_{12}^2 & \Delta_{13}^2 \\ \Delta_{12}^2 & \Delta_{22}^2 & \Delta_{23}^2 \\ \Delta_{13}^2 & \Delta_{23}^2 & \Delta_{33}^2 \end{pmatrix} + \begin{pmatrix} \tan\beta & -1 & -\frac{\nu_2}{x} \\ -1 & \cot\beta & -\frac{\nu_1}{x} \\ -\frac{\nu_2}{x} & -\frac{\nu_1}{x} & \frac{\nu_1\nu_2}{x^2} \end{pmatrix} \Delta_p^2. \tag{12}$$

The Δ_{ij}^2 and Δ_p^2 are given by

$$\begin{aligned} \Delta_p^2 &= \frac{3}{16\pi^2} h_t^2.(\lambda x).A_t.f(m_{t_1}^2, m_{t_2}^2), \\ \Delta_{11}^2 &= \frac{3}{8\pi^2} h_t^4 \nu_2^2.(\lambda x)^2.\left(\frac{A_t + \lambda x \cot\beta}{m_{t_2}^2 - m_{t_1}^2}\right)^2 g(m_{t_1}^2, m_{t_2}^2), \\ \Delta_{22}^2 &= \frac{3}{8\pi^2} h_t^4 \nu_2^2.\left(\ln\frac{m_{t_1}^2 m_{t_2}^2}{M_{SUSY}^4} + \frac{2A_t(A_t + \lambda x \cot\beta)}{m_{t_2}^2 - m_{t_1}^2} \ln\frac{m_{t_2}^2}{m_{t_1}^2}\right) \\ &+ \frac{3}{8\pi^2} h_t^4 \nu_2^2.\left(\frac{A_t(A_t + \lambda x \cot\beta)}{m_{t_2}^2 - m_{t_1}^2}\right)^2 g(m_{t_1}^2, m_{t_2}^2), \end{aligned}$$

‡In fact, this is only approximately true due to Higgs self–energy corrections; these are expected to be small for the lightest Higgs bosons [8].

$$
\begin{aligned}
\Delta_{33}^2 &= \frac{3}{8\pi^2} h_t^4 \nu_2^2 .(\lambda\nu_1)^2 . \left(\frac{A_t + \lambda x \cot\beta}{m_{t_2}^2 - m_{t_1}^2}\right)^2 g(m_{t_1}^2, m_{t_2}^2), \\
\Delta_{12}^2 &= \frac{3}{8\pi^2} h_t^4 \nu_2^2 .(\lambda x) . \left(\frac{A_t + \lambda x \cot\beta}{m_{t_2}^2 - m_{t_1}^2}\right) \left(\ln\frac{m_{t_2}^2}{m_{t_1}^2} + \frac{A_t(A_t + \lambda x \cot\beta)}{m_{t_2}^2 - m_{t_1}^2} g(m_{t_1}^2, m_{t_2}^2)\right), \\
\Delta_{13}^2 &= \frac{3}{8\pi^2} h_t^4 \nu_2^2 .(\lambda x).(\lambda\nu_1) . \left(\frac{A_t + \lambda x \cot\beta}{m_{t_2}^2 - m_{t_1}^2}\right)^2 g(m_{t_1}^2, m_{t_2}^2) \\
&- \frac{3}{8\pi^2} h_t^2 .(\lambda x).(\lambda\nu_1).f(m_{t_1}^2, m_{t_2}^2), \\
\Delta_{23}^2 &= \frac{\nu_1}{x} \Delta_{12}^2, \qquad (13)
\end{aligned}
$$

where the functions f and g are defined by

$$
\begin{aligned}
f(m_{t_1}^2, m_{t_2}^2) &= \frac{1}{m_{t_2}^2 - m_{t_1}^2} \left[m_{t_1}^2 \ln\frac{m_{t_1}^2}{M_{SUSY}^2} - m_{t_1}^2 - m_{t_2}^2 \ln\frac{m_{t_2}^2}{M_{SUSY}^2} + m_{t_2}^2 \right], \\
g(m_{t_1}^2, m_{t_2}^2) &= \frac{1}{m_{t_1}^2 - m_{t_2}^2} \left[(m_{t_1}^2 + m_{t_2}^2) \ln\frac{m_{t_2}^2}{m_{t_1}^2} + 2(m_{t_1}^2 - m_{t_2}^2) \right]. \qquad (14)
\end{aligned}
$$

If the squarks are degenerate in mass, then Eq.(10) implies that $A_t + \lambda x \cot\beta = 0$. Furthermore, if $m_{t_1} = m_{t_2} = M_{SUSY}$, then Eqs.(12-14) imply that the squark contribution to the CP-even mass–squared matrix vanishes. This is the limit of our previous analysis in Section 2 [18].

The CP-odd (pseudoscalar) mass–squared matrix is

$$
M_p^2 = \tilde{M}^2 + \delta\tilde{M}^2, \qquad (15)
$$

where $\tilde{M}^2$ is given in Eq.(7) and

$$
\delta\tilde{M}^2 = \begin{pmatrix} \tan\beta & 1 & \frac{\nu_2}{x} \\ 1 & \cot\beta & \frac{\nu_1}{x} \\ \frac{\nu_2}{x} & \frac{\nu_1}{x} & \frac{\nu_1\nu_2}{x^2} \end{pmatrix} \Delta_p^2. \qquad (16)
$$

Finally, the charged mass–squared matrix is

$$
M_c^2 = \begin{pmatrix} \tan\beta & 1 \\ 1 & \cot\beta \end{pmatrix} (m_4 x - \lambda_7 x^2 - \lambda_4 \nu_1 \nu_2 + \Delta_c^2), \qquad (17)
$$

where

$$
\Delta_c^2 = \frac{3}{16\pi^2} \sum_{m_a \in \{m_{t_1}, m_{t_2}, m_{b_1}\}} m_a^2 \left(\ln\frac{m_a^2}{M_{SUSY}^2} - 1\right) \frac{\partial^2 m_a^2}{\partial H_1^- \partial H_2^+}|_{vevs}, \qquad (18)
$$

and

$$
\begin{aligned}
\frac{\partial^2 m_{t_1}^2}{\partial H_1^- \partial H_2^+}|_{vevs} &= -\frac{h_t^4 \nu_2^2 .(\lambda x)^2 . \cot\beta}{(m_{t_1}^2 - m_{t_2}^2)(m_{t_1}^2 - m_{b_1}^2)} - \frac{h_t^2 .(\lambda x).A_t}{m_{t_1}^2 - m_{t_2}^2}, \\
\frac{\partial^2 m_{t_2}^2}{\partial H_1^- \partial H_2^+}|_{vevs} &= -\frac{h_t^4 \nu_2^2 .(\lambda x)^2 . \cot\beta}{(m_{t_2}^2 - m_{b_1}^2)(m_{t_2}^2 - m_{t_1}^2)} + \frac{h_t^2 .(\lambda x).A_t}{m_{t_1}^2 - m_{t_2}^2}, \\
\frac{\partial^2 m_{b_1}^2}{\partial H_1^- \partial H_2^+}|_{vevs} &= -\frac{h_t^4 \nu_2^2 .(\lambda x)^2 . \cot\beta}{(m_{b_1}^2 - m_{t_1}^2)(m_{b_1}^2 - m_{t_2}^2)}. \qquad (19)
\end{aligned}
$$

The bound on the lightest CP-even Higgs mass is a consequence of the fact that the minimum eigenvalue of M_s^2 is bounded by the minimum eigenvalue of the upper 2×2 submatrix of M_s^2. Using Eq. (17) to obtain the physical charged Higgs mass squared, m_c^2, in terms of m_4,

$$m_4 x - \lambda_7 x^2 = \frac{1}{2}(m_c^2 + \lambda_4 \nu^2) \sin 2\beta - \Delta_c^2, \tag{20}$$

we may eliminate m_4 from M_s^2 in favour of m_c^2, and thus write the upper 2×2 submatrix in the form

$$M_s'^2 = M'^2 + \delta M'^2, \tag{21}$$

where

$$M'^2 = \begin{pmatrix} 2\lambda_1 \nu_1^2 & 2(\lambda_3 + \lambda_4)\nu_1 \nu_2 \\ 2(\lambda_3 + \lambda_4)\nu_1 \nu_2 & 2\lambda_2 \nu_2^2 \end{pmatrix} + \begin{pmatrix} \tan\beta & -1 \\ -1 & \cot\beta \end{pmatrix} \frac{1}{2}(m_c^2 + \lambda_4 \nu^2) \sin 2\beta, \tag{22}$$

and

$$\delta M'^2 = \begin{pmatrix} \Delta_{11}^2 & \Delta_{12}^2 \\ \Delta_{12}^2 & \Delta_{22}^2 \end{pmatrix} + \begin{pmatrix} \tan\beta & -1 \\ -1 & \cot\beta \end{pmatrix} (\Delta_p^2 - \Delta_c^2). \tag{23}$$

It is clear from the form of the second term on the right–hand–side of Eq. (23) that the factor $\Delta_p^2 - \Delta_c^2$ does not change the bound at all, but merely serves, for a fixed $\tan\beta$, to shift the bound, when plotted as a function of m_c, to the left or right. We may then drop this term, since its presence may be re-parametrised by a shift in the free parameter m_c. In this way m_{b_1} is eliminated.

¿From Eq. (22) and Eq. (23) it is a simple matter to determine the shift in the minimum eigenvalue of M'^2 due to squark effects. Defining $A = (M'^2)_{11}$, $B = (M'^2)_{12}$ and $C = (M'^2)_{22}$, the shift is given by

$$\frac{1}{2}\left[\Delta_{11}^2 + \Delta_{22}^2 - \frac{(A-C)(\Delta_{11}^2 - \Delta_{22}^2) + 4B\Delta_{12}^2}{\sqrt{(A-C)^2 + 4B^2}}\right]. \tag{24}$$

Because this shift is a one loop effect, the couplings λ_i in A, B and C may be evaluated at any renormalisation point, since the difference between a coupling evaluated at two different renormalisation points is also a one loop effect, thus giving an overall error at the two loop level, which we neglect in our approximation.

In our previous analysis [18] we used triviality limits on the couplings h_t, λ and k [13, 14] to determine the bound on the lightest CP-even Higgs boson mass m_h^0 not including general squark effects, that is, with $\delta M'^2 \equiv 0$. Let the triviality limit on λ for a given h_t be λ_{max}. In table 1 we reproduce the values of h_t and λ_{max} which generated the bound m_h^0 of our previous analysis — all other values, for a given top quark mass, resulted in smaller values of the bound. It transpires that $k = 0$ in all cases. [§] In order to maximise the bound resulting from Eq. (22) we take the limit $m_c = \infty$, thus eliminating m_c as a parameter. As a function of m_c the bound approaches its maximum asymptotically as $m_c \to \infty$. The approach is rapid, with $m_c \sim 200$ GeV being a good approximation to $m_c = \infty$. Thus, the bound resulting

[§] Taking $k = 0$ gives rise to an axion in the physical spectrum. However, a small, non–zero value of k is sufficient to give the would–be axion a mass of several tens of GeV. Such a value of k does not invalidate our calculations.

Table 1: Lightest CP-even Higgs mass bound in the NMSSM including general squark effects. In row 1 is the top mass, m_t, in GeV; in rows 2 and 3 those values of $h_t(M_{SUSY})$ and $\lambda_{max}(M_{SUSY})$, which produce the bound in our previous analysis, this being in row 4, in GeV; in row 5 is the contribution to the bound resulting from a general squark mass spectrum rather than degenerate squarks with mass M_{SUSY}, in GeV; in row 6 is the lightest CP-even Higgs mass bound including general squark effects, in GeV; in rows 7 and 8 are those value of m_{t_1} and A_t, in GeV, which generate the bound in row 6; $m_{t_2} = 1$ TeV.

m_t	90	100	110	120	130	140	150	160	170	180	190
h_t	0.60	0.67	0.73	0.79	0.84	0.88	0.92	0.95	0.98	1.00	1.03
λ_{max}	0.87	0.85	0.83	0.81	0.79	0.77	0.74	0.71	0.67	0.63	0.50
m_h^0	145	143	140	137	135	132	129	126	124	123	126
δm_h	1	2	3	4	6	8	10	13	16	20	23
m_h	146	145	143	141	140	139	139	139	140	143	149
m_{t_1}	800	770	750	720	700	670	650	630	610	590	570
A_t	940	1000	1040	1130	1160	1270	1330	1410	1500	1630	1820

from the purely formal procedure of taking the limit $m_c = \infty$ is not unphysical [18]. Had this not been the case, then certainly we would have an upper bound, but one perhaps not capable of realisation in an actual spectrum.

We use the values of h_t and λ_{max} in table 1 to calculate the shift in the bound on the lightest CP-even Higgs boson mass–squared given by Eq. (24). The shift in the bound on the mass is denoted by δm_h. This is a hideously complicated function of many parameters, including the squark masses m_{t_1} and m_{t_2}, the soft SUSY breaking parameter A_t, the vev of the gauge singlet field x, and m_c. As a function of m_c the shift appears, numerically, to be maximised for m_c small, typically between 100 GeV and 200 GeV, with the difference between the shifts at these two points being less than 1 GeV. Thus, we set $m_c = 200$ GeV in the general squark contributions to the bound. Doing this enables us to retain the bound from our previous analysis derived from Eq. (22) in the limit $m_c = \infty$, and simply maximise the squark contributions separately and add them to our previous bound to yield the new bound $m_h = m_h^0 + \delta m_h$. The x dependence is not too strong, with the ratio $r = x/\nu$ typically taking values close to 10 in order to maximise the bound. Where this is not the case, the difference between the bound at its maximum, as a function of x, and that at $r = 10$ is less than 1 GeV. Thus, we set $r = 10$, this giving a reliable indication of the maximum shift. We calculate the shift for various ranges of squark masses and a range of values of A_t, and record the maximum value obtained. The squark masses are allowed to vary between 20 GeV and 1000 GeV in steps of 10 GeV. (1000 GeV is the upper limit since we take $M_{SUSY} = 1000$ GeV.) Given this restriction, the value of A_t which maximises the general squark contribution to the bound never exceeds 2 TeV. We impose the constraint $2h_t\nu_2(A_t + \lambda x \cot\beta) \leq |m_{t_1}^2 - m_{t_2}^2|$ which follows from the form of the squark mass matrix in Eq. (10).

In table 1 we show the bound on the lightest CP-even Higgs mass m_h including general squark effects (and m_h^0 for comparison). Table 1 also shows the values of m_{t_1} and A_t used to generate the bound. As m_t increases, m_{t_1} monotonically decreases, while A_t monotonically increases. The other squark mass, m_{t_2}, takes the value 1 TeV over the whole range of top quark masses — that one squark mass takes on its maximum permitted value to maximise the general squark contribution to the bound can be seen analytically. It will be noticed that these squark masses and A_t are somewhat larger than typically expected from GUT scenarios. However, we feel it wise not to restrict ourselves too excessively to particular prejudices regarding physics as yet unknown (though, of course, the notion of triviality does require the assumption of a SUSY desert up to the unification scale).

Our calculations suggest that the universal upper bound on the lightest CP-even Higgs mass is 149 GeV. Table 1 reveals that for large values of m_t squark effects may contribute up to $\delta m_h = 23$ GeV, but for small m_t the squark effects are small, as expected. We emphasise that we have not re-maximised the complete one loop corrected bound, but rather made the assumption that the values of h_t and λ_{max} which generated the bound of our previous analysis are not significantly modified by the inclusion of a general spectrum of squark masses.

4. Conclusions

We have combined an RG analysis of the Higgs sector of the NMSSM [18] with a calculation of general squark effects [19], and determined a bound on the mass of the lightest CP-even state of $m_h \leq 146, 139, 149\ GeV$ for $m_t = 90, 140, 190\ GeV$. Our calculations indicate that the effects of a general squark spectrum can be very significant for large top quark masses. One reason why, for large top quark masses, squark effects are important is that there exist finite, one–loop diagrams with vertices containing the uncontrolled soft SUSY breaking parameter A_t. These diagrams can give large contributions to the Higgs boson masses. We close by indicating that in the NMSSM there are similar diagrams involving loops of Higgs bosons, again with vertices containing soft SUSY breaking parameters, this time A_λ and A_k. The existence of these diagrams is directly due to the gauge singlet field N; they do not occur in the MSSM. We see no reason in principle why these new diagrams should not give rise to similarly large effects, except for obvious factors of 3 due to colour. We are currently in the process of estimating these effects [20].

Acknowledgements

This work was done in collaboration with Terry Elliott and Peter White, at the University of Southampton. I would like to thank the organisers of the workshop for their hospitality and the SERC for financial support.

References

1. H. P. Nilles, *Phys. Rep.***110** (1982) 1; H. E. Haber and G. L. Kane, *Phys. Rep.* **117** (1985) 75.
2. J. F. Gunion, H. E. Haber, G. L. Kane and S. Dawson, *The Higgs Hunter's Guide* (Addison-Wesley, Reading, MA, 1990).
3. J. Ellis, G. Ridolfi and F. Zwirner, *Phys. Lett.* **B257** (1991) 83; *ibid* **B262** (1991) 477; A. Brignole, J. Ellis, G. Ridolfi and F. Zwirner, *Phys. Lett.* **B271** (1991) 123.
4. H. Haber and R. Hempfling, *Phys. Rev. Lett.* **66** (1991) 1815.
5. Y. Okada, M, Yamaguchi and T. Yanagida, *Prog. Theor. Phys.* **85** (1991)1; *ibid Phys. Lett.* **B262** (1991) 54.
6. R. Barbieri, M. Frigeni and F. Caravaglios, *Phys. Lett.* **B258** (1991) 167.
7. J. L. Lopez and D. V. Nanopoulos, *Phys. Lett.* **B266** (1991) 397.
8. A. Brignole, *Phys. Lett.* **B277** (1992) 313; *ibid Phys. Lett.* **B281** (1992) 284.
9. K. Sasaki, M. Carena and C. E. M. Wagner, *Nucl. Phys.* **B381** (1992) 66.
10. P. Fayet, *Nucl. Phys.* **B90** (1975) 104.
11. J. Ellis, J. F. Gunion, H. E. Haber, L. Roszkowski and F. Zwirner, *Phys. Rev.* **D39** (1989) 844.
12. L. Durand and J. L. Lopez, *Phys. Lett.* **B217** (1989) 463; M. Drees, *Int. J. Mod. Phys.* **A4** (1989) 3635.
13. P. Binetruy and C. A. Savoy, *Phys. Lett.* **B277** (1992) 453.
14. J. R. Espinosa and M Quiros, *Phys. Lett.* **B279** (1992) 92.
15. U. Ellwanger and M. Rausch de Traubenberg, *Z. Phys.* **C53** (1992) 521; U. Ellwanger and M. Lindner, *Phys. Lett.* **B301** (1993) 365; U. Ellwanger, *Phys. Lett.* **B303** (1993) 271.
16. W. ter Veldhuis,*Purdue preprint* **PURD-TH-92-11**.
17. J. R. Espinosa and M. Quiros, *Phys. Lett.* **B302** (1993) 51.
18. T. Elliott, S. King and P. White, *Phys. Lett.* **B305** (1993) 71.
19. T. Elliott, S. King and P. White, *Southampton preprint* **SHEP-92/93-18**.
20. T. Elliott, S. King and P. White, *Southampton preprint* **SHEP-92/93-21**, in preparation.

The Search for Supersymmetry Anomalies–Does Supersymmetry Break Itself?

J. A. Dixon[1]
Center for Theoretical Physics
Physics Department
Texas A & M University
College Station, Texas 77843

Abstract

The established results concerning the BRS cohomology of supersymmetric theories in four space-time dimensions are briefly reviewed. The current status of knowledge concerning supersymmetry anomalies and the possibility that supersymmetry breaks itself through anomalies in local composite operators is then discussed. It turns out that the simplest allowable supersymmetry anomalies occur only in conjunction with the spontaneous breaking of gauge symmetry. A simple example of such a possible supersymmetry anomaly is presented.

1 Review and Discussion

Supersymmetry is currently one of the most popular ways to extend the standard model of strong, weak and electromagnetic interactions. It is also an essential ingredient of the superstring which is the only known candidate for a theory that might include gravity in a consistent way. However a major difficulty of supersymmetry is to explain why we do not observe it, assuming that it is really present. How and why is supersymmetry broken? It is possible that the considerations of the present paper may be relevant to that question.

It has been known for some time now that the BRS cohomology of $N = 1, D = 4$ rigid supersymmetry is very complicated [5][6][7]. All the non-trivial structure occurs in Lorentz non-trivial sectors, and was consequently not found in some investigations, for example [1], but see also [2][3].

This means that in principle there may be anomalies in these theories that violate supersymmetry. We know that anomalies usually play an important role in theories, so any examples of such anomalies would be interesting.

In this paper, I briefly touch on some recent [8][9] and also the older results concerning the BRS cohomology and then go on to some remarks concerning the current state of research on these questions [10][11].

[1]Supported in part by the U.S. Dept of Energy, under grant DE-FG05-91ER40633,
Email: dixon@phys.tamu.edu

The main problem now is that no supersymmetry anomalies have yet been found that correspond to this cohomology. I shall try to explain some of the fairly obvious reasons that such anomalies are not present in some simple cases, and then go on to exhibit a case which appears to require a detailed calculation to determine whether a supersymmetry anomaly is or is not present. Some of this material was not known at the time of the talk given at the conference, but it seems appropriate to include it here.

It is easy to write down the simplest examples where a supersymmetry anomaly could conceivably arise. The BRS cohomology of these theories indicates that there could be anomalies in the renormalization of composite operators (made from the elementary chiral superfields S of the theory) which are antichiral spinor superfields. These composite operators take forms such as:

$$\Psi_{1\alpha} = D^2[S_1 D_\alpha S_2] \tag{1}$$

$$\Psi_{2\alpha} = \overline{S}_1 D^2[S_2 D_\alpha S_3] \tag{2}$$

$$\Psi_{3\alpha} = \overline{S}_1 D^2[\overline{D}^2 \overline{S}_1 D_\alpha S_3] \tag{3}$$

One could add more chiral superfields S or more supercovariant derivatives of course. The main things to keep in mind are:

1. The expression for $\Psi_{1\alpha}$ should not vanish
2. It is frequently necessary to use more than one flavour of superfield S to prevent the expression from vanishing, because such expressions may be antisymmetric under interchange of flavour indices

To calculate the anomaly, one would couple such composite operators to the action with an elementary (i.e. not composite) antichiral spinor source superfield Φ^α. This means that one simply adds the following term to the usual action of the theory:

$$S_\Phi = \int d^6\overline{z}\ \Phi^\alpha \Psi_\alpha \tag{4}$$

where Ψ_α is some composite antichiral spinor superfield, examples of which are given above in (1-3).

Then the anomaly would appear in the form:

$$\delta\Gamma = m^k \int d^6\overline{z}\ \Phi^\alpha c_\alpha \overline{S}^n \tag{5}$$

where Γ is the one-particle irreducible generating functional, δ is the nilpotent BRS operator, $\int d^6\overline{z}$ is an integral over antichiral superspace, c_α is the constant ghost parameter of rigid supersymmetry, m^k is the mass parameter

m to some power k required by simple dimensional analysis, and $\overline{S}^n$ is the nth power of the antichiral superfield (this might include a sum over indices which distinguish different $\overline{S}$ superfields from each other).

To count masses we use the following assignments for the variables and the superfields (Notation defined below):

$$m = 1; \partial_\mu = 1; \theta_\alpha = \frac{-1}{2}; S = 1; \Phi_\alpha = \frac{1}{2}; \tag{6}$$

Now we define the component fields:

$$S = A + \theta^\alpha \psi_\alpha + \frac{1}{2}\theta^2 F \tag{7}$$

$$\Phi_\alpha = \phi_\alpha + W_{\alpha\dot\beta}\overline{\theta}^{\dot\beta} + \frac{1}{2}\overline{\theta}^2 \chi_\alpha \tag{8}$$

The dimensions of these coefficient fields are then:

$$D_\alpha = \frac{-1}{2}; A = 1; \psi_\alpha = \frac{3}{2}; F = 2;$$

$$\chi_\alpha = \frac{3}{2}; \phi_\alpha = \frac{1}{2}; W_{\alpha\dot\beta} = 1; \tag{9}$$

An examination of examples shows that elementary dimensional counting prevents the powers of m from working correctly to yield (5) whenever the only vertices of the diagram are chiral vertices involving only chiral fields. It should be possible to show this by a dimensional argument, but I have not taken the time to try to do that yet–at any rate it certainly seems to hold for a wealth of examples, one of which can be found in [4].

However when there is at least one gauge propagator in the diagram, the powers of m easily work out correctly to yield (5). But then one has to confront another problem, which is that one has to analyze the cohomology again in the presence of the gauge fields. This unsolved problem has been partially finessed in [10].

Another problem that is also unsolved is the problem of solving the full BRS cohomology of any supersymmetric theory including the sources that are necessary to formulate the full BRS identity. Essentially, this brings in the complication of ensuring that the BRS cohomology space is orthogonal to the equations of motion of the fields. I will also make here some new comments on this question, which has been the subject of my work over the summer [10].

When one tries to compute the full BRS cohomology of chiral matter coupled to gauge fields, the BRS operator becomes quite formidable, and I will not try to give all the details here. However we can discuss a sub-problem of interest without solving the entire problem, and that is what we do below.

Whenever one formulates a BRS identity in the manner pioneered by Zinn-Justin, it is necessary to also include sources $\tilde{f}_i$ for the variation of the fields f_i, and in the resulting 'full' BRS operator, these give rise to terms that involve the equations of motion of the corresponding fields. This turns out to be more or less equivalent to the Batalin-Vilkovisky quantization method. The essential point is that this eliminates from the cohomology space anything which vanishes by the equation of motion, i.e. anything which vanishes 'on-shell'. In our case this will eliminate all those objects in the cohomology space which involve superfields $\overline{S}$ which have mass terms in the action, as well as a number of higher order terms that are of no concern at present.

So we are now interested only in computing diagrams where the possible supersymmetry anomaly involves massless antichiral fields $\overline{S}$ in (5). But this raises another problem. The most promising simple case (see below) seems to involve a triangle diagram with the Φ^α superfield at one vertex, two chiral (or antichiral) superfields emerging from that vertex and the exchange of a vector superfield between these two lines. Now the mass counting implies that the anomaly has a higher power of mass than the composite operator from which it arises. The only way this can happen is if some of the interior lines are massive. Is there any way that interior massive lines can give rise to exterior massless lines while exchanging a vector superfield? The answer to this question is of course well known–this will happen if and only if the gauge symmetry is spontaneously broken. We will therefore assume that gauge symmetry *is* spontaneously broken and that supersymmetry *is not* spontaneously broken. Since we are looking for supersymmetry breaking through anomalies, it is reasonable to assume that it is not otherwise broken.

This combination is in fact very easy to achieve–as is well known, gauge symmetry breaking is natural and very easy to achieve in rigid supersymmetry, but spontaneous supersymmetry breaking can only be achieved with very contrived models, particularly if the gauge group is semisimple.

So now, if we want to examine the question of supersymmetry anomalies, we are forced to consider a supersymmetric gauge theory with spontaneous breaking of the gauge symmetry. But there are more conditions, at least for the supersymmetry anomalies that involve matter superfields. In order for the relevant diagrams to exist, we must have matter multiplets which break under the gauge breaking into a combination of massive and massless fields, so that a massive vector superfield can have a vertex with a massless and a massive chiral superfield.

This happens of course for the Higgs multiplet itself, but then the massless Goldstone supermultiplets do not contribute to the BRS cohomology space, as will be shown in the forthcoming paper [10]. We must have additional (non-Higgs) matter multiplets which break under the gauge breaking into a combination of massive and massless fields. There are many ways to do this, and an example is given below. Note that this happens also in the standard model, where the neutrino remains massless after spontaneous breaking of $SU(2) \times U(1)$ to $U(1)_{EM}$ simply because there is no right handed neutrino for it to form a mass with, (and because lepton conservation prevents the formation of a Majorana neutrino mass, in the minimal standard model at least). The relevant discussion of the standard model will also be the subject of a forthcoming paper [11].

So if we want to find a supersymmetry anomaly, we are driven to models with gauged supersymmetry and spontaneous breaking of the gauge symmetry through Higgs multiplets which develop a VEV in their 'A' components (but not their 'F' components–that would break supersymmetry). In addition these models must have matter which is massless at tree level, but which gets split into massive and massless components as a result of gauge breaking. These are the only models that have a chance of developing supersymmetry anomalies in some of their composite operators at the one loop level. Such models are of course very reminiscent of a supersymmetric version of the standard model of strong weak and electromagnetic interactions. It is just within the realm of possibility that these anomalies could account for the experimentally observed lack of supersymmetry in the world with no additional assumptions in the model at all–in which case we could say that supersymmetry breaks itself. But there is plenty of work to do before we can determine whether this notion is right. Even if the supersymmetry anomalies exist, considerable work will be necessary to deduce the form of the supersymmetry breaking they give rise to.

A rather interesting and new feature is that we can see that the particular 'soft' mass-dependent supersymmetry anomalies we are examining here, if there coefficients are non-zero, would give rise to a kind of supersymmetry breaking that is a function of the VEV that breaks the gauge symmetries, and which vanishes in the gauge symmetric limit. In addition, it has been conjectured [4] that such anomalies might also provide a natural mechanism whereby 'supersymmetry breaks itself', while at the same time retaining the cosmological constant at the zero value it naturally has in many unbroken supersymmetric theories. Clearly the spontaneous breaking of the gauge symmetry would not interfere with this feature.

2 A Simple Example

We consider a supersymmetric gauge theory based on the gauge group $SU(2)$ with matter in two vector multiplets and a singlet: $L^a : I = 1; H^a : I = 0; R : I = 0$. These 'a' indices transform with $i\epsilon^{abc}$ and take the values a=1,2,3. Since the 'a' indices are real and since δ_{ab} is an invariant tensor of $SU(2)$, there is no difference when we raise and lower these indices.

Without any good reason, we shall assume that the superpotential does not contain a mass term for the L field.

Since the 'Higgs field' H^a is in a real representation of the gauge group, it can have a mass term in the superpotential.

Now we assume the following form for the superpotential:

$$W = g_1 L^a H^a R + \frac{g_2 m}{2} H^a H^a + \frac{g_3}{4m}[H^a H^a]^2 \tag{10}$$

Note that renormalizability is not a property of this superpotential.

If $g_2 g_3 < 0$, the Higgs field will develop a VEV in its 'A' term that breaks the gauge symmetry down to $U(1)$ while leaving supersymmetry unbroken. The L and R fields develop no VEV. Let us denote components as follows:

$$L^a = A^a + \theta^\alpha \psi^a_\alpha + \frac{1}{2}\theta^2 F^a \tag{11}$$

$$H^a = B^a + mu^a + \theta^\alpha \phi^a_\alpha + \frac{1}{2}\theta^2 G^a \tag{12}$$

$$R = A + \theta^\alpha \psi_\alpha + \frac{1}{2}\theta^2 F \tag{13}$$

We distinguish $a = i, 3$ where $1 = 1, 2$. Here we have included a shift by the VEV:

$$< B^a >_{\text{before shift}} = \delta^{a3} m \sqrt{\frac{-g_2}{g_3}} \equiv \delta^{a3} mh \equiv mu^a \tag{14}$$

Then the 'F' term of the superpotential becomes:

$$W_F = \Big[g_1 L^a (B^a + \delta^{a3} mh) R + \frac{g_2 m}{2}(B^a + \delta^{a3} mh)(B^a + \delta^{a3} mh)$$

$$+ \frac{g_3}{4m}[(B^a + \delta^{a3} mh)(B^a + \delta^{a3} mh)]^2\Big]_F \tag{15}$$

In terms of components, this makes the following contribution to the action:

$$S_{\text{Chiral}} = \int d^4x W_F = \int d^4x \big\{ g_1 (mhA^3 F + mh\psi^{3\alpha}\psi_\alpha + mhF^3 A)$$

$$+ g_1 (A^a B^a F + \psi^a \phi^a A + F^a B^a A + A^a \phi^a \psi + A^a G^a A + \psi^a B^a \psi)$$

$$+ \text{terms involving H superfield only} \big\} \tag{16}$$

The essential point to note here is that there is no mass term like

$$m(A^i F + \psi^{i\alpha}\psi_\alpha + F^i A) \tag{17}$$

for $i = 1,2$ which would give a mass to the L^1 and L^2 superfields. They are massless after spontaneous breaking of the gauge symmetry.

The simplest composite operator (together with a source Φ_α that could develop a supersymmetry anomaly seems to be:

$$S_{\text{Composite}} = \int d^4x d^4\theta \left\{ \Phi^\alpha \overline{D}^{\dot\beta} [\overline{L}^a \overline{H}^a] \overline{D}_{\dot\beta} D_\alpha R \right\} \tag{18}$$

After translation of the Higgs field, we find the terms

$$S_{\text{Composite}} = mh \int d^4x \Big\{ \chi^\alpha \Big[(\sigma^\mu)^{\gamma\dot\beta} \partial_\mu \overline{A}^3 \sigma^\nu_{\alpha\dot\beta} \partial_\nu \psi_\gamma$$

$$+ \overline{\psi}^{3\dot\beta} (\sigma^\mu)_{\alpha\dot\beta} \partial_\mu F + \cdots \Big] + \cdots \Big\} \tag{19}$$

The form of the supersymmetry anomaly that we would like to calculate is

$$\delta\Gamma = m^4 \int d^4x d^2\overline{\theta} \Phi^\alpha c_\alpha \sum_{i=1,2} \overline{L}^i \overline{L}^i = m^4 \int d^4x \chi^\alpha c_\alpha \sum_{i=1,2} \overline{A}^i \overline{A}^i + \cdots \tag{20}$$

Clearly at this point there are innumerable examples. It may be that even more structure is needed before non-zero examples of supersymmetry anomalies can be found, or this may be non-zero itself. I have not yet tried to calculate this example, but at least it does not appear to be obviously zero, since it passes all the tests mentioned above.

3 Higher Spins

Now we turn to another topic, which is a fuller analysis of the cohomology of the BRS operator defined by the supersymmetry invariance of chiral multiplets of rigid $N = 1, D = 4$ supersymmetry. The new result here [8][9] is that this cohomology space contains potential anomalies in the renormalization of fermionic superfields with all half-integer spins. Formerly it had been shown that there were potential anomalies for fermionic superfields with spin $\frac{1}{2}$ only [7].

This may be very important for superstring theories, since such higher spin multiplets necessarily occur in all such theories.

The result is that there is an infinite set of states of the form

$$\mathcal{X}_{\alpha_1 \ldots \alpha_{k_n} \dot\beta_1 \ldots \dot\beta_{k_n+g}} = \text{Sym}_{\dot\beta_1 \ldots \dot\beta_{k_n+g}} \int d^4x \, d^2\theta \Big\{ \partial_{\alpha_1 \dot\beta_1} \ldots \partial_{\alpha_{k_1} \dot\beta_{k_1}} S_{a_1} \ldots$$

$$\ldots\partial_{\alpha_{k_{(n-1)}+1}\dot{\beta}_{k_{(n-1)}+1}}\ldots\partial_{\alpha_{k_n}\dot{\beta}_{k_n}} S_{a_n}\overline{c}_{\dot{\beta}_{k_n+1}}\overline{c}_{\dot{\beta}_{k_n+2}}\ldots\overline{c}_{\dot{\beta}_{k_n+g}}\Big\}. \tag{21}$$

in the cohomology space of the Wess-Zumino model. The corresponding complex conjugate expressions are obtained by converting dotted into undotted indices and vice versa, and $S \to \overline{S}$, $\overline{c} \to c$. By contraction and symmetrization of the undotted indices, we can decompose $\mathcal{X}$ into operators of the form

$$\mathcal{A}_{(a,b)} = \mathcal{A}_{(\alpha_1\alpha_2\cdots\alpha_a)(\dot{\beta}_1\dot{\beta}_2\cdots\dot{\beta}_b)}, \tag{22}$$

where $b = k_n + g$ and $k_n - a$ is even and greater than or equal to 0, since contractions always involve pairs of undotted indices. In particular, we are interested in polynomials with ghost charge $g = 1$, which correspond to anomalies. For these objects, we find $b - a$ is odd and positive. Therefore, the operators $\mathcal{A}$ are spinors.

These objects could appear as anomalies in the renormalization of composite operators with the same spin structure as the anomaly. To compute the anomalies of a given such composite operator, a term of the form

$$S_\Psi = \int d^4x\, d^4\theta \left[\Psi_{\alpha_1\ldots\alpha_a\dot{\beta}_1\ldots\dot{\beta}_b}\Phi^{\alpha_1\ldots\alpha_a\dot{\beta}_1\ldots\dot{\beta}_b}\right] \tag{23}$$

would be introduced into the action. Here Ψ is a composite operator with ghost charge zero and $\Phi^{\alpha_1\ldots\alpha_a\dot{\beta}_1\ldots\dot{\beta}_b}$ is a chiral source superfield. There is a matching between the indices of the anomaly and those of the anomalous operator, because both must couple to the source Φ. Now to compute the anomaly, some specific form for the composite operator Ψ would be chosen and then the one particle irreducible generating functional Γ including one vertex (23) would be calculated. If there is an anomaly, one would find that the supersymmetric variation of this part of Γ would be of the form

$$\delta\Gamma = \kappa \int d^4x\, d^2\theta \left[\mathcal{A}_{\alpha_1\ldots\alpha_a\dot{\beta}_1\ldots\dot{\beta}_b}\Phi^{\alpha_1\ldots\alpha_a\dot{\beta}_1\ldots\dot{\beta}_b}\right] \tag{24}$$

where κ is a calculable coefficient.

Because all possible anomalies have half-integer spin (see discussion of (22)), it follows that all operators which can be anomalous also have half-integer spin. Generally, the entire class of spinor operators in supersymmetric theories containing chiral matter can be anomalous.

We suspect that spontaneous breaking of the gauge symmetry will be necessary for non-zero computation of superanomalies in the higher spin cases just as in the spin one-half case.

Acknowledgments: I would like to thank my collaborators Ruben Minasian and Joachim Rahmfeld for many useful ideas, Ramzi Khuri for some very useful discussions about integrals, Heath Pois for many helpful remarks about phenomenological models of supersymmetry, Chris Pope for helping us to recognize the representations that arose in the solution of the higher spin problems and Mike Duff for his continued insistence that a supersymmetry anomaly be computed.

References

[1] Brandt, F., Dragon, N., Kreuzer, M.: Phys. Lett. **B231**, 263-270 (1989), Nucl. Phys. **B332**, 224-249 (1990), Nucl. Phys. **B332**, 250-260 (1990), Nucl. Phys. **B340**, 187-224 (1990)

[2] Brandt, F.: Lagrangedichten und Anomalien in vierdimensionalen supersymmetrischen Theorien. PhD-Thesis, Universität Hannover (1991)

[3] Brandt, F.: Lagrangians and anomaly candidates of $D = 1$, $N = 4$ rigid supersymmetry. Nucl. Phys. **B392**, 928 (1993)

[4] Dixon, J. A.: A natural mechanism for supersymmetry breaking with zero cosmological constant. CTP-TAMU-69/92, to be published in: Duff, M. J., Khuri, R. R., Ferrara, S. (eds.) From superstrings to supergravity. Workshop Proceedings, Erice 1992. Singapore, New Jersey, Hong Kong: World Scientific

[5] Dixon, J. A.: BRS cohomology of the chiral superfield. Commun. Math. Phys. **140**, 169-201 (1991)

[6] Dixon, J. A.: Supersymmetry is full of holes. Class. Quant. Grav. **7**, 1511-1521 (1990)

[7] Dixon, J. A.: Anomalies, Becchi-Rouet-Stora cohomology, and effective theories. Phys. Rev. Lett. **67**, 797-800 (1991)

[8] Dixon, J. A., Minasian, R.: BRS cohomology of the supertranslation in $D = 4$. Preprint CPT-TAMU-13/93 (hep-th@xxx/9304035)

[9] Dixon, J. A., Minasian, R., Rahmfeld, J.: The Complete BRS Cohomology of Supersymmetric Chiral Matter in $D = 4$. Preprint CTP-TAMU-20/93

[10] Dixon, J. A.: BRS Cohomology of Supersymmetric Gauge Theories with Spontaneous Breaking of Gauge Symmetry. Preprint CTP-TAMU-46/93

[11] Dixon, J. A.: Supersymmetry Anomalies and the Standard Model. Preprint CTP-TAMU-47/93

Supersymmetric Grand Unification

Deducing $\mathcal{L}_{fermion}$ at M_{GUT} using Low Energy Data

or

Towards a Theory of Fermion Masses

Stuart Raby * †
Department of Physics, The Ohio State University
Columbus, OH 43210

1 Introduction

In the last 20 years we have accumulated an enormous amount of data on elementary particles and their interactions. This data serves two purposes: to fix the phenomenological parameters of the Standard Model [SM] and to verify that the SM is an excellent description of nature. It is our goal to understand the origin of these many arbitrary parameters. In this talk we consider a supersymmetric [SUSY] SO(10) grand unified theory [GUT]. We present a straightforward procedure, incorporating a general operator analysis, which allows us to use low energy data to determine the fermionic sector of the theory at the GUT scale [1]. In what follows we first review the status quo, i.e. the low energy data [LED]. We then review the evidence for SUSY GUTs, and focus on the virtues of the particular group SO(10). Following a discussion of our dynamical principles, we present a general operator analysis for $\mathcal{L}_{fermion}$,i.e. the fermionic sector of the GUT theory. We argue that all fermion masses and mixing angles can be described with a minimum of 5 arbitrary parameters in the Yukawa sector of the theory at the GUT scale, M_G. Including the parameter $\tan\beta$, the ratio of Higgs vevs present in any SUSY theory, we thus obtain a 6 parameter description of fermion masses and mixing angles, leading to 8 predictions. Finally, we present preliminary results. These preliminary results are encouraging and eminently testable. Notwithstanding, the power of our analysis is in the paradigm whereby the use of LED, hand in hand with symmetry conditions, allows us determine the fermion sector of the theory at M_G.

---

*Talk presented at the HARC Workshop on Recent Advances in the Superworld, Houston, Texas, April 14-16, 1993.

†In collaboration with Greg Anderson, Department of Physics, The Ohio State University, Columbus, OH 43210; Savas Dimopoulos, Department of Physics, Stanford University, Stanford, CA 94305; Lawrence J. Hall, Department of Physics, University of California and Theoretical Physics Group, Lawrence Berkeley Laboratory, 1 Cyclotron Road, Berkeley, CA 94720 and Glenn Starkman, Department of Physics, University of Toronto, Toronto, Canada

2 Status Quo

There are 18 arbitrary parameters in the SM with 13 of these in the fermion sector of the theory.[‡] Let us briefly review the state of our knowledge of these 18 parameters since they play a central role in what follows. In Table 1 we list the 18 parameters of the SM along with their experimental and/or theoretical uncertainties.

Table 1. The 18 parameters of the Standard Model.

Parameter	Uncertainty	Comments
$\alpha, \sin^2\theta_W$	$< 1/2$ %	high accuracy
$\alpha_s(M_Z)$	$\sim 10\%$	less certain – scale dependent
m_e, m_μ, m_τ	$< 1/3\%$	high accuracy
m_c, m_b	$\sim 4\%$	less certain
$m_u/m_d, m_s/m_d$	?	chiral Lagrangian – ambiguous
$(m_u + m_d)/2$	?	QCD sum rules
$\lvert V_{cd}\rvert \approx \lvert V_{us}\rvert$	$\sim 1.5\%$	fair accuracy
$\lvert V_{cb}\rvert, \lvert V_{ub}/V_{cb}\rvert$	$\sim 20\%$	poorly known
m_t, m_H	$m_t = 150 \begin{matrix} +19 & +15 \\ -24 & -20 \end{matrix}$	$60 < m_H < 1000$ in SM
m_t, m_H	$m_t = 131 \begin{matrix} +23 & +5 \\ -28 & -5 \end{matrix}$	$60 < m_H < 150$ in MSSM
m_t	$m_t > 108$ GeV	Fermilab data
M_Z	$< .025\%$	high accuracy
J (Jarlskog invariant)	$\sim 30\%$	uncertainty in B_K

Some of these parameters are known to high accuracy, these include $\alpha, \sin^2\theta_W, m_e, m_\mu, m_\tau, M_Z$. A few are known with fair accuracy, $|V_{cd}|, m_c, m_b$. Finally, the following are poorly known: $\alpha_s(M_Z), m_u, m_d, m_s, |V_{cb}|, |V_{ub}/V_{cb}|, m_t, m_H$ and the Jarlskog invariant measure of CP violation, J.

Our knowledge of $\alpha_s(M_Z)$ suffers predominantly from theoretical uncertainties associated with renormalization scale ambiguities. The experimentally allowed range is now $\alpha_s(M_Z) = .118 \pm .007$ [3], where the errors probably underestimate the theoretical uncertainties.

Light quark masses are determined using chiral Lagrangians and QCD sum rules. In the chiral Lagrangian approach, light fermion masses are characterized by their transformation properties under the $SU(3)_L \times SU(3)_R$ chiral symmetry. The mass matrix, M, transforms as $(3, \bar{3})$. However, as shown by Kaplan and Manohar [4],

[‡]That is if we assume that neutrinos are massless. Incorporating neutrino masses leads to an additional 9 parameters, 3 masses and 6 mixing angles.

$M' = M + \kappa det(M) M^{\dagger -1}$ with κ an arbitrary constant, has the same transformation property. This leads to a large theoretical uncertainty in light quark masses.

The weak mixing angles $|V_{ub}|$ and $|V_{cb}|$ suffer from both large theoretical and/or experimental uncertainties. The theoretical uncertainties are evidenced by the different model dependent calculations of B decay. It is hoped that the application of heavy quark effective field theory to the exclusive semileptonic B decays will reduce these uncertainties, but this requires better statistics. The latest experimental result from $B \to D^* l\nu$ is [5] $|V_{cb}| = .050 \pm .008 \pm .007$ where the first error is from statistics and the second is from extrapolation uncertainties. The ratio $|V_{ub}/V_{cb}|$, determined from inclusive B decay, has large model dependences. Recently the experimental results from CLEO II changed significantly from previous measurements by both Argus and CLEO. The latest results give [5] $.038 \leq |V_{ub}/V_{cb}| \leq .097$. This is a factor of two smaller than previous results and the change was significantly larger than any of the experimental errors for any particular model. We will just have to wait and see if this settles down with more data.

The top and Higgs masses are only known from the radiative effects they have on electroweak parameters. The latest data, analyzed by Langacker [2], gives the results presented in Table 1, where MSSM denotes the minimal supersymmetric standard model. The difference between the SM and MSSM result is the requirement of a lighter Higgs boson in the MSSM.

Finally, J suffers from strong interaction uncertainties associated with the so-called bag constant, B_K. The value of J$\times B_K$ can be derived given the experimental value for ϵ_K, all quark masses, and the magnitudes of all Kobayashi-Maskawa [KM] elements. Thus, the uncertainty in J is limited by the uncertainty in B_K which is of order 30%.

I have briefly reviewed the status quo, since some of these parameters will be used as imputs to fix the fundamental parameters in the fermion mass matrices and the others test the theory. For example, the biggest uncertainties in our prediction of the top mass comes from the uncertainties in m_b and $\alpha_s(M_Z)$. Reducing these uncertainties would make this prediction much tighter.

3 SUSY GUTs and the virtues of SO(10)

• *SUSY GUT*

Given two parameters the fine structure constant, α_G, at the GUT scale, M_G, one determines the three low energy parameters, $\alpha, \sin^2\theta_W$ and α_s, all evaluated at some renormalization scale,μ, which for convenience we might choose to be M_Z. In actuality, we use the two best determined parameters, α and $\sin^2\theta_W$, to fix $\alpha_G \sim 1/25$, $M_G \sim 10^{16} GeV$ and predict (in a SUSY GUT) $\alpha_s(M_Z) = .125 \pm .002 \pm .009$ for a central value of $m_t = 138 GeV$. [2§] The errors take into account uncertainties in m_H, m_t

[§]**In a non-SUSY GUT the prediction is $\alpha_s(M_Z) \sim .07$ [2], which is inconsistent with the data.**

and estimates of threshold corrections at both M_G and the weak scale. This result is in remarkably good agreement with the data. For a heavier top, the central value for α_s decreases by several percent. This prediction assumes a supersymmetric desert, i.e. the only threshold between M_Z and M_G occurs below ~1 TeV due to the new spectrum of states encountered in the MSSM.

In any GUT, the number of fundamental parameters in the gauge sector of the theory decreases by one and only in a SUSY GUT does the resulting prediction agree with the low energy data [LED]. The powerful assumption of a SUSY desert has an equally important consequence. The LED becomes a window into physics at the GUT scale, i.e. measurements at the weak scale gives us information about the physics at M_G.

- *Virtues of SO(10)*

A single family of fermions fits into one irreducible representation — i.e. $16_i \supset \{u_i, d_i, e_i, \nu_i\}$ with $i = 1, 2, 3$ labelling the three families. We take the 3rd family to be the top family.

The two Higgs doublets required in the MSSM fit into one irreducible representation — i.e. $10 \supset \{H, H', H_3, H_3'\}$ where H, H' are weak doublets necessary for weak symmetry breaking and giving masses to quarks and leptons and H_3, H_3' are color triplet Higgs which must get mass of order M_G to avoid rapid proton decay. We will return to this point shortly.

In order to accomplish the GUT scale symmetry breaking we must have additional representations including $\{45, 16, \overline{16}, \cdots\}$, where, for example, a 16 and $\overline{16}$ vev can break SO(10) to SU(5) and the 45 vev can then break SU(5) to SU(3) × SU(2) × U(1). This may happen at the same scale, $M_G << M_P$, or at two separate scales, where the first occurs at a scale v_{10} such that $v_5 = M_G << v_{10} << M_P$.

We have not considered other possible representations which may be relevant for GUT symmetry breaking, such as 54, 126, etc. We shall now assume that only the 45 plays a crucial role in the generation of fermion masses. It is thus necessary to elaborate the possible directions the 45 vev may point in the two dimensional space of U(1) subgroups of SO(10) which commute with SU(3) × SU(2) × U(1). Although there are only two orthogonal directions in this space, we nevertheless consider the following 4 possible symmetry breaking vevs —

$$\begin{aligned} \langle 45 \rangle_X &= v_{10} e^{i\alpha_X} \mathbf{X} \\ \langle 45 \rangle_Y &= v_5 e^{i\alpha_Y} \mathbf{Y} \end{aligned} \tag{1}$$

$$\begin{aligned} \langle 45 \rangle_{B-L} &= v_5 e^{i\alpha_{(B-L)}} (\mathbf{B} - \mathbf{L}) \\ \langle 45 \rangle_{T_{3R}} &= v_5 e^{i\alpha_{T_{3R}}} \mathbf{T_{3R}} \end{aligned} \tag{2}$$

where we have explicitly represented them as two groups of two orthogonal vevs. We consider all four since two of these vevs (one in each group) are well motivated.

X, in Eq. (1), is the U(1) direction which leaves SU(5) invariant. This is why we have taken the magnitude of the vev to be v_{10}, whereas all the others are taken to be v_5 since they do not commute with SU(5).

B − L, in Eq. (2), just measures baryon number minus lepton number. It can play a crucial role in splitting the weak doublet and color triplet Higgs multiplets, i.e. solving the hierarchy problem. The Higgs doublets carry zero B-L whereas the triplets have non-zero B-L. Thus if the Higgs in the 10 gets mass by coupling to this 45, only the color triplets will acquire mass at the scale M_G. Hence, this vev is expected to be a necessary ingredient in any complete SO(10) model which also solves the hierarchy problem.

4 Dynamic Principles

Let us now discuss the dynamical principles which guide us towards a theory of fermion masses.

0. At zeroth order, we work in the context of a SUSY GUT with the MSSM below M_G.

1. We use SO(10) as the GUT symmetry with three families of fermions $\{16_i \;\; i = 1,2,3\}$ and the minimal electroweak Higgs content in one 10. Using SO(10) symmetry relations allows us to reduce the number of fundamental parameters.

2. We will assume that there are family symmetries which enforce zeros of the mass matrix, although we will not specify these symmetries at this time. As we will make clear shortly, these symmetries will be realized at the level of the fundamental theory defined at M_P.

3. Only the third generation obtains mass via a dimension 4 operator. The fermionic sector of the Lagrangian thus contains the term $\mathcal{L}_f \supset A\ O_{33} \equiv A\ \ 16_3\ 10\ 16_3$. This term gives mass to t, b and τ. It results in the symmetry relation — $\lambda_t = \lambda_b = \lambda_\tau \equiv A$ at M_G. This relation has been studied before by Ananthanarayan, Lazarides and Shafi [ALS] [6] and using m_b and m_τ as input it leads to reasonable results for m_t and $\tan\beta$.

4. All other masses come from operators with dimension > 4. As a consequence, the family hierarchy will be related to the ratio of scales above M_G. We will show shortly how to understand the higher dimension operators in terms of an effective field theory at M_G, obtained by integrating out states with mass $> M_G$.

5. **[Predictivity requirement]** We demand the minimal set of effective fermion mass operators at M_G consistent with the **LED**.

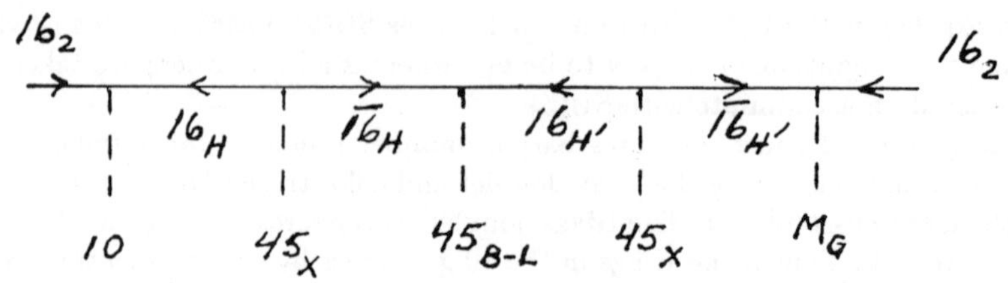

Figure 1: Tree diagram which leads to effective operator when massive states are integrated out.

Let us now consider the general **operator basis for fermion masses**. Let $\mathcal{L}_{fermion}$ include operators of the form

$$\mathrm{O_{ij}} = \mathbf{16_i}\ (\cdots)_n\ \mathbf{10}\ (\cdots)_m\ \mathbf{16_j} \tag{3}$$

where

$$(\cdots)_n = \frac{M_G^k\ 45_{k+1}\cdots 45_n}{M_P^l\ 45_X^{n-l}}$$

and the 45 vevs in the numerator can be in any of the 4 directions, $\mathbf{X}, \mathbf{Y}, \mathbf{B}-\mathbf{L}, \mathbf{T_{3R}}$ discussed earlier.

We said that such operators can be considered as the result of integrating out states with mass $> M_G$. For example, you can convince yourself that an operator of the form $O_{22} = 16_2\ 10\ \frac{45_{B-L}\ M_G}{45_X^2}\ 16_2$ is generated by the tree graph of Figure 1, assuming $v_{10} >> v_5 \sim M_G$. Note it is at this level in the fundamental theory at M_P that additional family symmetries are needed to enforce zeros in the mass matrix. It is also trivial to evaluate the Clebsch-Gordon coefficients associated with any particular operator since the matrices $X, Y, B-L, T_{3R}$ are diagonal. Their eigenvalues on the fermion states are given in Table 2.

5 Operator analysis

Our goal is to find the *minimal* set of fermion mass operators consistent with the LED. With any given operator set one can evaluate the fermion mass matrices for up and down quarks and charged leptons. One obtains relations between mixing angles and ratios of fermion masses which can be compared with the data. It is easy to show, however, without any detailed calculations that the minimal operator set consistent with the LED is given by

$$\begin{aligned} \mathcal{L}_{fermion} \supset\ & O_{33} + O_{23} + O_{22} + O_{12} \quad ---\text{“22” texture} \\ & \text{or} \\ & O_{33} + O_{23} + O'_{23} + O_{12} \quad ---\text{“23'” texture} \end{aligned} \tag{4}$$

It is clear that at least 3 operators are needed to give non-vanishing mass to all charged fermions, i.e. $det(m_a) \neq 0$ for $a = u, d, e$. That the operators must be in

Table 2. Quantum numbers of the four 45 vevs on fermion states. Note, if u denotes a left-handed up quark, then $\bar{u}$ denotes a left-handed charge conjugate up quark.

	X	Y	B − L	T_{3R}
u	1	1/3	1	0
$\bar{u}$	1	-4/3	-1	-1/2
d	1	1/3	1	0
$\bar{d}$	-3	2/3	-1	1/2
e	-3	-1	-3	0
$\bar{e}$	1	2	3	1/2
ν	-3	-1	-3	0
$\bar{\nu}$	5	0	3	-1/2

the [33, 23 and 12] slots is not as obvious but is not difficult to show. It is then easy to show that 4 operators are required in order to have CP violation. This is because, with only 3 SO(10) invariant operators, we can redefine the phases of the three 16s of fermions to remove the three arbitrary phases. With one more operator, there is one additional phase which cannot be removed. A corollary of this observation is that this minimal operator set results in just 5 arbitrary parameters in the Yukawa matrices of all fermions, 4 magnitudes and one phase. This is the minimal parameter set which can be obtained without solving the remaining problems of the fermion mass hierarchy, one overall real mixing angle and a CP violating phase. We should point out however that the problem of understanding the fermion mass hierarchy and mixing has been rephrased as the problem of understanding the hierarchy of scales above M_G. Moreover given any particular operator set which fits the low energy data we would be obliged at some later time to construct the complete GUT theory, including symmetries which forbid additional operators and a consistent description of symmetry breaking scales. We leave this problem for future analysis.

For now we shall describe the detailed analysis of the "22" texture.¶ Models with "22" texture give the following Yukawa matrices at M_G –

$$\lambda_{\mathbf{a}} = \begin{pmatrix} 0 & z_a\, C & 0 \\ z'_a\, C & y_a\, E\, e^{i\phi} & x_a\, B \\ 0 & x'_a\, B & A \end{pmatrix} \tag{5}$$

¶As of this writing, we have found no models with "23'" texture which fit the LED.

with the subscript $a = \{u, d, e\}$. The constants $x_a, x'_a, y_a, z_a, z'_a$ are Clebschs which can be determined once the 3 operators (O_{23}, O_{22}, O_{12}) are specified. Recall, we have taken $O_{33} = A\ 16_3\ 10\ 16_3$, which is why the Clebsch in the 33 term is independent of a. Finally, combining the Yukawa matrices with the Higgs vevs to find the fermion mass matrices we have 6 arbitrary parameters given by A, B, C, E, ϕ and $\tan\beta$ describing 14 observables. We thus obtain 8 predictions. We shall use the best known parameters, $e, \mu, \tau, c, b, |V_{cd}|$, as input to fix the 6 unknowns. We then predict the values of $u, d, s, t, \tan\beta, |V_{cb}|, |V_{ub}|$ and J.

We now show how (within the context of our dynamic principles) we can use the LED to guide us towards *the theory of fermion masses* at M_G. We search for all operators (O_{23}, O_{22}, O_{12}) with dimension ≤ 10 which fit the data. Using a coarse-grained analysis, it is easy to show that just 9 models (with one caveat) may fit the data. We now describe this analysis.

5.1 *3rd generation fit* — O_{33}

Using the values of α and $\sin^2\theta_W$ evaluated at M_Z, we obtain α_G, M_G and $\alpha_s(M_Z)$. There is a theoretical uncertainty in this prediction due to unknown threshold corrections at both M_G and the weak scale. There is also a 10% experimental uncertainty in $\alpha_s(M_Z)$. Using arbitrary threshold corrections we can obtain any experimentally allowed value of α_s. We thus allow for all values of $\alpha_s(M_Z) = .12 \pm .01$ self-consistently by introducing arbitrary threshold corrections.

The analysis for the third generation follows —

$$m_b = \frac{v}{\sqrt{2}}\ A\ \cos\beta\ \left(\frac{\eta_b}{S_b}\right) \tag{6}$$

$$m_\tau = \frac{v}{\sqrt{2}}\ A\ \cos\beta\ \left(\frac{\eta_\tau}{S_\tau}\right) \tag{7}$$

$$m_t = \frac{v}{\sqrt{2}}\ A\ \sin\beta\left(\frac{1}{S_t}\right) \tag{8}$$

where $v = 246GeV$ and the terms in parentheses are renormalization group factors which are implicit functions of both A and $\alpha_s(M_Z)$. The numerator takes into account renormalization from M_Z to m_b or m_τ, and the denominator takes into account the running from M_G to M_Z. We use two loop renormalization group equations.

Using $m_b = 4.25 \pm .1GeV$ and $m_\tau = 1.7841GeV$ (these are running masses $m(m)$) as input for a given value of $\alpha_s = .118$(for example) we obtain A from the relation

$$\frac{m_b}{m_\tau} = \left(\frac{\eta_b}{\eta_\tau}\frac{S_\tau}{S_b}\right)(\alpha_s(M_Z), A).$$

Plugging this value of A into the expression for m_τ we then obtain $\tan\beta$. We find $\tan\beta \sim 59$ for $m_b = 4.34GeV$. Now using the values of A and $\tan\beta$ in the expression for m_t we find $m_t(pole) \sim 188GeV$. In general, we find values of $\tan\beta = 54 \pm 5$

and $m_t = 185 \pm 15GeV$. Both m_t and $\tan\beta$ increase for increasing values of α_s or decreasing values of m_b. Thus $m_t = 170GeV$ and $\tan\beta = 50$ is obtained for $\alpha_s(M_Z) = .110$, $m_b = 4.35GeV$. Similar results have been obtained previously by ALS [6]. Note they find lower values of m_t since they allow for values of α_s which are lower than those presently admissable by the data.

5.2 *2nd generation* — O_{22}

Let us now consider the 2nd generation. We have 4 relations which must be satisfied by the LED.

$$|V_{cb}| \approx |x_u - x_d| \; \frac{B}{A} \sim 1/20 \tag{9}$$

$$\frac{m_\mu}{m_\tau} \approx |y_e \; \frac{E}{A} \; e^{i\phi} - x_e \; x'_e \frac{B^2}{A^2}| \sim 1/17 \tag{10}$$

$$\frac{m_s}{m_b} \approx |y_d \; \frac{E}{A} \; e^{i\phi} - x_d \; x'_d \frac{B^2}{A^2}| \sim 1/25 \tag{11}$$

$$\frac{m_c}{m_t} \approx |y_u \; \frac{E}{A} \; e^{i\phi} - x_u \; x'_u \frac{B^2}{A^2}| \sim 10^{-2} \tag{12}$$

We have written these equations using the parameters at M_G, ignoring for the moment small renormalization group corrections. Using the first relation we see that the ratio $B/A \sim 1/10$, assuming Clebschs of order 1. The 2nd and 3rd relations thus require $E/A \sim 1/10$. The last relation then requires that the Clebsch, $y_u << 1$. Finally, the relation $m_s = m_\mu/3$ at M_G, first suggested by Georgi-Jarlskog [7], must be incorporated, since it is in good agreement with the LED. We thus conclude that the Clebschs, y_a,(including RG corrections) should approximately be in the ratio[||]

$$y_u : y_d : y_e =< 1/3 : 1 : 3.$$

The Clebschs, y_a, are derived from the operator O_{22}. We have searched over all dimension 5 and 6 operators to find solutions to the above Clebsch ratios. We find 6 solutions —

$$16_2 \; (45_X) \; 10 \; (\frac{45_{B-L}}{45_X}) \; 16_2 \tag{13}$$

$$16_2 \; (\frac{1}{45_X}) \; 10 \; (45_{B-L}) \; 16_2$$

$$16_2 \; (45_X) \; 10 \; (45_{B-L}) \; 16_2$$

$$16_2 \; 10 \; (\frac{45_{B-L}}{45_X}) \; 16_2$$

[||] If the Clebschs are not of order 1 as we assumed, then it is possible that the 4 relations may be satisfied with some fine-tuning and a completely different ratio of Clebschs. We have not pursued this possibility further. This is our one caveat.

$$16_2\ 10\ (45_X\ 45_{B-L})\ 16_2$$
$$16_2\ 10\ (\frac{45_{B-L}}{45_X^2})\ 16_2$$

However, note that *all* solutions give the *same* ratio of Clebschs —

$$y_u : y_d : y_e = 0 : 1 : 3. \tag{14}$$

5.3 *1st generation* — O_{12}

We can now show that the operator O_{12} is *unique*. The first two generations satisfy the relations —

$$\frac{m_d}{m_s} \approx 9\ \frac{z_d\ z'_d}{z_e\ z'_e}\ \frac{m_e}{m_\mu} \left(\frac{\eta_\mu\ \eta_d}{\eta_e\ \eta_s}\right) \tag{15}$$

$$\frac{m_u}{m_d} \approx \frac{z_u\ z'_u}{z_d\ z'_d}\ \frac{m_s}{m_c}\ \tan^2\beta\ \left(\frac{S_d}{S_u}\right)^2 \tag{16}$$

$$|V_{cd}| = \left|\sqrt{\frac{z_d}{z'_d}}\sqrt{\frac{m_d}{m_s}} - \sqrt{\frac{z_u}{z'_u}}\sqrt{\frac{m_u}{m_c}}\ e^{-i\phi}\right| \tag{17}$$

The first relation is satisfied *if*

$$z_d\ z'_d \approx z_e\ z'_e.$$

This relation is satisfied if the Clebsch is derived from an SU(5) invariant vev, i.e. 45_X. Since $\tan\beta$ is large, the second relation requires

$$\frac{z_u\ z'_u}{z_d\ z'_d} \approx (\frac{1}{3})^{6 \text{ or } 7}.$$

Finally, the last relation requires

$$z_d \approx z'_d.$$

The unique operator which satisfies the above 3 relations is

$$O_{12} = 16_1\ (\frac{45_X}{M_P})^3\ 10\ (\frac{45_X}{M_P})^3\ 16_2. \tag{18}$$

5.4 O_{23}

We have now determined, using simple arguments, all but one of the operators. The charged fermion mass matrices are given by —

$$U = \begin{pmatrix} 0 & C & 0 \\ C & 0 & x_u\ B \\ 0 & x'_u\ B & A \end{pmatrix} \tag{19}$$

$$D = \begin{pmatrix} 0 & -27\ C & 0 \\ -27\ C & E\ e^{i\phi} & x_d\ B \\ 0 & x'_d\ B & A \end{pmatrix} \tag{20}$$

$$E = \begin{pmatrix} 0 & -27\ C & 0 \\ -27\ C & 3\ E\ e^{i\phi} & x_e\ B \\ 0 & x'_e\ B & A \end{pmatrix} \tag{21}$$

With this form for the mass matrices, the KM element V_{cb} satisfies the relation —

$$|V_{cb}| \approx \chi \sqrt{\frac{m_c}{m_t}} \sqrt{\frac{S_u}{\eta_c\ S_t\ S}} \sim .058\ \chi. \tag{22}$$

where $\chi \equiv \frac{|x_u - x_d|}{\sqrt{|x_u x'_u|}}$ and the last term results from using central values of the input parameters. Experimentally, we have an upper bound on $|V_{cb}| \leq .054$. We thus require that the function of Clebschs, χ, satisfy $\chi < 1$.

We have searched over all dimension 5 and 6 operators for $\chi < 1$. We find 9 possible models with only 3 different values of $\chi = \frac{2}{3}, \frac{5}{6}$, or $\frac{8}{9}$. The 9 models are listed below.

$$\chi = 2/3$$

$$1 \quad 16_2\ (45_Y)\ 10\ (\frac{1}{45_X})\ 16_3 \tag{23}$$

$$2 \quad 16_2\ (45_Y)\ 10\ (\frac{45_{B-L}}{45_X})\ 16_3 \tag{24}$$

$$3 \quad 16_2\ (\frac{45_Y}{45_X})\ 10\ (\frac{1}{45_X})\ 16_3 \tag{25}$$

$$4 \quad 16_2\ (\frac{45_Y}{45_X})\ 10\ (\frac{45_{B-L}}{45_X})\ 16_3 \tag{26}$$

$$\chi = 5/6$$

$$5 \quad 16_2\ (45_Y)\ 10\ (\frac{45_Y}{45_X})\ 16_3 \tag{27}$$

$$6 \quad 16_2\ (\frac{45_Y}{45_X})\ 10\ (\frac{45_Y}{45_X})\ 16_3 \tag{28}$$

$$\chi = 8/9$$

$$7 \quad 16_2\ 10\ (\frac{1}{45_X^2})\ 16_3 \tag{29}$$

$$8 \quad 16_2\ 10\ (\frac{45_{B-L}}{45_X^2})\ 16_3 \tag{30}$$

$$9 \quad 16_2\ 10\ (\frac{45_{B-L}}{45_X})^2\ 16_3 \tag{31}$$

This is as far as we can get with our coarse-grained search. We must now take the 9 distinct models and test the predictions. We are presently in the midst of a complete renormalization group analysis, obtaining predictions as a function of the input parameters, $e, \mu, \tau, c, b, |V_{cd}|$ and $\alpha_s(M_Z)$. For now we present some preliminary results.

6 Preliminary results

The results in Table 3 are for inputs $m_c = 1.23 GeV, m_b = 4.34 GeV, |V_{cd}| = .221$ and $\alpha_s(M_Z) = .118$ for 4 different models. The model numbers are defined in Eqs. (23 - 31)

Table 3. Results for models 3, 6, 8, and 9.

	3	6	8	9
$m_d[MeV]$	6.8	6.9	6.9	7.3
m_u/m_d	.82	.79	.80	.79
m_s/m_d	24.8	24.1	24.9	22.4
m_t	188	188	188	188
$\tan\beta$	59	59	59	59
$\|V_{cb}\|$	.039	.049	.052	.052
$\|V_{ub}/V_{cb}\|$	.065	.064	.066	.068
$J \times 10^5$	1.8	2.9	3.3	3.7

Note, $|V_{cb}|$ and J are sensitive to the value of χ and so can distinguish between the 3 types of models. With better data, $|V_{cb}|$ will distinguish between these choices. At present we can use ϵ_K, the CP violating parameter in K decay, to distinguish these models through its dependence on J. We find that models with $\chi = 2/3$ tend to give too little CP violation. Finally, we see that m_s/m_d and m_u/m_d will also be useful in constraining the models. Complete results will be presented in an upcoming publication. [1]

7 Conclusion

We have presented a straightforward method for trying to understand the origin of fermion masses. There are four main ingredients —

- The assumption of a SUSY GUT with its SUSY DESERT implies that the Low Energy Data becomes a window into the physics at the GUT scale.

- The assumption of symmetries (SO(10) plus family symmetries) with a general operator analysis allows us to reduce the number of fundamental parameters in the theory.
- With the above assumptions, we obtain predictive theories of fermion masses. Given 6 inputs, e, μ, τ, c, b, $|V_{cd}|$, we find 8 predictions for $u, d, s, t, \tan\beta$, $|V_{cb}|$, $|V_{ub}|$ and J.
- This method allows us to use the LED to systematically find the theory of fermion masses at M_G.

A final note: there is no apriori reason to believe that this scheme should work, *but* if it does work, i.e. if our results agree with the LED, then perhaps we will have learned something about the physics at the GUT scale. For example, consider the set of possible models 1 – 9 in Eqs. (23 - 31). Each one has the vev 45_X in the denominator. This can only occur if this vev gives mass $>> M_G$ to some fermion. When this fermion is integrated out in the effective theory at M_G, the vev then appears in the denominator of these higher dimension operators. Thus fermion masses and mixing angles at low energies may tell us about the hierarchy of symmetry breaking scales above M_G.

References

[1] G.W. Anderson, S. Dimopoulos, L.J. Hall, S. Raby, and G. Starkman, in preparation.

[2] P. Langacker, Pennsylvania preprint UPR-0555T (1993).

[3] Bethke, Proceedings of the International Conference on High Energy Physics, Dallas, TX, August 1992.

[4] D. Kaplan and A. Manohar, *Phys. Rev. Lett.* **56**, 2004 (1986); H. Leutwyler, *Nucl. Phys.* **B337**, 108 (1990).

[5] P. Drell, Proceedings of the International Conference on High Energy Physics, Dallas, TX, August 1992.

[6] B. Ananthanarayan, G. Lazarides and Q. Shafi, *Phys. Rev.***D44** 1613 (1991); H. Arason et al., *Phys. Rev. Lett.* **67** 2933 (1991); S. Kelley, J.Lopez and D. Nanopoulos, *Phys. Lett.* **B274** 387 (1992). See also the earlier work of M. Olechowski and S. Pokorski, Phys. Lett. B214, 393 (1988).

[7] H. Georgi and C. Jarlskog, *Phys. Lett.* **86B**, 297 (1979).

Basic Results and Experimental Predictions from SUSY GUTs

JORGE L. LOPEZ[(a),(b)], D. V. NANOPOULOS[(a),(b)], and A. ZICHICHI[(c)1]

[(a)] *Center for Theoretical Physics, Department of Physics, Texas A&M University College Station, TX 77843–4242, USA*
[(b)] *Astroparticle Physics Group, Houston Advanced Research Center (HARC) Mitchell Campus, The Woodlands, TX 77381, USA*
[(c)] *CERN, 1211 Geneva 23, Switzerland*

ABSTRACT

The most basic results of Super Grand unification are reviewed as well as their consequences, including experimental predictions for existing facilities. One of these results is that supersymmetric particles can be at the verge of discovery, lurking around the corner at present and near future facilities. Experimentally viable models must satisfy *all* available experimental data. A recent realization is that several important indirect constraints can be obtained by calculating one-loop supersymmetric contributions to rare Standard Model processes, such as $b \to s\gamma$, $(g-2)_\mu$, and the one-loop electroweak radiative corrections. Finally, we emphasize that experimentally testable models must be as simple as theoretical constraints allow. An archetypal example is the $SU(5) \times U(1)$ supergravity model which can be described in terms of only three parameters, one of which is the soon-to-be determined top-quark mass. A further reduction to only one parameter (plus the top-quark mass) is in fact possible, making the experimental predictions of $SU(5) \times U(1)$ supergravity very sharp.

1 Introduction

The LEP e^+e^- collider at CERN has been in operation since 1989 with a center-of-mass energy around the Z-boson mass ($M_Z \approx 91\,\mathrm{GeV}$). During this time a vast amount of data has been collected on many different decay modes of the Z. Analyses of the totality of the data in the context of the Standard Model of electroweak interactions show no deviations from expectations, which have been tested at the level of one-loop and for some quantities two-loop corrections [1]. In particular, the weak mixing angle ($\sin^2\theta_W(M_Z)$) and the QCD running coupling ($\alpha_3(M_Z)$) have been determined with unprecedented accuracy. This picture-perfect agreement between theory and experiment has had several consequences. One of these is the immediate rejection of large classes of models which predicted new physics below the electroweak scale. This

[1] Conference speaker.

is the case even for a particularly well motivated extension of the Standard Model, namely supersymmetry, whose phenomenological predictions need as input the lower limits on the masses of the superpartners of the known matter and gauge particles.

The lesson from LEP is therefore how to reconcile the successes of the Standard Model with the compelling reasons for the existence of new physics beyond it. An obvious shortcoming of the Standard Model is the ad-hoc nature of the many parameters involved: the quark and lepton masses, the quark mixing angles, and the CP violating phase. However, even if explanations for these parameters could be found within physics not so far from the electroweak scale, a problem would remain when extrapolating the theory to much larger energies. This so-called *gauge hierarchy problem* is manifest in theories with elementary scalar fields, such as the Higgs boson in the Standard Model. The idea is that radiative corrections to the Higgs-boson mass become extremely large if the theory also contains very massive particles, as is the case in unified theories, or any theory which attempts to incorporate gravity in a common framework. This is a "problem" for the theory, as opposed to a terminal disease, in that there is a grossly undesirable solution which involves fine-tuning parameters to large numbers of decimal places, order-by-order in perturbation theory.

Particle physics seems to always find solutions to its problems by invoking the existence of larger symmetries. The proposed solutions to the above two problems of the Standard Model are no different: unified models and supersymmetry, respectively. The former postulate that the Standard Model is actually embedded in a larger group structure which includes $SU(3)_C \times SU(2)_L \times U(1)_Y$ but which is only fully manifest at very large energies, where new degrees of freedom are excited and the symmetry group is effectively enlarged to *e.g.*, $SU(5), SU(5)\times U(1), SO(10)$. In the larger theory relations among the parameters arise, which help explain some of the regularities observed in the Standard Model parameters, such as the quantization of the electric charge and some relations among the quark and lepton masses. In fact, the existence of three light generations of quarks and leptons had already been predicted in this context in studies of the m_b/m_τ ratio [2]. An immediate consequence of the larger symmetry is that the gauge couplings of the Standard Model are "unified", *i.e.*, take the same value at the unification scale (M_U), even though they are measured to be different numbers at the weak scale. By-products of great experimental interest are the new interactions among the quarks and leptons which may be induced by the heavy degrees of freedom. Among these, the most model-independent one leads to the prediction of proton decay, which must occur at a rather slow rate ($\tau_p \gtrsim 10^{32}\,$y) to avoid conflict with experimental limits.

Supersymmetry solves the gauge hierarchy problem by predicting the existence of partners for the ordinary particles which differ by half-a-unit of spin, *e.g.*, electron($\frac{1}{2}$)-selectron(0), photon(1)-photino($\frac{1}{2}$). The radiative corrections to the Higgs-boson mass are now tamed by the fact that the partners of the heavy particles have an opposite effect on the correction. However, since supersymmetry cannot be an unbroken symmetry (otherwise the sparticles would have the same mass as their Standard Model partners and no such sparticles have been observed) the cancellation is not perfect, but up to the typical mass splitting of particles and their partners. If this splitting can be predicted to be no larger than $\sim 1\,\text{TeV}$, then the gauge hierarchy problem will remain solved. In fact, it is this further requirement which demands that we enlarge the symmetry even more, by making supersymmetry a local symmetry called *supergravity*. In this class of theories the mass splittings can be explicitly computed in terms of very few parameters. This ability is of great experimental interest, since a large class of (*e.g.*, collider and rare) low-energy processes can be calculated in terms of few parameters, providing close knit correlations among the several experimental predictions, which may otherwise appear arbitrary.

We will consider two such supergravity models: the minimal $SU(5)$ [3] and $SU(5)\times U(1)$ [4]. Conceptually, the $SU(5)$ model is a typical (and the simplest) grand unified supersymmetric model which predicts the existence of proton decay at a rate which should be on the verge of being observed. In fact, its non-supersymmetric ancestor, the Georgi-Glashow $SU(5)$ model [5], was ruled out experimentally on the basis of its incorrect (too short) prediction for the proton lifetime. Theoretically, this model also suffers from the gauge hierarchy problem, which is present in the absence of supersymmetry. In the supersymmetric $SU(5)$ model, the unification of gauge couplings occurs at a scale $M_U \sim 10^{16}\,\text{GeV}$, and the proton lifetime problem of non-supersymmetric $SU(5)$ is naturally solved. As we now well know, unification of the gauge couplings also does not work in non-supersymmetric $SU(5)$, while it works very well in its supersymmetric counterpart. The goodness of the prediction for the m_b/m_τ relation alluded to above is also maintained with the addition of supersymmetry [2].

The $SU(5) \times U(1)$ supergravity model is best motivated in the context of superstrings, where it is found that $SU(5)\times U(1)$ (in contrast with $SU(5)$) is relatively easy to obtain. In this model the unification of couplings should occur at the string scale ($M_U \sim 10^{18}\,\text{GeV}$). This fact appears to require a non-minimal spectrum of particles at intermediate scales, in contrast with the $SU(5)$ model where the "big desert" picture is assumed to hold. Furthermore, the $SU(5) \times U(1)$ supergravity model allows a further reduction of the unknown parameters of the theory, making its predictions very sharp.

2 Convergence of gauge couplings: geometry versus physics

The three gauge couplings of the Standard Model, as measured at the Z-boson mass scale, can be expressed in terms of: α_e, $\sin^2\theta_W$, and α_3. Experimentally we know that (see *e.g.*, [6]),

$$\alpha_e^{-1}(M_Z) = 127.9 \pm 0.1, \quad (1)$$

$$\alpha_3(M_Z) = 0.120 \pm 0.010, \quad (2)$$

$$\sin^2\theta_W(M_Z) = 0.2324 \pm 0.0006. \quad (3)$$

The $U(1)_Y$ and $SU(2)_L$ gauge couplings are related to this by $\alpha_1 = \frac{5}{3}(\alpha_e/\cos^2\theta_W)$ and $\alpha_2 = (\alpha_e/\sin^2\theta_W)$. In numbers:

$$\text{At} \quad Q = M_Z \qquad \left\{ \begin{array}{lcr} \alpha_1^{-1} & = & 58.91 \pm 0.16 \\ \alpha_2^{-1} & = & 29.72 \pm 0.080 \\ \alpha_3^{-1} & = & 8.33 \pm 0.69 \end{array} \right. \tag{4}$$

The three gauge couplings evolve with increasing values of the scale Q in a logarithmic fashion, and may become equal at some higher scale, signaling the possible presence of a larger gauge group. However, this need not be the case: the three gauge couplings may meet and then depart again. Conceptually, the presence of a unified group is essential in the discussion of unification of couplings. In this case, the newly excited degrees of freedom will be such that all three couplings will evolve together for scales $Q > M_U$, and one can then speak of a unified coupling. (For a recent review and extensive references to the current literature see *e.g.*, Ref. [7].)

The running of the gauge couplings is prescribed by a set of first-order differential equations: the renormalization group equations (RGEs) for the gauge couplings. In general, there is one such equation for each dynamical variable in the theory (*i.e.*, for each gauge coupling, Yukawa coupling, and sparticle mass). These equations give the rate of change of the gauge coupling as the scale Q is varied. This rate of change is proportional to (some power of) the gauge coupling itself, and the coefficient of proportionality is called the *beta function.* The beta functions encode the spectrum of the theory, and how the various gauge couplings influence the running of each other (a higher-order effect). Assuming that all supersymmetric particles have a common mass M_{SUSY}, the RGEs are:

$$\frac{d\alpha_i^{-1}}{dt} = -\frac{b_i}{2\pi} - \sum_{j=1}^{3} \frac{b_{ij}\alpha_j}{8\pi^2}. \tag{5}$$

where $t = \ln(Q/M_U)$, with Q the running scale and M_U the unification mass. The one-loop (b_i) and two-loop (b_{ij}) beta functions are given by

$$b_i = \left(\tfrac{33}{5}, 1, -3\right), \tag{6}$$

$$b_{ij} = \begin{pmatrix} \frac{199}{25} & \frac{27}{5} & \frac{88}{5} \\ \frac{9}{5} & 25 & 24 \\ \frac{11}{5} & 9 & 14 \end{pmatrix}. \tag{7}$$

These equations are valid from $Q = M_{SUSY}$ up to $Q = M_U$. For $M_Z < Q < M_{SUSY}$ an analogous set of equations holds, but with beta functions which reflect the non-supersymmetric nature of the theory (*i.e.*, with all the sparticles decoupled),

$$b'_i = \left(\tfrac{41}{10}, -\tfrac{19}{6}, -7\right), \tag{8}$$

$$b'_{ij} = \begin{pmatrix} \frac{199}{50} & \frac{27}{10} & \frac{44}{5} \\ \frac{9}{10} & \frac{35}{6} & 12 \\ \frac{11}{10} & \frac{9}{2} & -26 \end{pmatrix}. \tag{9}$$

The non-supersymmetric equations are supplemented with the initial conditions given in Eq. (4).

If the above is all the physics which is incorporated in the study of the convergence of the gauge couplings, then it is easy to see that the couplings will always meet at some scale M_U, provided that M_{SUSY} is tuned appropriately [8, 9, 10, 11]. This is a simple consequence of euclidean geometry, as can be seen from Eq. (5). Neglecting the higher-order terms, we see that as a function of t, α_i^{-1} are just straight lines. In fact, the slope of these lines changes at $Q = M_{SUSY}$, where the beta functions change. The convergence of three straight lines with a change in slope is then guaranteed by euclidean geometry, as long as the point where the slope changes is tuned appropriately. What is non-trivial about the convergence of the couplings is that with the initial conditions given in Eq. (4), the change in slope needs to be $M_{SUSY} \sim 1\,\mathrm{TeV}$ [10].

This prediction for the likely scale of the supersymmetric spectrum (*i.e.*, $M_{SUSY} \sim 1\,\mathrm{TeV}$ [10]) is in fact *incorrect* [12]. The reason is simple: the physics at the unification scale, which is used to predict the value of M_{SUSY}, has been ignored completely. In fact, such a geometrical picture of convergence of the gauge couplings is physically inconsistent, since for scales $Q > M_U$ the gauge couplings will depart again. One must consider a unified theory to be assured that the couplings will remain unified. This entails the study of a new kind of effect, namely the influence on the running of the gauge couplings of the degrees of freedom which are excited near the unification scale (*i.e.*, the *heavy threshold effects*) [8]. In fact, the whole concept of a single unification point needs to be abandoned. The upshot of all this is that the theoretical uncertainties on the values of the parameters describing the heavy GUT particles are such that the above prediction for M_{SUSY} is washed out completely [12]. Furthermore, the insertion of a realistic spectrum of sparticles at low energies (as opposed to an unrealistic common M_{SUSY} mass) blurs the issue some more. *Thus, it is perfectly possible to obtain acceptable unification, with supersymmetric particle masses as low as experimentally allowed.*

A related point is that LEP data do not uniquely demonstrate that the gauge couplings must unify at a scale $M_U \sim 10^{16}\,\mathrm{GeV}$. This is probably the simplest conclusion one could draw. However, this conclusion is easily altered by for example considering models with particles at intermediate scales, *i.e.*, by populating the "big desert". In fact, one such simple modification allows the gauge couplings to converge at the string scale $M_U \sim 10^{18}\,\mathrm{GeV}$ instead [4].

3 Constraints from unification

The convergence of the gauge couplings implies that given α_e and α_3, one is able to compute the values of $\sin^2\theta_W$, the unification scale M_U, and the unified coupling α_U. In lowest-order approximation (*i.e.*, neglecting all GUT thresholds, two-loop effects, and taking $M_{SUSY} = M_Z$) one obtains

$$\begin{aligned}
\ln\frac{M_U}{M_Z} &= \frac{\pi}{10}\left(\frac{1}{\alpha_e} - \frac{8}{3\alpha_3}\right), & (10)\\
\frac{\alpha_e}{\alpha_U} &= \frac{3}{20}\left(1 + \frac{4\alpha_e}{\alpha_3}\right), & (11)\\
\sin^2\theta_W &= 0.2 + \frac{7\alpha_e}{15\alpha_3} \quad . & (12)
\end{aligned}$$

These equations provide a rough approximation to the actual values obtained when all effects are included. Nonetheless, they embody the most important dependences on the input parameters. In Fig. 1 (from Ref. [11]) we show the relation between M_U and α_3 for various values of $\sin^2\theta_W$. One can observe that:

$$\begin{aligned}
\alpha_3 \uparrow \quad &\Rightarrow \quad M_U \uparrow \quad \text{for fixed } \sin^2\theta_W & (13)\\
\sin^2\theta_W \uparrow \quad &\Rightarrow \quad M_U \uparrow \quad \text{for fixed } \alpha_3 & (14)\\
\alpha_3 \uparrow \quad &\Rightarrow \quad \sin^2\theta_W \downarrow \quad \text{for fixed } M_U & (15)
\end{aligned}$$

These are the most important systematic correlations, which are not really affected by the neglected effects. These correlations are evident in Eqs. (10–12) and also in Fig. 1. In this figure we also show the lower bound on M_{GUT} which follows from the proton decay constraint. Clearly a lower bound on $\alpha_3(M_Z)$ results, which allows the world-average value. It is also possible to establish an anticorrelation between M_{SUSY} and M_U, *i.e.*, as the supersymmetric spectrum gets heavier, the unification mass decreases. This is shown in Fig. 2. As we have emphasized, realistically speaking M_{SUSY} is not a very meaningful parameter since the supersymmetric spectrum could span order of magnitude in mass. Nonetheless, a similar effect is observed in more realistic situations.

4 The origin of M_{SUSY} and the need for local supersymmetry

The calculations which we have described so far, attempted to determine the value of M_{SUSY}, or more properly the supersymmetric particle spectrum, by fitting the spectrum to obtain the "best possible" unification picture. This program did not succeed because of the large inherent uncertainties in the physics at the GUT scale. Nevertheless, for a given GUT model, it should be possible to compute the GUT threshold effects and attempt the "best fit" procedure to deduce the corresponding light supersymmetric spectrum. This picture is not very satisfying since we would like to know why the supersymmetric spectrum would be the way the fit required it to be. In other words, the real question is: what determines the values of the sparticle masses? And why should these be below ~ 1 TeV, so that the gauge hierarchy problem is not re-introduced?

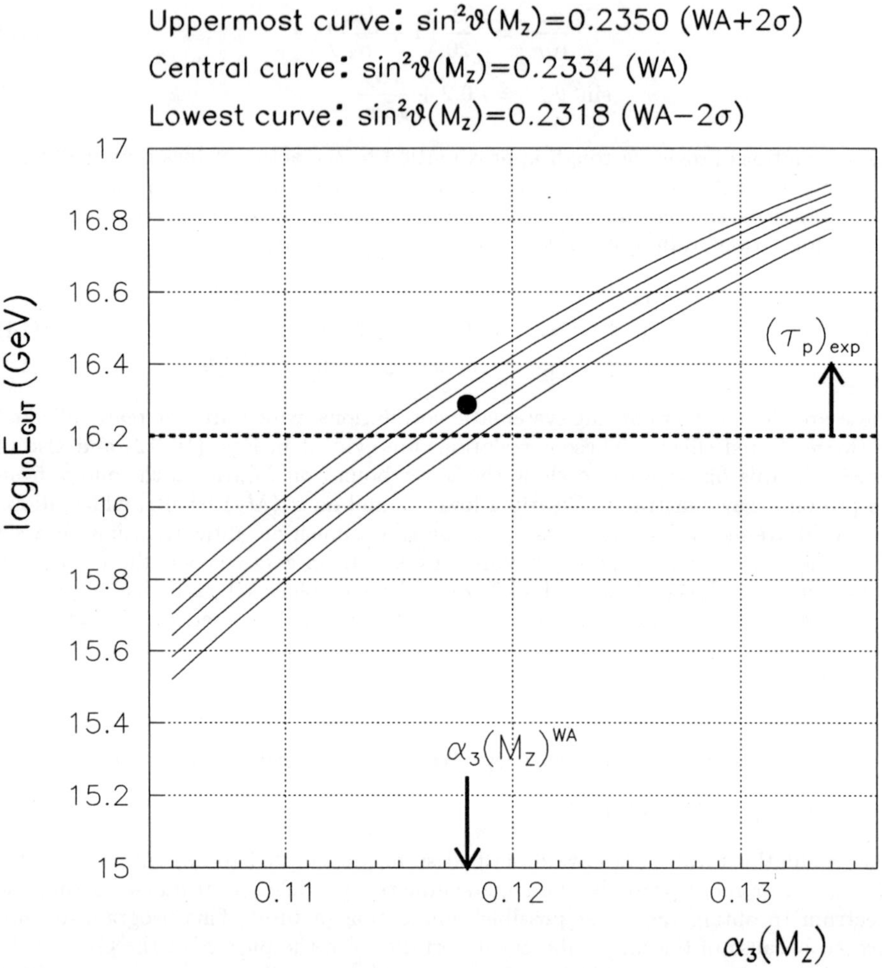

Figure 1: The unification scale M_{GUT} versus $\alpha_3(M_Z)$ for various values of $\sin^2\theta_W(M_Z)$ within $\pm 2\sigma$ of the world-average value. Also indicated is the lower bound on M_{GUT} from the lower limit on the proton lifetime.

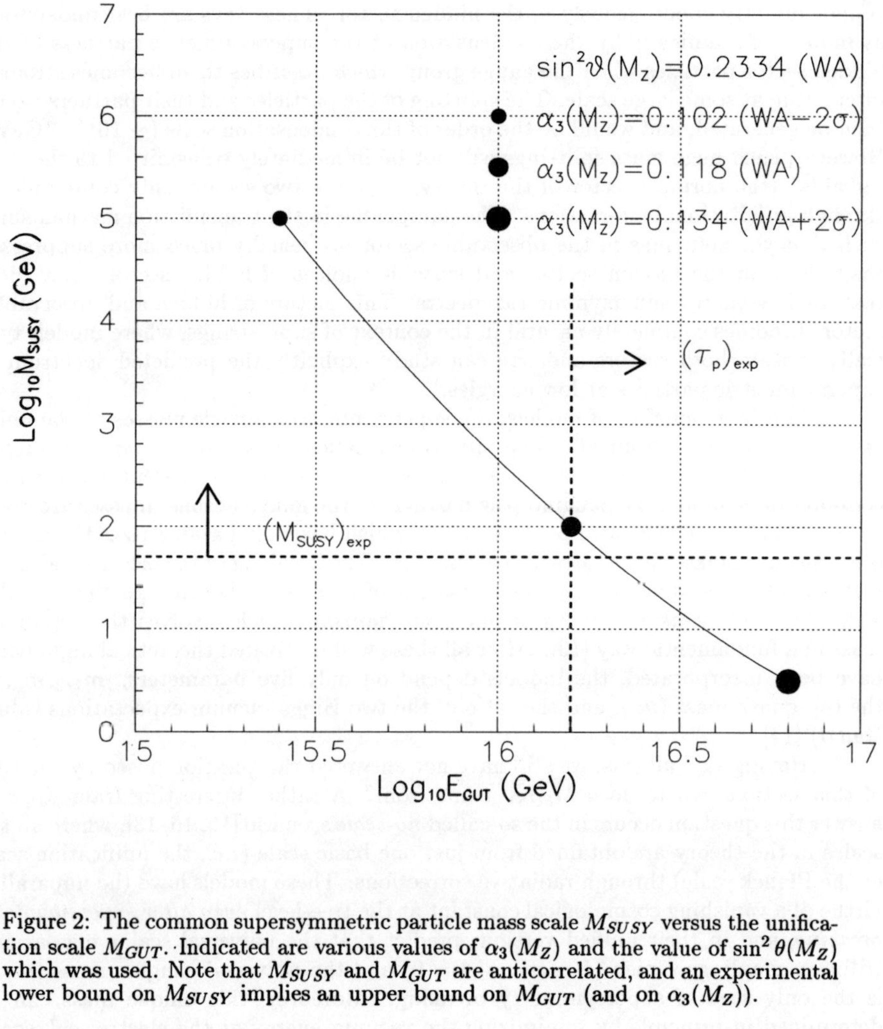

Figure 2: The common supersymmetric particle mass scale M_{SUSY} versus the unification scale M_{GUT}. Indicated are various values of $\alpha_3(M_Z)$ and the value of $\sin^2\theta(M_Z)$ which was used. Note that M_{SUSY} and M_{GUT} are anticorrelated, and an experimental lower bound on M_{SUSY} implies an upper bound on M_{GUT} (and on $\alpha_3(M_Z)$).

As mentioned in Sec. 1, considering a theory with supergravity (instead of global supersymmetry) provides the means to compute the masses of the sparticles [3]. This framework assumes that supersymmetry breaking occurs in a "hidden sector" of the theory, where "gravitational particles" (those introduced when the supersymmetry was made local) may grow vacuum expectation values (vevs) which break supersymmetry spontaneously in the hidden sector. These vevs are best understood as induced dynamically by the condensation of the supersymmetric partners of the hidden sector particles when the gauge group which describes them becomes strongly interacting at some large scale. The splitting of the particles and their partners would then be generated, and would be the order of the condensation scale ($\sim 10^{12-16}\,\mathrm{GeV}$). However, such huge mass splittings will not be immediately transmitted to the "observable" (the normal) sector of the theory, since the two sectors only communicate through gravitational interactions. The dampening in the transmission mechanism is such that the splittings in the observable sector are usually much more suppressed than those in the hidden sector, and suitable choices of hidden sectors may yield realistic low-energy supersymmetric spectra. This picture of hidden and observables sectors becomes completely natural in the context of superstrings, where models typically contain both sectors and one can study explicitly the predicted spectrum of supersymmetric particles at low energies.

In a large number of models, the supersymmetric particle masses at the unification scale are also "unified". This situation is called *universal soft-supersymmetry-breaking*, and the masses of all scalar partners (*e.g.*, squarks and sleptons) take the common value of m_0, the gaugino (the partners of the gauge bosons) masses are given by $m_{1/2}$, and there is a third parameter (A) which basically parametrizes the mixing of stop-squark mass eigenstates at low energies. The breaking of the electroweak symmetry is obtained dynamically in the context of these models, through the so-called *radiative electroweak symmetry breaking mechanism*, which involves the top-quark mass in a fundamental way [13]. After all these well motivated theoretical ingredients have been incorporated, the models depend on only five parameters: $m_{1/2}, m_0, A$, the top-quark mass (m_t), and the ratio of the two Higgs vacuum expectations values ($\tan\beta$) [14].

Having said all this, we still have not answered the question posed by the title of this section: where does M_{SUSY} come from? A rather interesting framework to answer this question occurs in the so-called *no-scale* scenario [15, 16, 13], where all the scales in the theory are obtained from just one basic scale (*i.e.*, the unification scale or the Planck scale) through radiative corrections. These models have the unparalled virtue of a vanishing cosmological constant at the tree-level *even after supersymmetry breaking*, and in their unified versions predict that the universal scalar masses and trilinear couplings vanish (*i.e.*, $m_0 = A = 0$) and the universal gaugino mass ($m_{1/2}$) is the only seed of supersymmetry breaking. Moreover, this unique mass can be determined in principle by minimizing the vacuum energy at the electroweak scale. The generic result is $m_{1/2} \sim M_Z$ [15, 16], in agreement with theoretical prejudices (*i.e.*, "naturalness"). Furthermore, no-scale supergravity is obtained in the infrared limit of superstrings [17].

More generally, in generic supergravity models the five-dimensional parameter space is constrained by phenomenological requirements, such as sparticle and Higgs-boson masses not in conflict with present experimental lower bounds, a sufficiently long proton lifetime, a sufficiently old Universe (a cosmological constraint on the amount of dark matter in the Universe today), various indirect constraints from well measured rare processes, etc. In addition, further theoretical constraints can be imposed which give m_0 and A as functions of $m_{1/2}$, and thus reduce the dimension of the parameter space down to just three. In what follows we will focus on some specific supergravity models which are so constrained that precise experimental predictions can be made.

5 Specific Models: SU(5) and SU(5)xU(1)

The two supergravity models that we have chosen belong to the class of models we just described. However, these models are even more predictive that the run-of-the-mill supergravity model because further well motivated theoretical assumptions are made.

5.1 The minimal SU(5) supergravity model

In this model the gauge group is $SU(5)$, which completely contains the Standard Model as a subgroup. This implies that all three Standard Model gauge couplings should evolve from low energies and become equal at (and above) the unification scale M_U (up to heavy threshold corrections). This scale has some dependence on the uncertainties on the various parameters; one usually obtains $M_U \sim 10^{16}\,\mathrm{GeV}$. The Standard Model particles and their superpartners are assigned to the $\mathbf{\bar{5}}$ and $\mathbf{10}$ representations:

$$\bar{\mathbf{5}} = \{d^c, L = \left(\begin{array}{c} e \\ \nu_e \end{array} \right)\}, \qquad \mathbf{10} = \{Q = \left(\begin{array}{c} u \\ d \end{array} \right), u^c, e^c\}. \tag{16}$$

The heavy GUT particles which are excited near the unification scale include the **24** representation of gauge bosons (and gauginos) which contains the twelve Standard Model gauge bosons (8 gluons, $W^\pm, Z, \gamma$) plus twelve heavy (charged and colored) gauge bosons which mediate the $SU(5)$ gauge interactions. There is also a **24** representation of Higgs bosons (and their superpartners), and the vacuum expectation value of the neutral component of this set effects the $SU(5) \to SU(3)_C \times SU(2)_L \times U(1)_Y$ gauge symmetry breaking. The minimal $SU(5)$ spectrum also includes a pair of pentaplet representations ($\mathbf{5},\mathbf{\bar{5}}$) which contain the two Higgs doublets of the low-energy supersymmetric theory, but also a pair of colored Higgs triplets.

Perhaps the most decisive property of this model is the prediction for the proton lifetime. Proton decay can be mediated in two ways: by exchange of heavy gauge or Higgs bosons (dimension-six operators), or by the exchange of heavy Higgsinos (dimension-five operators) [18, 19]. The dimension-six operators predict a proton lifetime proportional to M_U^4 or M_H^4 respectively, and the expected values of these mass scales make this contribution rather small. On the other hand, the dimension-five operators entail a proton lifetime proportional to $M_{\widetilde{H}}^2$, which requires $M_{\widetilde{H}} > M_U$ and some strong constraints on the supersymmetric spectrum (if it is to be light enough to be observable). In fact, the three-dimensional soft-supersymmetry-breaking parameter space $(m_{1/2}, m_0, A)$ is constrained in such a way that the squarks and sleptons should be heavy, while the neutralinos and charginos could be light [20].

Another important constraint on the parameter space of the minimal $SU(5)$ supergravity model is provided by cosmological considerations. The models we consider include a discrete symmetry called *R-parity* which has value +1 for the ordinary particles, and −1 for the supersymmetric particles. This implies that supersymmetric particles must always be created or destroyed in pairs, *i.e.*, at each vertex there are always none, two, or four supersymmetric particles. In particular, the *lightest supersymmetric particle* (LSP) must be stable in these models. Astrophysical considerations require this particle to be neutral and colorless [21]. The only two candidates are the lightest neutralino (a linear combination of the weakly interacting neutral sparticles: photino, zino, two Higgsinos) and the sneutrino. It turns out that it is the neutralino which is the lightest one. Since the neutralinos are stable, they must have pair-annihilated away quite efficiently in the early Universe, otherwise their present relic abundance would be too large, and the Hubble parameter would correspond to a Universe younger than the oldest known stars.

The cosmological constraint can be usually satisfied as long as the supersymmetric particles are not too heavy. However, the proton decay constraint requires heavy squarks and sleptons and thus the constraint is relevant. In fact, the only particles which mediate pair-annihilation efficiently are the Z-boson and the lightest Higgs boson, which are light. In practice, the neutralino mass must be near half the masses of these particles so that resonant s-channel annihilation occurs [22]. The resulting effect on the parameter space of the model is quite severe [23].

5.2 The SU(5)xU(1) supergravity model

The $SU(5) \times U(1)$ ("flipped $SU(5)$") gauge group [24] differs from $SU(5)$ in several ways. The Standard Model particles are also assigned to the $\bar{\mathbf{5}}$ and $\mathbf{10}$ representations, but in a "flipped" way (*i.e.*, $u^c \leftrightarrow d^c$, $e^c \leftrightarrow \nu^c$) relative to the minimal $SU(5)$ case,

$$\bar{\mathbf{5}} = \{u^c, L = \begin{pmatrix} e \\ \nu_e \end{pmatrix}\}, \qquad \mathbf{10} = \{Q = \begin{pmatrix} u \\ d \end{pmatrix}, d^c, \nu^c\}, \qquad \mathbf{1} = e^c. \tag{17}$$

Note that a right-handed neutrino ν^c has appeared naturally. This leads to an automatic *see-saw* mechanism to generate small neutrino masses in this model [25]. The heavy GUT spectrum of fields includes the **24** representation of gauge boson plus the $U(1)$ gauge boson of $SU(5) \times U(1)$. Unlike $SU(5)$, the symmetry breaking Higgs fields are contained in the $\mathbf{10}, \overline{\mathbf{10}}$ representations (which have neutral components because of the "flipped" assignment). This property is central to the appeal of $SU(5) \times U(1)$

as a paradigm of a string model [26], since larger Higgs representations (like the **24**) are not easily obtainable in string model building. There is also a pair of Higgs pentaplets $\mathbf{5},\overline{\mathbf{5}}$ which contain the two Higgs doublets of the low-energy theory, and the heavy Higgs triplets.

Proton decay is a concern that is easily dismissed in the typical $SU(5) \times U(1)$ models: the dimension-six proton decay operators are small as usual, while the perilous dimension-five operators are strongly suppressed. The latter is a direct consequence of the $SU(5)\times U(1)$ symmetry as it applies to the solution of the *doublet-triplet splitting problem.* The Higgs triplets are not only heavy but also cannot mediate the dangerous proton decay diagram [25, 27].

The model we have studied is most simply understood in the context of string model building when the gauge couplings of the Standard Model unify at the string scale $M_U \sim 10^{18}\,\mathrm{GeV}$ [28]. This entails the addition of another $\mathbf{10}, \overline{\mathbf{10}}$ representations to the heavy GUT spectrum, otherwise the gauge couplings would unify as in the minimal $SU(5)$ model. (Specific string models with this property have also been constructed [29].)

For the soft-supersymmetry-breaking parameters we have used two string-inspired scenarios, which allow one to determine two of the parameters as functions of the third one. We have considered,

(i) the *no-scale* model: $m_0 = A = 0$

(ii) the *dilaton* model: $m_0 = \frac{1}{\sqrt{3}} m_{1/2}\,, \quad A = -m_{1/2}$ [30].

Note that the dimension of the parameter space is now reduced to just three: m_t, $\tan\beta$, and $m_{1/2}$, where $m_{1/2} \propto m_{\tilde{g}}$. For both these scenarios there is a yet more restrictive possibility which we call the *strict no-scale* and the *special dilaton* scenarios. A further requirement allows one to deduce the value of $\tan\beta$ as a function of m_t and $m_{1/2}$, yielding highly predictive two-parameter models.

With these choices for the supersymmetry breaking parameters, one determines that the cosmological constraint does not restrict the models in any way [28, 31]. In fact, these two models offer a natural explanation for the required amount of dark matter in the Universe.

6 Experimental Predictions

As remarked above, in a generic supergravity model one expects a large degree of correlation among the predictions for, *e.g.*, the sparticle and Higgs masses and the rates for the various experimental processes of interest. These generic models are further constrained by the non-observation of any sparticle or Higgs boson. Interestingly enough, such five-parameter models are still not very predictive in that many possible trends of correlations are possible. In contrast, in the two specific supergravity models described in Sec. 5, the correlations become sharp and of much more experimental interest.

Without imposing any further constraints on the models, besides the present collider bounds on sparticle and Higgs masses and the proton decay and cosmological constraints as applicable, we have calculated rates for the following processes at the indicated facilities: (DI stands for deep-inelastic)

$$\begin{array}{ll}
\text{Tevatron [32]} & p\bar{p} \to \chi_1^{\pm}\chi_2^0, \quad \chi_1^{\pm} \to \chi_1^0 l^{\pm}\nu_l, \quad \chi_2^0 \to \chi_1^0 l^+ l^-, \quad l = e,\mu \quad \text{“trileptons”} \\
\text{LEPI [33]} & e^+e^- \to Zh \to f\bar{f}h \\
\text{LEP II [34]} & \left\{ \begin{array}{l}
e^+e^- \to \chi_1^+\chi_1^-, \quad \chi_1^+ \to \chi_1^0 q\bar{q}', \quad \chi_1^- \to \chi_1^0 l^-\nu_l, \; l = e,\mu \quad \text{“mixed”} \\
e^+e^- \to \chi_1^+\chi_1^-, \quad \chi_1^+ \to \chi_1^0 l^+\bar{\nu}_l, \quad \chi_1^- \to \chi_1^0 l^-\nu_l, \; l = e,\mu \; \text{“dilepton”} \\
e^+e^- \to \tilde{l}_R^+\tilde{l}_R^-, \quad \tilde{l}_R^+ \to \chi_1^0 l^+, \quad \tilde{l}_R^- \to \chi_1^0 l^-, \quad l = e,\mu,\tau \quad \text{“dilepton”} \\
e^+e^- \to Z^*h \to f\bar{f}h, \quad h \to \chi_1^0\chi_1^0, b\bar{b}, c\bar{c}, \tau^+\tau^-, gg
\end{array} \right. \\
\text{HERA [35]} & \left\{ \begin{array}{l}
e^-p \to \tilde{e}_R^-\chi_1^0 + p, \quad \tilde{e}_R^- \to \chi_1^0 e^- \quad \text{“elastic selectron – neutralino”} \\
e^-p \to \tilde{\nu}_e\chi_1^- + p, \quad \chi_1^- \to \chi_1^0 e^-\bar{\nu}_e \quad \text{“elastic sneutrino – chargino”} \\
e^-p \to \tilde{e}_R^-\chi_1^0 + X, \quad \tilde{e}_R^- \to \chi_1^0 e^- \quad \text{“DI selectron – neutralino”} \\
e^-p \to \tilde{\nu}_e\chi_1^- + X, \quad \chi_1^- \to \chi_1^0 e^-\bar{\nu}_e \quad \text{“DI sneutrino – chargino”}
\end{array} \right.
\end{array}$$

All the above processes are kinematically accessible in the $SU(5) \times U(1)$ models. In the minimal $SU(5)$ model the sleptons are heavy and neither the slepton pair production at LEPII nor any of the indicated processes at HERA are allowed.

We now give a sample of the actual results obtained for the most important processes listed above. These are shown in Figures 3,4,5,6.

More recently it has been realized that indirect experimental constraints on the $SU(5) \times U(1)$ models exist and can be quite significant. These constraints come generally from three sources: (i) the experimentally allowed range for $B(b \to s\gamma)$ as recently determined by the CLEO Collaboration [36]; (ii) the long-standing value for the anomalous magnetic moment of the muon [37]; and (iii) the precision LEP measurements of the electroweak parameters in the form of the allowed range for the ϵ_1 parameter [38]. (Only the last constraint is relevant in the minimal $SU(5)$ supergravity model because of its relatively heavier spectrum.)

The first two constraints exclude regions of the parameter space of the $SU(5) \times U(1)$ models which span all allowed values of the chargino mass, when viewing the parameter space in the $(m_{\chi_1^\pm}, \tan\beta)$ plane for fixed m_t. On the other hand, the ϵ_1 constraint basically implies an upper bound on m_t: $m_t \lesssim 165\,\mathrm{GeV}$, unless the

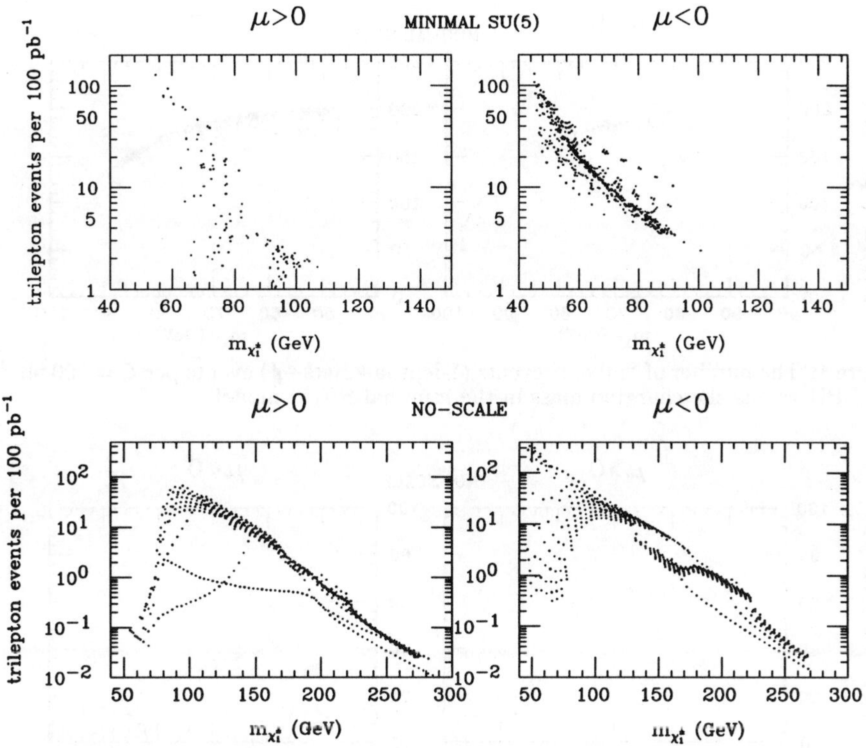

Figure 3: The number of trilepton events at the Tevatron per 100 pb^{-1} in the minimal $SU(5)$ model and the no-scale $SU(5) \times U(1)$ model (for $m_t = 130$ GeV). Note that with 200 pb^{-1} and 60% detection efficiency it should be possible to probe basically all of the parameter space of the minimal $SU(5)$ model, and probe chargino masses as high as 175 GeV in the no-scale model.

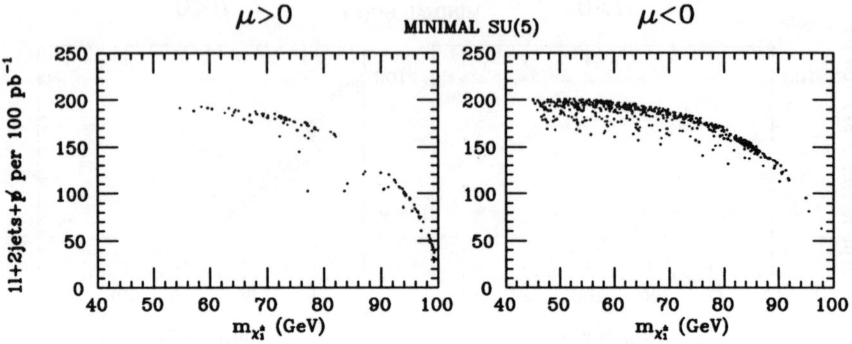

Figure 4: The number of "mixed" events (1-lepton+2jets+$\not{p}$) events per $\mathcal{L} = 100\,\text{pb}^{-1}$ at LEPII versus the chargino mass in the minimal $SU(5)$ model.

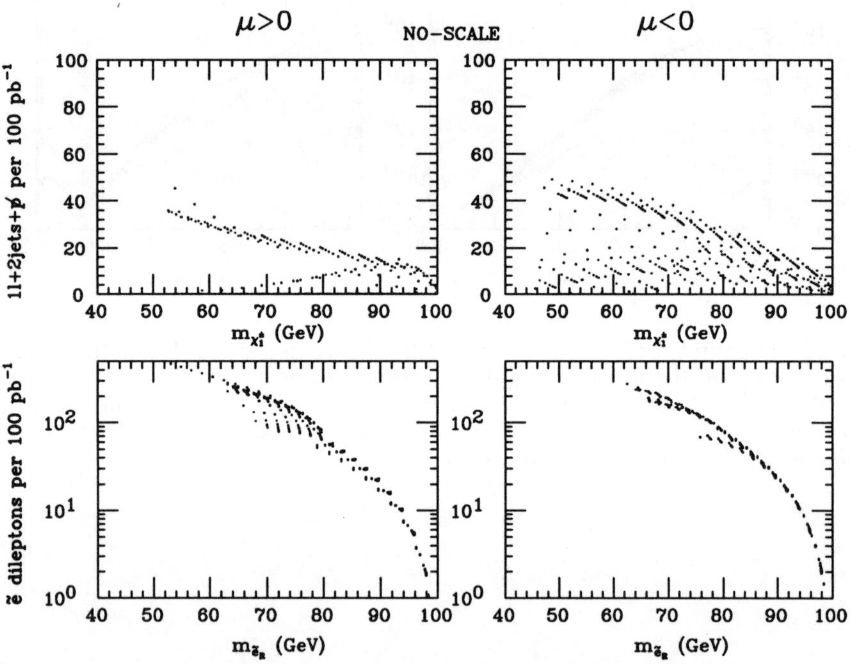

Figure 5: The number of "mixed" events (1-lepton+2jets+$\not{p}$) events per $\mathcal{L} = 100\,\text{pb}^{-1}$ at LEPII versus the chargino mass in the no-scale model (top row). Also shown (bottom row) are the number of di-electron events per $\mathcal{L} = 100\,\text{pb}^{-1}$ from selectron pair production versus the lightest selectron mass.

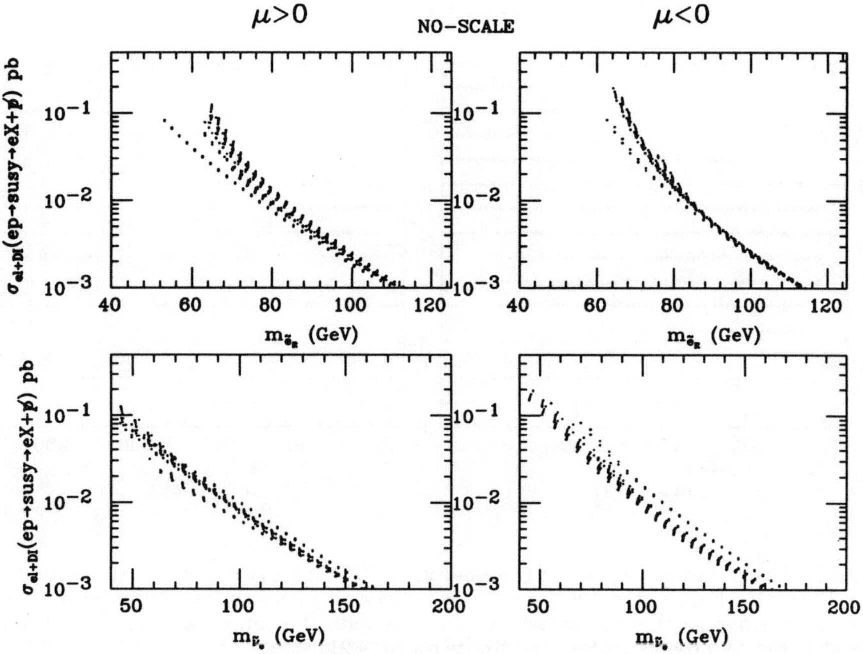

Figure 6: The elastic plus deep-inelastic total supersymmetric cross section at HERA ($ep \to$ susy $\to eX + \not{p}$) in the no-scale model versus the lightest selectron mass ($m_{\tilde{e}_R}$) and the sneutrino mass ($m_{\tilde{\nu}}$). The short- and long-term limits of sensitivity are expected to be 10^{-2} pb and 10^{-3} pb respectively.

chargino is very light ($m_{\chi_1^\pm} \lesssim 70\,\mathrm{GeV}$) in which case the upper bound on m_t can be relaxed up to $m_t \lesssim 180\,\mathrm{GeV}$. As an example of the effect of the constraints, in Fig. 7 we show the parameter space for $m_t = 130\,\mathrm{GeV}$ (when the ϵ_1 constraint is not restrictive) for the no-scale model [39]. Clearly the region of parameter space accessible to LEPII searches ($m_{\chi_1^\pm} \lesssim 100\,\mathrm{GeV}$) has become quite constrained.

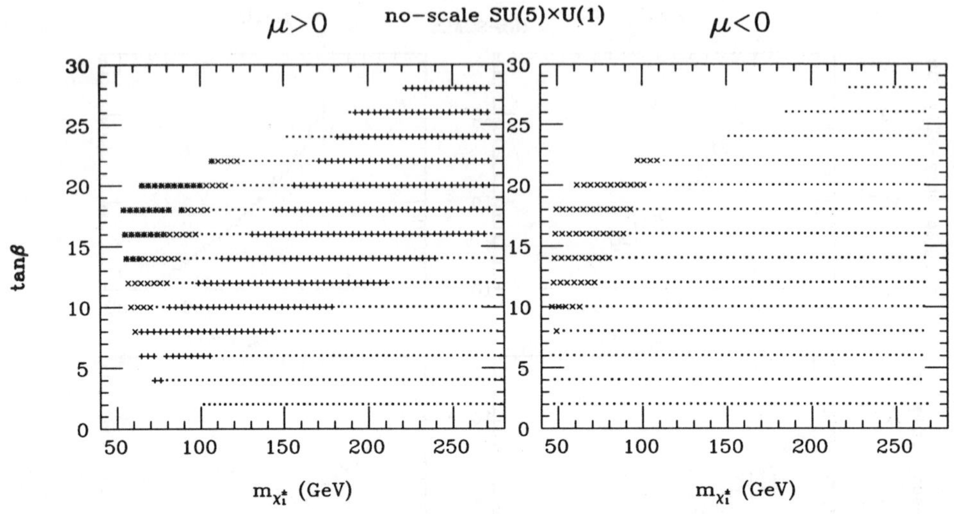

Figure 7: The parameter space of the no-scale $SU(5) \times U(1)$ supergravity model for $m_t = 130\,\mathrm{GeV}$. Points denoted by periods satisfy all presently known experimental constraints, whereas those denoted by pluses violate the limits on $B(b \to s\gamma)$ and those denoted by crosses violate the limits on $(g-2)_\mu^{susy}$.

7 Conclusions

In sum, we have presented a simplified tour of supersymmetric unified theories which hopefully will allow non-experts on the field to get acquainted with such a topical subject. One of our goals was to show that supersymmetric particles can well be "around the corner" and at the verge of discovery at present and near future facilities, such as the Tevatron, HERA, and LEPII. It is also apparent that there are many experimental constraints on supersymmetric unified models which need to be consistently imposed to speak about experimentally viable models. Finally, these experimantally testable models should be as simple as the best theoretical motivations allow. There is little to be learned from generic models which can predict anything – since predicting anything is akin to predicting nothing.

Acknowledgements

This work has been supported in part by DOE grant DE-FG05-91-ER-40633.

References

[1] For a recent review see *e.g.*, G. Altarelli, CERN-TH.6867/93.

[2] A. J. Buras, J. Ellis, M. K. Gaillard, D. V. Nanopoulos, Nucl. Phys. B **135** (1978) 66; D. V. Nanopoulos and D. A. Ross, Nucl. Phys. B **157** (1979) 273, Phys. Lett. B **108** (1982) 351, and Phys. Lett. B **118** (1982) 99.

[3] For reviews see *e.g.*, R. Arnowitt and P. Nath, *Applied N=1 Supergravity* (World Scientific, Singapore 1983); H. P. Nilles, Phys. Rep. **110** (1984) 1.

[4] For a recent review see J. L. Lopez, D. V. Nanopoulos, and A. Zichichi, CERN-TH.6926/93 and Texas A & M University preprint CTP-TAMU-33/93.

[5] H. Georgi and S. Glashow, Phys. Rev. Lett. **32** (1974) 438.

[6] See *e.g.*, P. Langacker and N. Polonsky, Phys. Rev. D **47** (1993) 4028.

[7] J. L. Lopez, Erice 93 Subnuclear School Lecture, Texas A & M University preprint CTP-TAMU-42/93.

[8] J. Ellis, S. Kelley, and D. V. Nanopoulos, Phys. Lett. B **249** (1990) 441; Phys. Lett. B **260** (1991) 131; Nucl. Phys. B **373** (1992) 55.; Phys. Lett. B **287** (1992) 95.

[9] P. Langacker and M.-X. Luo, Phys. Rev. D **44** (1991) 817.

[10] U. Amaldi, W. de Boer, and H. Fürstenau, Phys. Lett. B **260** (1991) 447.

[11] F. Anselmo, L. Cifarelli, A. Peterman, and A. Zichichi, Nuovo Cim. **104A** (1991) 1817 and **105A** (1992) 1025.

[12] F. Anselmo, L. Cifarelli, A. Peterman, and A. Zichichi, Nuovo Cim. **105A** (1992) 581; F. Anselmo, L. Cifarelli, and A. Zichichi, Nuovo Cim. **105A** (1992) 1357.

[13] For a review see A. B. Lahanas and D. V. Nanopoulos, Phys. Rep. **145** (1987) 1.

[14] See *e.g.*, S. Kelley, J. L. Lopez, D. V. Nanopoulos, H. Pois, and K. Yuan, Nucl. Phys. B **398** (1993) 3.

[15] J. Ellis, A. Lahanas, D. V. Nanopoulos, and K. Tamvakis, Phys. Lett. B **134** (1984) 429.

[16] J. Ellis, C. Kounnas, and D. V. Nanopoulos, Nucl. Phys. B **241** (1984) 406, Nucl. Phys. B **247** (1984) 373.

[17] E. Witten, Phys. Lett. B **155** (1985) 151.

[18] S. Weinberg, Phys. Rev. D **26** (1982) 287; N. Sakai and T. Yanagida, Nucl. Phys. B **197** (1982) 533.

[19] J. Ellis, D. V. Nanopoulos, and S. Rudaz, Nucl. Phys. B **202** (1982) 43; B. Campbell, J. Ellis, and D. V. Nanopoulos, Phys. Lett. B **141** (1984) 229; K. Enqvist, A. Masiero, and D. V. Nanopoulos, Phys. Lett. B **156** (1985) 209; P. Nath, A. Chamseddine, and R. Arnowitt, Phys. Rev. D **32** (1985) 2348; P. Nath and R. Arnowitt, Phys. Rev. D **38** (1988) 1479.

[20] R. Arnowitt and P. Nath, Phys. Rev. Lett. **69** (1992) 725; P. Nath and R. Arnowitt, Phys. Lett. B **287** (1992) 89.

[21] J. Ellis, J. S. Hagelin, D. V. Nanopoulos, K. A. Olive and M. Srednicki, Nucl. Phys. B **238** (1984) 453.

[22] J. L. Lopez, D. V. Nanopoulos, and A. Zichichi, Phys. Lett. B **291** (1992) 255.

[23] J. L. Lopez, D. V. Nanopoulos, and H. Pois, Phys. Rev. D **47** (1993) 2468; R. Arnowitt and P. Nath, Phys. Lett. B **299** (1993) 58 and **307** (1993) 403(E); P. Nath and R. Arnowitt, Phys. Rev. Lett. **70** (1993) 3696; J. L. Lopez, D. V. Nanopoulos, and K. Yuan, Phys. Rev. D **48** (1993) 2766.

[24] S. Barr, Phys. Lett. B **112** (1982) 219, Phys. Rev. D **40** (1989) 2457; J. Derendinger, J. Kim, and D. V. Nanopoulos, Phys. Lett. B **139** (1984) 170.

[25] I. Antoniadis, J. Ellis, J. Hagelin, and D. V. Nanopoulos, Phys. Lett. B **194** (1987) 231.

[26] I. Antoniadis, J. Ellis, J. Hagelin, and D. V. Nanopoulos, Phys. Lett. B **231** (1989) 65.

[27] J. Ellis, J. Hagelin, S. Kelley, and D. V. Nanopoulos, Nucl. Phys. B **311** (1988/89) 1.

[28] J. L. Lopez, D. V. Nanopoulos, and A. Zichichi, CERN-TH.6667/92, Texas A & M University preprint CTP-TAMU-68/92 (to appear in Phys. Rev. D).

[29] J. L. Lopez, D. V. Nanopoulos, and K. Yuan, Nucl. Phys. B **399** (1993) 654.

[30] V. Kaplunovsky and J. Louis, Phys. Lett. B **306** (1993) 269.

[31] J. L. Lopez, D. V. Nanopoulos, and A. Zichichi, CERN-TH.6903/93, Texas A & M University preprint CTP-TAMU-31/93.

[32] J. L. Lopez, D. V. Nanopoulos, X. Wang, and A. Zichichi, Phys. Rev. D **48** (1993) 2062.

[33] J. L. Lopez, D. V. Nanopoulos, H. Pois, X. Wang, and A. Zichichi, Phys. Lett. B **306** (1993) 73.

[34] J. L. Lopez, D. V. Nanopoulos, H. Pois, X. Wang, and A. Zichichi, Phys. Rev. D **48** (1993) 4062.

[35] J. L. Lopez, D. V. Nanopoulos, X. Wang, and A. Zichichi, Phys. Rev. D **48** (1993) 4029.

[36] J. L. Lopez, D. V. Nanopoulos, and G. Park, Phys. Rev. D **48** (1993) R974; J. L. Lopez, D. V. Nanopoulos, G. Park, and A. Zichichi, Texas A & M University preprint CTP-TAMU-40/93 (to appear in Phys. Rev. D).

[37] J. L. Lopez, D. V. Nanopoulos, and X. Wang, Texas A & M University preprint CTP-TAMU-44/93 (to appear in Phys. Rev. D).

[38] J. L. Lopez, D. V. Nanopoulos, G. Park, H. Pois, and K. Yuan, Phys. Rev. D **48** (1993) 3297.

[39] J. L. Lopez, D. V. Nanopoulos, G. Park, X. Wang, and A. Zichichi, in preparation.

RENORMALIZATION GROUP STUDY OF THE MINIMAL SUPERSYMMETRIC STANDARD MODEL: NO SCALE MODELS

PIERRE RAMOND
Institute for Fundamental Theory and Physics Department,
University of Florida, 215 Williamson Hall
Gainesville, Florida 32611, U.S.A.

ABSTRACT

We present partial numerical results in the Minimal Supersymmetric Standard Model with soft breaking of supersymmetry, and radiative breaking of the electroweak symmetry. We impose the additional relation $m_b = m_\tau$ at the GUT scale. For the special case of the strict no-scale model, in which global supersymmetry breaking arises solely from soft gaugino masses, we find that M_t is expected to be lighter than ~ 128 GeV. A higher upper bound for M_t of ~ 138 is predicted, if M_b is at the lower end of its experimental uncertainty.

1. Introduction

In this paper, I report some preliminary results of work in progress, done in collaboration with E. Piard and D. Castaño. In the last few years, it has become apparent, using the ever increasing accuracy in the measurement of the strong coupling, that Supersymmetry (SUSY) affords an elegant means to achieve gauge coupling unification [15, 16, 17] at scales consistent with Grand Unified Theories (GUTs) [18, 19, 21, 22]. Whereas in the Standard Model (SM) the three gauge couplings unify "two by two" forming the "GUT triangle," in the simplest Minimal SUSY Extension of the SM (MSSM) these gauge couplings spectacularly unify at a point (within the experimental errors in their values). Given that the scale of unification in these models is generally above the lower bound set by proton decay, these supersymmetric grand unified theories (SUSY-GUTs) have regained increasing interest. Constraints from the Yukawa sectors of such models have also yielded interesting predictions for various low energy parameters including the top quark mass [23, 35, 13, 14].

Crucial to these analyses has been the use of the Renormalization Group (RG) in extrapolating from the GUT scale to experimental scales. Here, we use of our previous RG analysis for the standard model[30], including the supersymmetric two loop β functions for the MSSM. Also soft symmetry breaking (SSB) terms are added. These lead to a nondegenerate superparticle spectrum as well as to the radiative breaking of the electroweak symmetry. A brief discussion of the effective one loop potential is presented. The boundary conditions at the unification scale in these minimal low energy supergravity models are discussed. Next the numerical procedure employed is described. The treatment of thresholds and the "special" form of the β functions needed is discussed next. Similar analyses have appeared in the literature [8, 12, 9, 11].

In the following, we only concern ourselves with the simplest schemes which have a chance of reproducing the data. If these are indeed those chosen by nature, they have some very distinct signatures. The top quark, although allowed by the standard model to be as heavy as 200 GeV, comes out much lighter. The lightest Higgs particle also comes out light, so light in fact that it may be very hard to detect. Another feature is that these models all predict the existence of a stable particle, the LSP, whose abundance can be calculated, based on the standard cosmological scenario. LSPs are typically too light to bring Ω to one, the preferred value of theorists.

2. Minimal Supersymmetric Standard Model

In the minimal supersymmetric extension of the standard model, every particle has a supersymmetric partner, with spin differing by a half [24]. Supersymmetry requires a second Higgs field with opposite hypercharge to the first as the superpotential cannot contain both a field and its complex conjugate. The second Higgs is also needed for both chiral $U(1)$ and $SU(2)$ global anomaly cancellation and to give its sector a mass. For renormalizable theories, the superpotential can have at most degree three interactions. The superpotential for the MSSM is (suppressing the $SU(2)$ and Weyl metrics)

$$W = \hat{\overline{u}}\mathbf{Y}_u\hat{\Phi}_u\hat{Q} + \hat{\overline{d}}\mathbf{Y}_d\hat{\Phi}_d\hat{Q} + \hat{\overline{e}}\mathbf{Y}_e\hat{\Phi}_d\hat{L} + \mu\hat{\Phi}_u\hat{\Phi}_d \ . \tag{1}$$

where the hat indicates a chiral superfield and the overline denotes a left handed CP conjugate of a right handed field, $\overline{\psi} = i\sigma_2\psi_R^*$. The usual Yukawa interactions are accompanied by new Yukawa interactions among the scalar quarks and leptons and the Higgsinos in the supersymmetric Lagrangian. There are also new gauge Yukawa interactions involving the gauginos. The new purely scalar interactions form the scalar potential which is positive definite in supersymmetric theories. The scalar potential will be discussed in a subsequent section. A remarkable aspect of supersymmetry is that all these new interactions require no new couplings. Table I displays the $SU(3) \times SU(2) \times U(1)$ quantum numbers of the chiral (all left handed) and vector superfields of the MSSM.

The last term is put in by hand to avoid the exact Peccei-Quinn (PQ) symmetry that the superpotential would otherwise exhibit. Although there are other ways to treat the PQ symmetry, including producing an axion, we leave the other possibilities to a future investigation.

Table 1: Quantum Numbers

	$\hat{Q}$	$\hat{\bar{u}}$	$\hat{\bar{d}}$	$\hat{L}$	$\hat{\bar{e}}$	$\hat{\Phi}_u$	$\hat{\Phi}_d$	$\hat{g}^A$	$\hat{W}^a$	$\hat{B}$
$U(1)$	$+\frac{1}{6}$	$-\frac{2}{3}$	$+\frac{1}{3}$	$-\frac{1}{2}$	$+1$	$+\frac{1}{2}$	$-\frac{1}{2}$	0	0	0
$SU(2)$	**2**	**1**	**1**	**2**	**1**	**2**	**2**	**1**	**3**	**1**
$SU(3)$	**3**	$\overline{\mathbf{3}}$	$\overline{\mathbf{3}}$	**1**	**1**	**1**	**1**	**8**	**1**	**1**

3. Minimal Low Energy Supergravity Model

Since no super particles have been observed experimentally, supersymmetry, if truly present in nature, must be broken. One way to accomplish this breaking is to add to the Lagrangian soft breaking terms.

The most general soft symmetry breaking potential for the MSSM can be written (including gaugino mass terms)

$$\begin{aligned} V_{soft} &= m^2_{\Phi_u}\Phi_u^\dagger\Phi_u + m^2_{\Phi_d}\Phi_d^\dagger\Phi_d + B\mu(\Phi_u\Phi_d + h.c.) \\ &\quad + \sum_i (\, m^2_{Q_i}\tilde{Q}^\dagger\tilde{Q} + m^2_{L_i}\tilde{L}^\dagger\tilde{L} + m^2_{u_i}\tilde{\bar{u}}^\dagger\tilde{\bar{u}} + m^2_{d_i}\tilde{\bar{d}}^\dagger\tilde{\bar{d}} + m^2_{e_i}\tilde{\bar{e}}^\dagger\tilde{\bar{e}}\,) \\ &\quad + \sum_{i,j} (\, A_u^{ij} Y_u^{ij}\tilde{\bar{u}}_i\Phi_u\tilde{Q}_j + A_d^{ij} Y_d^{ij}\tilde{\bar{d}}_i\Phi_d\tilde{Q}_j + A_e^{ij} Y_e^{ij}\tilde{\bar{e}}_i\Phi_d\tilde{L}_j + h.c.\,)\;, \end{aligned} \tag{2}$$

$$V_{gaugino} = \frac{1}{2}\sum_{l=1}^{3} M_l\lambda_l\lambda_l + h.c.\;, \tag{3}$$

where $V_{gaugino}$ is the Majorana mass terms for the gaugino fields, λ_l (suppressing the group index), corresponding to $U(1)$, $SU(2)$, and $SU(3)$, respectively. In the above, there are in principle sixty-three soft symmetry breaking parameters. This is hardly desirable to explain the standard model which already has eighteen parameters! It is clear that we need a further organizing principle. While this may be done by invoking extra symmetries such as flavor blindness (blandness), one attractive possibility is to couple the standard model to $N = 1$ supergravity (SUGRA). In the minimal low energy supergravity model, supersymmetry is explicitly broken by induced soft breaking terms of this form but with fewer parameters[10].

We will assume that the spontaneous breaking of the local N=1 supersymmetry is communicated to the "visible" sector by weak gravitational interactions from some "hidden" sector. This spontaneous symmetry breaking of supergravity manifests itself at low energy as explicit soft breaking terms of global supersymmetry.

The SUGRA Lagrangian is characterized by two arbitrary functions of the fields, a real function (the Kähler potential) that determines the kinetic terms of the chiral superfields, and an analytic function, transforming as the symmetric product of the adjoint representation of the gauge group, that determines the kinetic terms of the gauge fields. In models with minimal kinetic terms for the chiral superfields, this leads to a common (gravitino) mass, m_0, for all the scalars of the model. The presence of non-minimal gauge kinetic terms implies non-zero masses, M_l, for the gauginos at the GUT scale, M_X. By further assuming gauge coupling unification, we can take the three gaugino masses to be equal. Furthermore, the trilinear soft couplings A_u^{ij}, A_d^{ij}, and A_e^{ij} are all equal to a common value A_0. With minimal chiral kinetic terms, the bilinear soft coupling B_0 is related to A_0 as $B_0 = A_0 - m_0$. This scenario has obvious, desirable features. First, it is very predictive since it has a few parameters accounting for thirty-one new masses. Second, the universal nature of the squark and slepton masses at M_X helps to avoid the appearance of unwanted flavor changing neutral current (FCNC) effects. In fact, one could argue that the absence of FCNCs hints at a universal mass for the scalars.

All of these couplings will evolve to different values under the renormalization group. The complete scalar potential now appears as

$$V = V_F + V_D + V_{soft} \;, \tag{4}$$

where V_F contains the potential contributions from the F-terms

$$\begin{aligned} V_F \;=\; & |\tilde{\bar{u}}\mathbf{Y}_u\tilde{Q} + \mu\Phi_d|^2 + |\tilde{\bar{d}}\mathbf{Y}_d\tilde{Q} + \tilde{\bar{e}}\mathbf{Y}_e\tilde{L} + \mu\Phi_u|^2 \\ & + |\mathbf{Y}_u\tilde{Q}\Phi_u|^2 + |\mathbf{Y}_d\tilde{Q}\Phi_d + \mathbf{Y}_e\tilde{L}\Phi_d|^2 \\ & + |\tilde{\bar{u}}\mathbf{Y}_u\Phi_u + \tilde{\bar{d}}\mathbf{Y}_d\Phi_d|^2 + |\tilde{\bar{e}}\mathbf{Y}_e\Phi_d|^2 \;, \end{aligned} \tag{5}$$

and V_D contains the potential contributions from the D-terms

$$\begin{aligned} V_D \;=\; & \frac{g'^2}{2}\Big(\,\frac{1}{6}\tilde{Q}^\dagger\tilde{Q} - \frac{2}{3}\tilde{\bar{u}}^\dagger\tilde{\bar{u}} + \frac{1}{3}\tilde{\bar{d}}^\dagger\tilde{\bar{d}} - \frac{1}{2}\tilde{L}^\dagger\tilde{L} + \tilde{\bar{e}}^\dagger\tilde{\bar{e}} + \frac{1}{2}\Phi_u^\dagger\Phi_u - \frac{1}{2}\Phi_d^\dagger\Phi_d\,\Big)^2 \\ & + \frac{g_2^2}{8}\Big(\,\tilde{Q}^\dagger\vec{\tau}\tilde{Q} + \tilde{L}^\dagger\vec{\tau}\tilde{L} + \Phi_u^\dagger\vec{\tau}\Phi_u + \Phi_d^\dagger\vec{\tau}\Phi_d\,\Big)^2 \\ & + \frac{g_3^2}{8}\Big(\,\tilde{Q}^\dagger\vec{\lambda}\tilde{Q} - \tilde{\bar{u}}^\dagger\vec{\lambda}^*\tilde{\bar{u}} - \tilde{\bar{d}}^\dagger\vec{\lambda}^*\tilde{\bar{d}}\,\Big)^2 \;, \end{aligned} \tag{6}$$

where $\vec{\tau} = (\tau_1, \tau_2, \tau_3)$ are the $SU(2)$ Pauli matrices and $\vec{\lambda} = (\lambda_1, \ldots, \lambda_8)$ are the Gell-Mann matrices. In general, one must impose constraints on the parameters to avoid charge and color breaking minima in the scalar potential. Some *necessary* constraints have been formulated, such as

$$A_U^2 \;<\; 3(m_Q^2 + m_u^2 + m_{\Phi_u}^2) \;, \tag{7}$$
$$A_D^2 \;<\; 3(m_Q^2 + m_d^2 + m_{\Phi_d}^2) \;, \tag{8}$$
$$A_E^2 \;<\; 3(m_L^2 + m_e^2 + m_{\Phi_d}^2) \;. \tag{9}$$

However, these relations are in general neither *sufficient* nor indeed always *necessary* [25]. Their derivation involves very specific assumptions about the spontaneous symmetry breaking.

4. Radiative Electroweak Breaking

An appealing feature of the models we are considering is that they can lead to the breaking of the electroweak symmetry radiatively [1, 2, 3, 4]. The one loop effective Higgs potential in these models can be expressed as the sum of the tree level potential plus a correction coming from the sum of all one loop diagrams with external lines having zero momenta

$$V_{1-loop}(\Lambda) = V_{tree}(\Lambda) + \Delta V_1(\Lambda) \ . \tag{10}$$

The one loop correction is given by

$$\begin{aligned} \Delta V_1(\Lambda) &= \frac{1}{64\pi^2}\mathrm{Str}\{\mathcal{M}^4(\ln\frac{\mathcal{M}^2}{\Lambda^2} - \frac{3}{2})\} \\ &= \frac{1}{64\pi^2}\sum_p(-1)^{2s_p}(2s_p+1)m_p^4(\ln\frac{m_p^2}{\Lambda^2} - \frac{3}{2}) \ , \end{aligned} \tag{11}$$

where $\mathcal{M}^2$ is the field dependent squared mass matrix of the model and m_p is the eigenvalue mass of the pth particle of spin s_p. The tree level part of the potential is

$$\begin{aligned} V_{tree}(\Lambda) &= m_1^2(\Lambda)\Phi_d^\dagger\Phi_d + m_2^2(\Lambda)\Phi_u^\dagger\Phi_u + m_3^2(\Lambda)(\ \Phi_u\Phi_d + h.c.\ ) \\ &+ \frac{g'^2(\Lambda)}{8}(\ \Phi_u^\dagger\Phi_u - \Phi_d^\dagger\Phi_d\ )^2 + \frac{g_2^2(\Lambda)}{8}(\ \Phi_u^\dagger\vec{\tau}\Phi_u + \Phi_d^\dagger\vec{\tau}\Phi_d\ )^2 \end{aligned} \tag{12}$$

where

$$m_1^2(\Lambda) = m_{\Phi_d}^2(\Lambda) + \mu^2(\Lambda) \ , \tag{13}$$

$$m_2^2(\Lambda) = m_{\Phi_u}^2(\Lambda) + \mu^2(\Lambda) \ , \tag{14}$$

$$m_3^2(\Lambda) = B(\Lambda)\mu(\Lambda) \ . \tag{15}$$

The parameters of the potential are taken as running ones, that is, they vary with scale according to the renormalization group. The logarithmic term in the one loop correction is important in making $V_{1-loop}(\Lambda)$ independent of Λ to this order (up to non-field dependent terms).

Given the low energy scale of electroweak breaking, we must use the renormalization group to evolve the parameters of the potential to a convenient scale such as M_Z (where the experimental values of the gauge couplings are usually cited) thereby making this leading log approximation valid. The exact scale is not critical as long as it is in the electroweak range. If we define,

$$\overline{m}_i^2 = m_i^2 + \frac{\partial\Delta V_1}{\partial v_i^2} \ , \tag{16}$$

with $v_1 = v_d$, $v_2 = v_u$ and

$$\frac{\partial\Delta V_1}{\partial v_i^2} = \frac{1}{32\pi^2}\sum_p(-1)^{2s_p}(2s_p+1)m_p^2(\ln\frac{m_p^2}{\Lambda^2} - 1)\frac{\partial m_p^2}{\partial v_i^2} \ , \tag{17}$$

then minimization of the potential yields the following two conditions among its parameters

$$\frac{1}{2}m_Z^2 = \frac{\overline{m}_1^2 - \overline{m}_2^2\tan^2\beta}{\tan^2\beta - 1}, \tag{18}$$

where $m_Z^2 = (g'^2 + g_2^2)v^2/2$, $v^2 = v_u^2 + v_d^2$, and

$$B\mu = \frac{1}{2}(\overline{m}_1^2 + \overline{m}_2^2)\sin 2\beta \ , \tag{19}$$

where $\tan\beta = v_u/v_d$.

Although results based on the tree level potential cannot always be trusted, one can still use it to get some idea under what conditions electroweak breaking occurs. The renormalization group evolution of $m_{\Phi_u}^2$ can be such that it turns negative at low energies, if the top quark mass is large enough, whereas $m_{\Phi_d}^2$ runs positive. From Eq. (12), the scale at which breaking occurs is set by the condition

$$m_1^2(\Lambda_b)\ m_2^2(\Lambda_b) - m_3^4(\Lambda_b) = 0 \ . \tag{20}$$

If the free parameters are adjusted properly, then the correct value of the Z^0 mass ($M_Z = 91.17$ GeV) can be achieved.

In the tree level analysis, there is another critical scale that must be considered. It is evident from Eq. (12) that the potential becomes unbounded from below along the equal field (neutral component) direction, if

$$m_1^2(\Lambda_s) + m_2^2(\Lambda_s) < 2m_3^2(\Lambda_s) \ . \tag{21}$$

Since $m_1^2 m_2^2 - m_3^4 \geq 0$ implies $m_1^2 + m_2^2 \geq 2m_3^2$, condition (21) can only occur at scales lower than condition (20), so $\Lambda_s < \Lambda_b$. From this analysis, one concludes that the tree level vacuum expectation values (VEVs) of the scalar fields obtained by minimizing the potential are zero above Λ_b, and grow to infinity as one approaches Λ_s where the potential becomes unbounded from below. It follows that the appropriate scale at which to minimize the tree level potential and evaluate the VEVs is critical. This scale must be such that the one loop corrections of the effective potential may safely be neglected. Only at such a scale can the tree level results be trusted. However, there is more than one scale involved, and therefore, it is difficult if not impossible to find a scale at which all logarithms may be neglected. Indeed, the use of tree level minimization conditions to compute the VEVs at an arbitrary scale (e.g., $\Lambda = M_Z$) leads to incorrect conclusions about the regions of parameter space that yield consistent electroweak breaking scenarios [26]. When ΔV_1 is included, however, the value of Λ is not critical as long as it is in the neighborhood of M_Z.

Ref. [26] gives a prescription for arriving at a scale ($\hat{\Lambda}$) at which the tree level and the one loop effective potential results for the VEVs agree. Three qualitatively different cases are considered. Following Ref. [26], we let M_{SUSY} parametrize the superparticle thresholds, then the cases can be characterized by the orderings: (a) $M_{SUSY} < \Lambda_s < \Lambda_b$, (b) $\Lambda_s < M_{SUSY} < \Lambda_b$, and (c) $\Lambda_s < \Lambda_b < M_{SUSY}$. In each case, the prescription is to take $\hat{\Lambda} = \max\{M_{SUSY}, \Lambda_s\}$. Two cases deserve special mention. Case (a) cannot be handled using the tree level analysis because $v_u, v_d \to \infty$ near Λ_s. Fortunately, phenomenological bounds rule this case out anyway. In case (c), there is actually no electroweak breaking. For scales below M_{SUSY}, the superparticles have decoupled and the effective theory is not supersymmetric. Therefore, the running mass parameters of the potential freeze into their values at $\Lambda = M_{SUSY}$ at which scale there is no electroweak breaking. Finally, it must be emphasized that the apparent violent behavior of the VEVs with scale in the tree level analysis is an artifact of the approximation. The only physical potential is the full effective potential, and it either breaks electroweak symmetry or not. If it does, then the scalar fields have non-zero VEVs, and these VEVs are non-zero over all scales varying according to the anomalous dimension of their respective scalar fields (in the Landau gauge).

In this paper, we do not rely on the tree level analysis, rather we incorporate the one loop corrections. We include the dominant contributions from the third generation, that is, those of the top and stop, bottom and sbottom, and tau and stau [28, 29]. The one loop effective potential is constant against the renormalization group to this order around the electroweak scale. We choose the Z^0 mass as the scale at which to evaluate the minimization conditions. Eqs. (18) and (19) can be written

$$\mu^2(M_Z) = \frac{\overline{m}^2_{\Phi_d} - \overline{m}^2_{\Phi_u}\tan^2\beta}{\tan^2\beta - 1} - \frac{1}{2}m_Z^2 , \tag{22}$$

$$B(M_Z) = \frac{\overline{m}_1^2 + \overline{m}_2^2)\sin 2\beta}{2\mu(M_Z)} , \tag{23}$$

where $\overline{m}^2_{\Phi_{u,d}} = m^2_{\Phi_{u,d}} + \partial\Delta V_1/\partial v^2_{u,d}$ and used to solve for $\mu(M_Z)$ and $B(M_Z)$ given the value of all the relevant parameters at M_Z. We note that the form of Eq. (22) does not fix the sign of μ, and a choice for its sign must be made (μ is multiplicatively renormalized). The right hand sides of these equations implicitly involve the VEV at M_Z. In a consistent scenario it would have the value of $v(M_Z) = 174.1$ GeV. If the parameters are such that $\mu^2 < 0$, then the scenario is inconsistent and the electroweak symmetry fails to be broken.

5. Boundary Conditions at M_X

In this paper, as in the previous one [30] we work in the modified minimal subtraction scheme ($\overline{\text{MS}}$) of renormalization. The parameters of the Lagrangian are not in general equal to any corresponding physical constant. For example, in the case of masses, except for those of the bottom and top quark (see [30]), all other physical masses, M, will be determined from their corresponding running masses by the relation

$$M = m(\Lambda)|_{\Lambda=M} . \tag{24}$$

This equation is easily solved in the course of an integration of the RGEs for the different masses by noting the scale at which it is valid.

Because SUGRA models make simplifying predictions about the soft parameters at the unification scale, we initiate the evolution of the renormalization group equations at this scale. It has been demonstrated that the introduction of supersymmetry leads to gauge coupling unification at approximately $\sim 10^{16}$ GeV. Therefore we take $M_X = 10^{16}$ GeV, and evolve down to 1 GeV, the conventional scale at which the running quark masses are given [32].

Table 2: Models

	A_0	m_0	$m_{1/2}$	B_0
Strict No-scale	0	0	free	free
No-scale	0	0	free	0
String Inspired	0	free	free	0

At the unification scale, M_X, all the scalars will have a common mass

$$\begin{aligned} m_{Q_i}(M_X) &= m_{u_i}(M_X) = m_{d_i}(M_X) = m_{L_i}(M_X) \\ &= m_{e_i}(M_X) = m_{\Phi_u}(M_X) = m_{\Phi_d}(M_X) = m_0 \ , \end{aligned} \tag{25}$$

as will the gauginos

$$M_1(M_X) = M_2(M_X) = M_3(M_X) = m_{1/2} \ . \tag{26}$$

The prefactors of the trilinear soft scalar terms are equal at M_X

$$A_u^{ij}(M_X) = A_d^{ij}(M_X) = A_e^{ij}(M_X) = A_0 \ . \tag{27}$$

Also we define the bilinear soft scalar coupling and the mixing mass at M_X by $B(M_X) \equiv B_0$, and $\mu(M_X) \equiv \mu_0$.

Furthermore, to constrain the parameter space, we will take the bottom and tau masses equal at M_X

$$m_b(M_X) = m_\tau(M_X) \ . \tag{28}$$

This being the best motivated mass relation in supersymmetric grand unified theories [34].

The present analysis restricts itself to three subclasses of soft symmetry breaking models. These have various soft parameters equal to zero at M_X. The first class of models follows from the no-scale model [33] and has $A_0 = m_0 = B_0 = 0$ (strict no-scale model). In these models, only gaugino masses provide global supersymmetry breaking. The second class is the less constraining no-scale case that has only $A_0 = m_0 = 0$. A third class we consider with $A_0 = B_0 = 0$ comes from string derived models. Table II lists these three possibilities. In both strict no-scale and string inspired cases, we must have $B_0 = 0$. However, since for us B_0 is an output rather than an input variable, $B_0 = 0$ results must be inferred from its behavior upon varying other input parameters.

6. Numerical Procedure

In this work as in Ref. [30], we will use routines, based on the "shooting" method, to solve systems of nonlinear equations. The method involves making a guess for the solution, then assessing its merits based on how well the equations are satisfied, given some tolerance. The process is optimized and iterated until the routine converges on a solution.

As discussed previously, we start out runs at M_X at which scale we can make simplifying assumptions about the soft breaking terms based on various SUGRA models. This requires that we use the solution routines to consistently find the M_X values of all known low energy parameters such as lepton and quark masses and mixing angles and gauge couplings. This amounts to solving for sixteen unknowns (nine masses, three angles and a phase, and three gauge couplings). Alternatively, we could start our runs at M_Z or 1 GeV; however, this now requires solving for sixty-three unknowns (the values of the soft breaking terms at low energy) that must evolve to just four different values at M_X. The efficiency of the former method is obvious.

There are seven free parameters in the model we consider. These are A_0, B_0, m_0, $m_{1/2}$, μ_0, $\tan\beta$, and m_t. The two minimization constraints (18) and (19) reduce this set to five, which are taken to be A_0, m_0, $m_{1/2}$, $\tan\beta$, and m_t. In the present framework, B_0 and μ_0 will be determined using the numerical solutions routines in conjunction with the minimization of the one loop effective potential at M_Z in the process of evolving from M_X to 1 GeV. Minimization at M_Z will give $B(M_Z)$ and $\mu(M_Z)$. To arrive at B_0 and μ_0 (their corresponding values at M_X), we employ the solution routine as follows. A guess for B_0 and μ_0 is made at M_X and then the parameters of the model are run to M_Z at which scale the evolved value of B is compared to the minimization output value for B at M_Z. The same is done for μ. If the compared values agree to some set accuracy, then B_0 and μ_0 are the required values. Other analyses that also extract $B(M_Z)$ and $\mu(M_Z)$ simply evolve these two parameters via their renormalization group equations back to M_X to find B_0 and μ_0 relying on their near decoupling from the full set of RGEs. We note that the sign of μ is not determined from the minimization procedure, thus we must make a choice for it. To constrain the parameter space further, the bottom quark and tau lepton masses will be taken equal at M_X. This equality is a characteristic of many SUSY-GUTs. This constrains the model to four free parameters, A_0, m_0, $m_{1/2}$, and $\tan\beta$. Demanding that $m_b(M_X) = m_\tau(M_X)$ and achieving the correct physical masses for the bottom quark and tau lepton fixes the mass of the top quark which affects the evolution of the bottom Yukawa significantly. We shall assume gauge coupling unification, an assumption which appears reasonable when one considers SUSY models with SUSY breaking scales $\lesssim 10$ TeV.

In a complete treatment, the solution routines would be used to find the precise (similar) values of α_1, α_2, and α_3 at M_X that will evolve to the experimentally known values at M_Z, however this increases the CPU time considerably. We shall therefore sacrifice some precision in their M_Z values by taking them exactly equal at M_X. This is already a theoretical oversimplification since one does not expect the gauge couplings to be exactly equal due to threshold effects at the GUT scale. We find that for all cases we have studied, the common value $\alpha_1^{-1}(M_X) = \alpha_2^{-1}(M_X) = \alpha_3^{-1}(M_X) = 25.31$ leads to errors no bigger than 1%, 5%, and 10% in $\alpha_1(M_Z)$, $\alpha_2(M_Z)$, and $\alpha_3(M_Z)$, respectively. This is not so bad considering that the (combined experimental and theoretical) errors on $\alpha_3(M_Z)$ from some processes can be as large as 10% [30].

It is well known that there is a fine tuning problem inherent in the radiatively induced electroweak models. For certain values of the parameters, the top quark mass must be tuned to an "unnaturally" high degree of accuracy to achieve the correct value of M_Z. This problem is generally handled by rejecting models that require "too much" tuning. The amount of tuning is usually defined quite arbitrarily. The usual procedure is to define fine tuning parameters

$$c_i = \left| \frac{x_i^2}{M_Z^2} \frac{\partial M_Z^2}{\partial x_i^2} \right| , \tag{29}$$

where x_i are parameters of the theory such as m_0, $m_{1/2}$, μ, or m_t. One then demands that the c_i be less than some chosen value that is typically taken to be 10.

We have analyzed to some extent the differences in using the tree level *vs.* one loop effective potential. The basis for the "theoretical" fine tuning problem can be seen, if one makes some simplifying assumptions, in the dependence of M_W on the top quark Yukawa coupling, y_t, [31]

$$M_W \sim M_X e^{-1/y_t^2} . \tag{30}$$

We remark that this fine tuning problem is exacerbated if one uses only the tree level analysis of the potential. The vacuum expectation value coming from the minimization conditions of the tree level potential changes rapidly from 0 to infinity over the interval (Λ_s, Λ_b). Using the prescription of Ref.[26] for the scale $\hat{\Lambda}$ at which to adequately minimize the tree level potential, to extract $v(\hat{\Lambda})$, and thereby to arrive at a value for M_Z, one finds that although a small variation in $y_t(M_X)$ may lead to a small variation in Λ_b, the steepness in the tree level VEV can lead to a large variation in the value of $v(\hat{\Lambda})$ and therefore in M_Z. Hence, in the tree level analysis, solutions which may be within the bounds of the "theoretical" fine tuning may nevertheless display a fine tuning aspect because of this "tree level" fine tuning of $\hat{\Lambda}$. However, our use of the one loop effective potential (i.e., including ΔV_1) stabilizes the VEV around the M_Z scale and this particular fine tuning goes away. The true VEVs depend on scale through wave function renormalization effects which are never large as can be seen from the form of the renormalization group equations for the VEVs.

In this analysis, we shall also reject solutions based on fine tuning considerations; however, our method differs somewhat from the usual one in that it is incorporated in the solution routine described above. The routine is an iterative one which determines the convergence properties of the solution. Very slow convergence reflects an inherent fine tuning. Therefore, if the convergence is too slow, we will reject the solution. Effectively we are rejecting any solution which the computer cannot pinpoint within an allotted number of iterations.

Given values for A_0, m_0, $m_{1/2}$, $\tan\beta$, and $\text{sign}(\mu)$, the solution routines search for the values of $v(M_X)$, $m_{b,\tau}(M_X)$, $m_t(M_X)$, B_0, and μ_0. The process by which B_0 and μ_0 are found was described above. The remaining three parameters are determined similarly. The routine makes a guess for $v(M_X)$, $m_{b,\tau}(M_X)$, and $m_t(M_X)$, then the full renormalization group equations are evolved to 1 GeV calculating superparticle threshold masses in the process and minimizing the one loop effective potential at M_Z. The merits of the guess for $v(M_X)$, $m_{b,\tau}(M_X)$, and $m_t(M_X)$ is assessed by comparing the resulting values of M_Z, $m_\tau(1 \text{ GeV})$, and $m_b(1 \text{ GeV})$ with the expected ones. The process is iterated until the correct values are achieved to within some tolerance.

7. Thresholds

In the minimal low energy supergravity model being considered, the super particle spectrum is no longer degenerate as in the simple global supersymmetry model in which all the super particles are given a common mass, M_{SUSY}. In the simple case, one makes one course correction in the renormalization group evolution at M_{SUSY}. In the model with soft symmetry breaking, the nondegenerate spectrum should lead to various course corrections at the super particle mass thresholds. To this end, the renormalization group β functions must be cast in a new form which makes the implementation of the thresholds effects (albeit naive) evident. Since the $\overline{\text{MS}}$ renormalization group equations are mass independent, particle thresholds must be handled using the decoupling theorem [36], and each super particle mass has associated with it a boundary between two effective theories. Above a particular mass threshold the associated particle is present in the effective theory, below the threshold the particle is absent.

The simplest way to incorporate this is to (naively) treat the thresholds as steps in the particle content of the renormalization group β functions. This method is not always entirely adequate. For example, in the case of the $SU(2)$ gauge coupling there will be scales in the integration process at which there are effectively a half integer number of doublets using this method. We believe, nevertheless, that this method does yield the correct, general behavior of the evolution. It is a simple means of implementing the smearing effects of the non-degenerate super particle spectrum. The determination of the spectrum of masses is done without iteration as is common in other analyses. Our method deduces the physical masses by solving the equation $m(\Lambda) = \Lambda$ for each superparticle in the process of evolving from M_X to 1 GeV. The usual iterative method requires several runs to find a consistent solution.

8. Analysis

The tremendous computing task involved in analyzing the full parameter space of the soft symmetry breaking models, using the methods described as designed, would be far too time consuming given the computing facilities available to us. Therefore, in the following analysis, some simplifications will be made in the procedural method. First, only the heaviest family of quarks and leptons will have non-zero mass. This will cut down on the CPU time required for the solution routines to consistently find their values at M_X. Second, as stated previously, the value of the strong coupling at M_Z will be allowed to vary from its central value of .113 by at most 10%. This translates into a similar error in the bottom quark mass. Third, the allotted number of Runge-Kutta steps, involved in numerically integrating the renormalization group equations, will be cut down to ~ 100.

Our method involves four input parameters A_0, m_0, $m_{1/2}$, $\tan\beta$ (and the sign(μ)). The output is B_0, μ_0, M_t, M_h, and all the masses of the extra particles associated with the MSSM. Efficient use of CPU time required that we proceed as follows. For a given model, our initial exploration of the parameter space was performed in a coarse grained fashion. $\tan\beta$ was most commonly coarse grained as 2, 5, and 10, with only some runs involving higher values (e.g., 15, 20). The other three input parameters were varied in steps of 50 and 100 GeV. Values larger than ~ 500 GeV were rarely ever used. We subsequently narrowed down on the allowed hyperenvelope by fine graining around the edges of the expected region (based on the coarse graining results).

Our raw data consists of those runs which satisfy the following two criteria. The first is consistent electroweak breaking; that is, the correct value of M_Z is achieved from the minimization of the one loop effective potential with $\mu^2(M_Z) > 0$. The second criterion is no fine tuning, as implemented in our method (see Section 6). The solution routines employed return a numbered code representing the convergence properties of the solution which we use to screen the runs.

The raw data is then progressively filtered based on three physical constraints. R-parity is a discrete symmetry which distinguishes particles from superparticles by assigning a +1 to particles and a −1 to the superparticles. Conserved R-parity requires the existence of a stable lightest supersymmetric particle (LSP). Astrophysical considerations indicate that the LSP must be neutral and colorless. Cosmological considerations based on the LSPs contribution to the density of the universe indicate that it must have a mass less than ~ 200 GeV [5, 6, 7]. First, points in parameter space that lead to LSPs other than neutralinos with masses less than ~ 200 GeV are cut. Second, flavor changing neutral current bounds are used to reject runs in which intergenerational splitting of squark and slepton masses is too large. Third, experimental limits on the masses of the superparticles are used as another criterion to reject runs.

9. Results

In the following, we discuss the results of our analysis of the three classes of models listed in Table II. Because we have performed a coarse grained study, our results should only be considered qualitatively valid. Based on these, we hope to be able to ascertain the general trends in the data, and make some general predictions about the feasibility of the models considered.

Because the GUT inspired constraint Eq. (28) is enforced in this analysis, the results will depend on the mass of the bottom quark. Namely, lower bottom quark masses require larger values of the top quark mass to satisfy this relation. Most results will be reported for the case $m_b(1\text{ GeV}) = 6.00$ GeV, but lower mass (5.70 GeV) and higher mass (6.33 GeV) cases were also studied. The running value of 6.00 GeV for $m_b(1\text{ GeV})$ corresponds to a physical bottom mass $M_b = 4.85 \pm .15$ GeV, with the uncertainty coming from the error in the strong coupling, as discussed above.

9.1. No-scale Case

In all cases considered, the mass of the LSP, when it is a neutralino, is observed to be correlated with the value of $m_{1/2}$. Therefore, we find that $m_{1/2}$ cannot be taken too large ($\lesssim$ 400 GeV). In the no-scale case, we present plots of M_t vs. $m_{1/2}$ for three values of m_b(1 GeV) and containing all points satisfying the various criteria outlined in Section 8. Figure 1 is such a plot for m_b(1 GeV) = 6.00 GeV. The "right edge" of the envelope is defined by points whose neutralino LSPs are just slightly heavier than the lightest charged superparticle (usually a $\tilde{\tau}_R$ for us). The "left edge" defines the threshold of consistent electroweak breaking. The top and bottom edges are set by the requirement that Eq. (28) hold. The range of $\tan\beta$ considered leads to definite lower and upper bounds on M_t [23, 35, 13, 14].

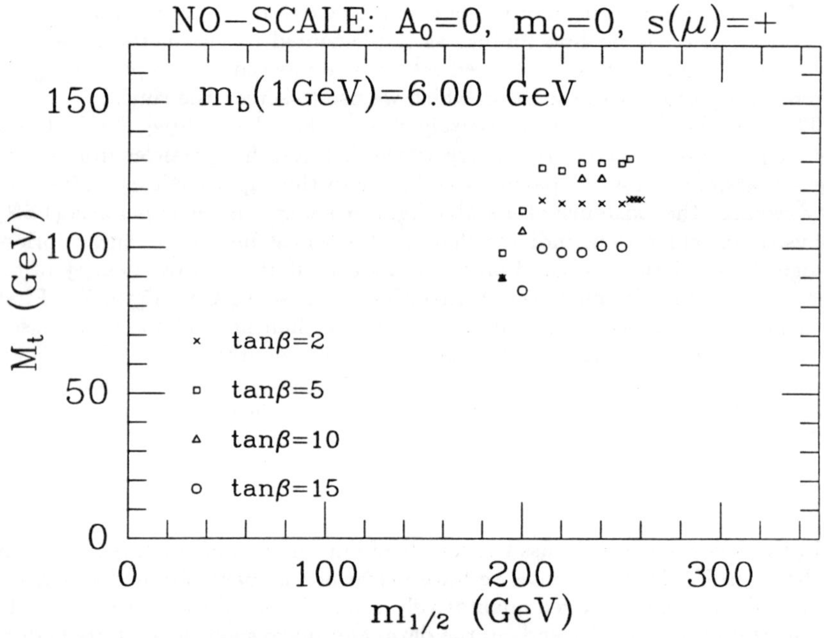

Figure 1. Plot of M_t vs. $m_{1/2}$ for $\tan\beta = 2, 5, 10, 15$ and m_b(1 GeV) = 6.00 GeV.

In Figs. 2 and 3, we display similar plots with lower (m_b(1 GeV) = 5.70 GeV) and higher (m_b(1 GeV) = 6.33 GeV) bottom quark masses. The previously noted dependence of the data on M_b is evident from these figures.

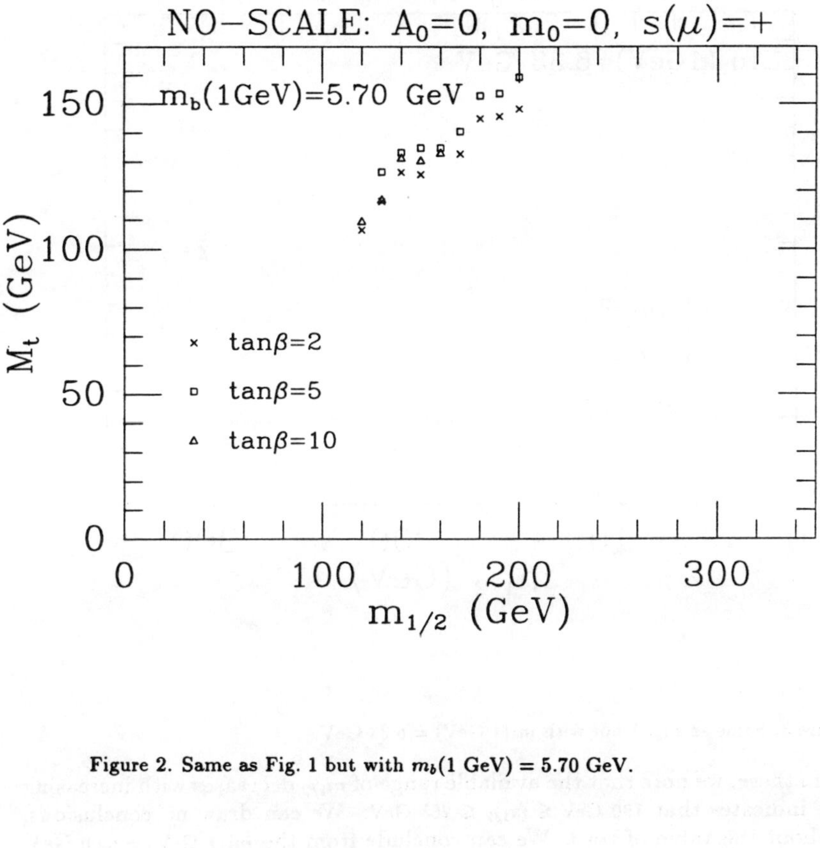

Figure 2. Same as Fig. 1 but with m_b(1 GeV) = 5.70 GeV.

In Fig. 3, we change the value of the bottom quark mass to illustrate the dependence of our results on its value.

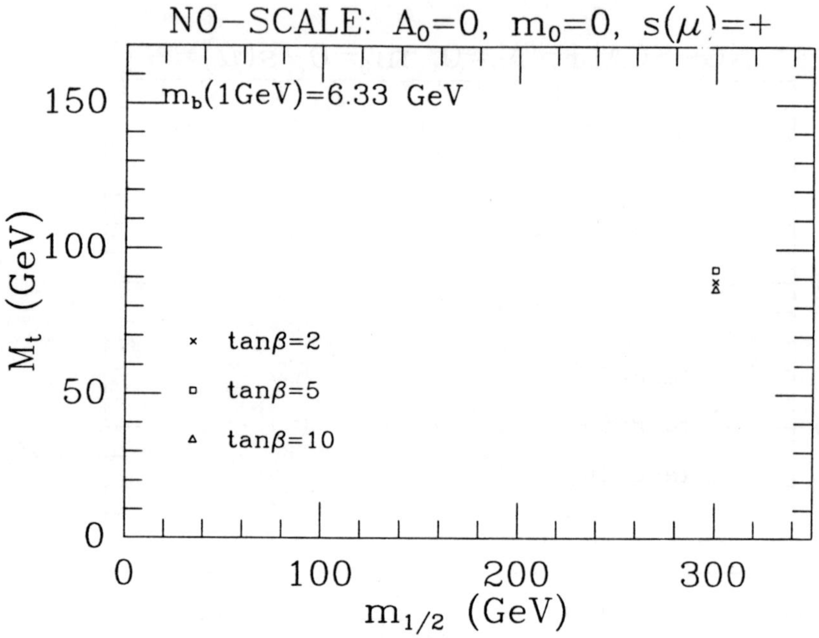

Figure 3. Same as Fig. 1 but with $m_b(1 \text{ GeV}) = 6.33$ GeV.

From these, we note that the available range of $m_{1/2}$ decreases with increasing M_b. Fig. 1 indicates that 190 GeV $\lesssim m_{1/2} \lesssim$ 265 GeV. We can draw no conclusions, however, about the value of $\tan\beta$. We can conclude from the $m_b(1 \text{ GeV}) = 6.00$ GeV case that $85 \lesssim M_t \lesssim 132$ GeV. From similar plots involving M_h, we conclude that $25 \lesssim M_h \lesssim 78$ GeV.

In Fig. 4, we display the familiar dependence of M_t on $\tan\beta$ for a particular value of $m_{1/2}$ in the allowed range.

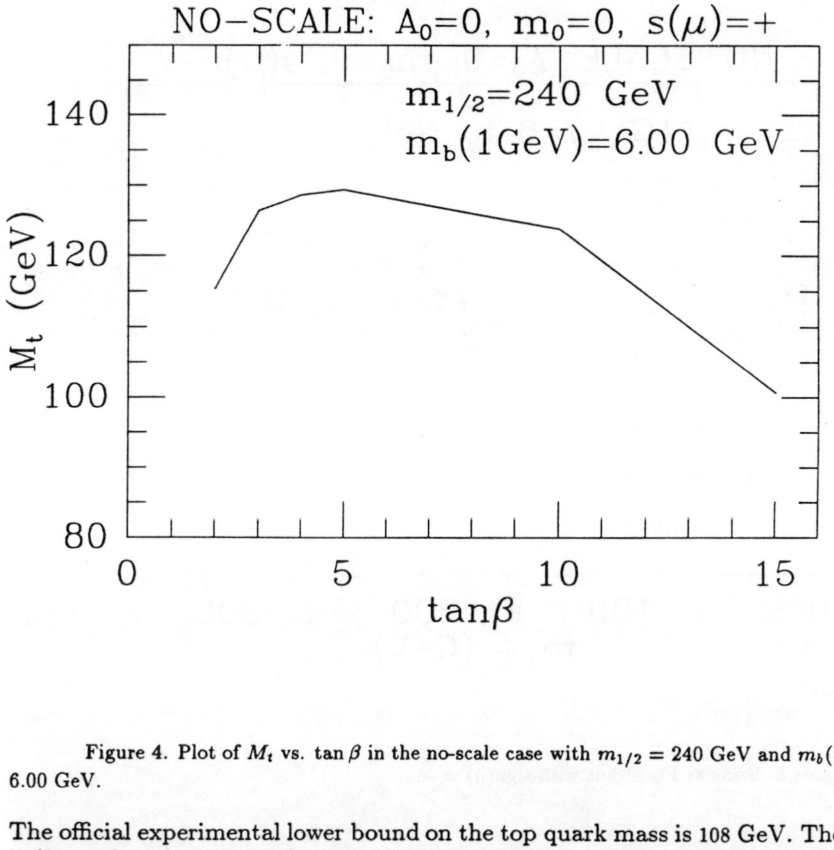

Figure 4. Plot of M_t vs. $\tan\beta$ in the no-scale case with $m_{1/2} = 240$ GeV and $m_b(1\text{ GeV}) =$ 6.00 GeV.

The official experimental lower bound on the top quark mass is 108 GeV. The figures indicate that the top quark cannot have a mass greater than ~ 132 GeV in this model, if $m_b(1\text{ GeV}) = 6.00$ GeV. This upper bound is raised to ~ 160 GeV if $m_b(1\text{ GeV}) = 5.70$ GeV, and the model is ruled out, if $m_b(1\text{ GeV}) = 6.33$ GeV.

Figure 5 is similar to Fig. 1, but we have chosen sign of μ negative. The allowed region is displaced down with respect to the positive μ case with the upper bound on M_t now 113 GeV and very close to the experimental limit.

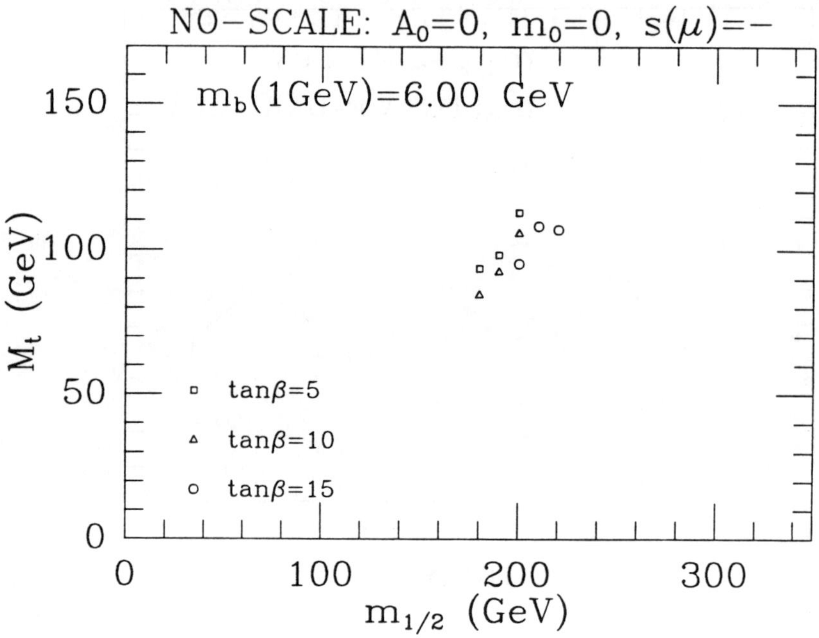

Figure 5. Same as Fig. 1 but with sign(μ) = −.

9.2. Strict No-scale Case

The results of the strict no-scale case must be interpolated from the no-scale results, because B_0 is not an input parameter in our procedural method. Therefore, we plot in Fig. 6 M_t vs. B_0 and deduce the M_t bounds from slicing the data along $B_0 = 0$. Admittedly, the relatively small number of points makes the perimeter of the region unclear in some areas. We get approximately $90 \lesssim M_t \lesssim 127$ GeV. For M_h, we find $40 \lesssim M_h \lesssim 78$ GeV with an uncertainty in the lower bound of ~ 10 GeV due to lack of definition in the lower end of the envelope. Inspection of the data indicates that $3.5 \lesssim \tan\beta \lesssim 9$. Thus, it appears that $\tan\beta$ cannot be too small or too large to accommodate the strict no-scale case.

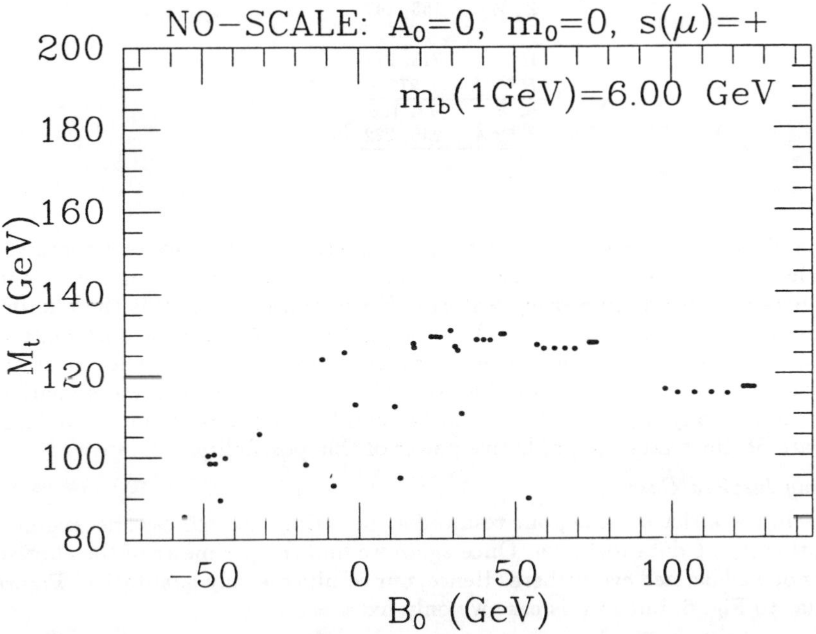

Figure 6. Plot of M_t vs. B_0 for points with $A_0 = m_0 = 0$ used to interpolate to $B_0 = 0$ and deduce the M_t range in this case.

Table 3: Strict No-scale Scenario

A_0	0
m_0	0
$m_{1/2}$	240
μ_0	171
$\tan\beta$	8.3
sign(μ)	+
M_t	126
$\tilde{d}$	473 – 506
$\tilde{u}$	390 – 532
$\tilde{e}$	94 – 176
$\tilde{\nu}_L$	180
$\tilde{\gamma}$	92
$\tilde{Z}$, $\tilde{W}$	153, 147
$\tilde{g}$	557
$\tilde{H}^0$	213, 271
$\tilde{H}^{\pm}$	270
h, H	77, 158
$H^{\pm}$, A	235, 222

Finally, in Table 3, we display the spectrum of superparticle masses for a representative strict no-scale scenario with $\tan\beta = 8.3$ and $m_{1/2} = 240$ GeV. The masses are in GeVs. There are several interesting features. For instance, the LSP is the photino in this case and has a mass of 92 GeV. It is too light to account for the dark matter. The top quark mass is just above the experimental limit at 126 GeV, and the Higgs boson mass is 77 GeV, which underlines one of the signatures of these scenarios: the Higgs is typically light, so that it will be found in two photon modes at SSC. At any rate, it illustrates the predictive power of this possibility.

9.3. String Inspired Case

As in the strict no-scale, our results for the string inspired case necessitates interpolating $A_0 = 0$ data to $B_0 = 0$. Once again we find the perimeter of the allowed region is not well defined everywhere. Hence, our results are only qualitative. Figure 7 is similar to Fig. 6, but in this case we only fix $A_0 = 0$.

The strict no-scale case is a special case of the string inspired one, therefore the 127 GeV top quark mass upper bound is not expected to decrease but rather to increase in this case. Slicing along $B_0 = 0$ yields $85 \lesssim M_t \lesssim 140$ GeV. Similarly, for the light Higgs we get $40 \lesssim M_h \lesssim 80$ GeV. The data indicates in this case, as in the strict no-scale case, that there is a lower bound on $\tan\beta$ of ~ 3.

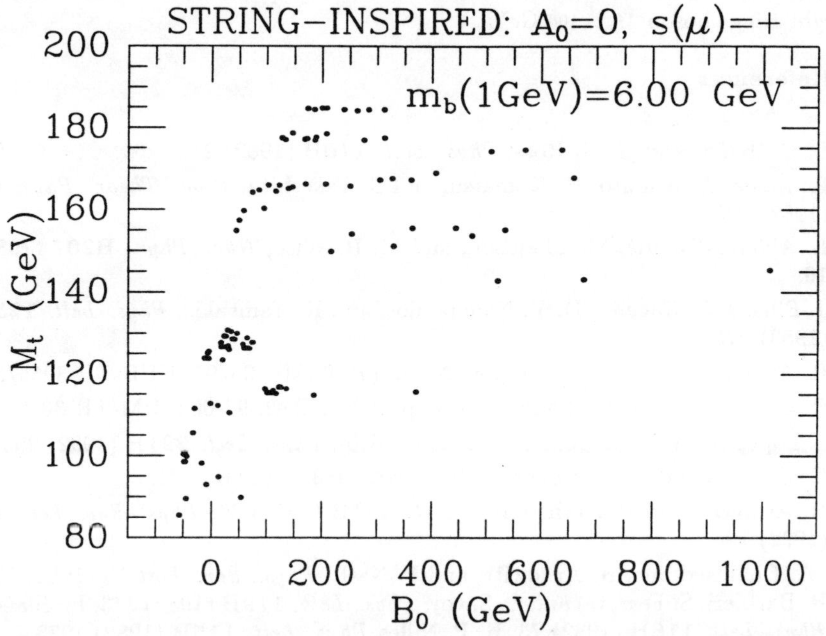

Figure 7. Same as Fig. 6 but for points with $A_0 = 0$ only.

10. Conclusions

Minimal low energy supergravity models were considered. It is quite remarkable that for as simple a breaking of supersymmetry as that offered by the strict no-scale model, we can reproduce the standard model results, including the appealing feature of automatic breaking of the electro-weak symmetry. The importance of the value of the top quark mass to these schemes cannot be understated, since we found in our study of specific models, rather restrictive upper bounds for the top quark mass. No-scale models in which only gaugino masses provide global supersymmetry breaking yield top quarks with masses less than ~ 127 GeV. The results are sensitive to the value of the bottom quark mass. Lower bottom quark masses, within the experimental uncertainty, lead to higher top quark upper bounds. In these models, the ratio of vacuum expectation values of the two Higgs fields is expected to be larger than $\sim 70°$.

Although the perimeter of the allowed regions were often fuzzy, we could, nevertheless, draw some general conclusions from our results. For all our runs, with no restrictions on the soft terms, we find for the top quark $M_t \lesssim 186$ GeV and for the light Higgs boson $M_h \lesssim 100$ GeV .

11. References

1. L. E. Ibáñez and G. G. Ross, *Phys. Lett.* **110B** (1982) 215.
2. K. Inoue, A. Kakuto, H. Komatsu, and S. Takeshita, *Prog. Theor. Phys.* **68** (1982) 927.
3. L. Alvarez-Gaumé, M. Claudson, and M. B. Wise, *Nuc. Phys.* **B207** (1982) 16.
4. J. Ellis, J. S. Hagelin, D. V. Nanopoulos, and K. Tamvakis, *Phys. Lett.* **125B** (1983) 275.
5. R. Arnowitt and P. Nath, Report No. CTP-TAMU-65/92, NUB-TH-3055/92.
6. R. G. Roberts and L. Roszkowski, Report No. RAL-93-003, UM-TH-33/92.
7. J. Lopez, D. V. Nanopoulos, and A. Zichichi, *Phys. Lett.* **291B** (1992) 255.
8. G. G. Ross and R. G. Roberts, *Nuc. Phys.* **B377** (1991) 571.
9. R. Arnowitt and P. Nath, *Phys. Lett.* **287B** (1992) 89; *Phys. Rev. Lett.* **69** (1992) 725.
10. A. Chamseddine, R. Arnowitt, and P. Nath, *Phys. Rev. Lett.***49** (1982) 970. R. Barbieri, S. Ferrara, and C. Savoy, *Phys. Lett.* **119B** (1982) 343; L. Ibañez, *Phys. Lett.* **118B** (1982) 73; H. P. Nilles *Phys. Lett.* **115B** (1982) 1973.
11. P. Nath and R. Arnowitt, *Phys. Lett.* **289B** (1992) 368.
12. S. Kelly, J. Lopez, H. Pois, D. V. Nanopoulos, and K. Yuan, *Phys. Lett.* **273B** (1991) 423.
13. B. Ananthanarayan, G. Lazarides, and Q. Shafi, *Phys. Rev.* **D44** (1991) 1613.
14. A. Giveon, L.J. Hall, and U. Sarid, *Phys. Lett.* **271B** (1991) 138.
15. U. Amaldi, W. de Boer, and H. Fürstenau, *Phys. Lett.* **260B** (1991) 447.
16. J. Ellis, S. Kelley, and D. Nanopoulos, *Phys. Lett.* **260B** (1991) 131.
17. P. Langacker and M. Luo, *Phys. Rev.* **D44** (1991) 817.
18. J. C. Pati and A. Salam, *Phys. Rev. D* **10** (1974) 275.
19. H. Georgi and S. Glashow, *Phys. Rev. Lett.* **32** (1974) 438.
20. H. Georgi, in *Particles and Fields-1974*, edited by C. E. Carlson, AIP Con-

ference Proceedings No. 23 (American Institute of Physics, New York, 1975) p. 575.

21. H. Fritzsch and P. Minkowski, *Ann. Phys. NY* **93** (1975) 193.
22. F. Gürsey, P. Ramond, and P. Sikivie, *Phys. Lett.* **60B** (1975) 177.
23. H. Arason, D. J. Castaño, B. Keszthelyi, S. Mikaelian, E. J. Piard, P. Ramond, and B. D. Wright, *Phys. Rev. Lett.* **67** (1991) 2933.
24. For a review see H. P. Nilles, *Phys. Reports* **110C** (1984) 1 and references therein.
25. J. F. Gunion, H. E. Haber, and M. Sher, *Nuc. Phys.* **B306** (1988) 1.
26. G. Gamberini, G. Ridolfi, and F. Zwirner, *Nuc. Phys.* **B331** (1990) 331.
27. Y. Okada, M. Yamaguchi, and T. Yanagida, *Prog. Theor. Phys.* **85** (1991) 1.
28. J. Ellis, G. Ridolfi, and F. Zwirner, *Phys. Lett.* **262B** (1991) 477.
29. M. Drees and M.M. Nojiri, *Phys. Rev.* **D45** (1992) 2482.
30. H. Arason, D. J. Castaño, B. Keszthelyi, S. Mikaelian, E. J. Piard, P. Ramond, and B. D. Wright, *Phys. Rev.* **D46** (1992) 3945.
31. A. B. Lahanas and D. V. Nanopoulos, *Phys. Rep.* **145** (1987) 1.
32. J. Gasser and H. Leutwyler, *Phys. Rep.* **87** (1982) 77.
33. See Ref. [31] and references therein.
34. H. Arason, D. J. Castaño, E. J. Piard, and P. Ramond, *Phys. Rev.* **D47** (1992) 232.
35. S. Kelley, J. L. Lopez, and D. V. Nanopoulos, *Phys. Lett.* **274B** (1992) 387.
36. T. W. Appelquist and J. Carrazzone, *Phys. Rev.* **D11** (1975) 2856.
37. N. K. Falck, *Zeit. für Phys.* **C30** (1986) 247.
38. M. E. Machacek and M. T. Vaughn, *Nuc. Phys.* **B222** (1983) 83; *Nuc. Phys.* **B236** (1984) 221; *Nuc. Phys.* **B249** (1985) 70.
39. H. E. Haber and R. Hempfling, *Phys. Rev. Lett.* **66** (1991) 1815.
40. J. Ellis, G. Ridolfi, and F. Zwirner, *Phys. Lett.* **257B** (1991) 83.

A FERMION MASS MATRIX ANSATZ FOR THE FLIPPED SU(5) MODEL

George K. Leontaris
Physics Department, Ioannina University
Ioannina, GR-451 10, Greece

ABSTRACT

A fermion mass matrix ansatz is proposed in the context of Grand Unified Supersymmetric Theories (GUTs). The fermion mass matrices are evolved down to the electroweak scale by solving the renormalization group equations for the gauge and Yukawa couplings. Eight inputs are introduced at the GUT scale to predict the 13 arbitrary parameters of the Standard model. The constraints imposed by the charged fermion data are used to make predictions in the neutrino sector. In particular, the neutrino mass matrix is worked out in the case of the flipped SU(5) model and it is found that the *generalized* see-saw mechanism which occurs naturally in this model can provide a solution to the solar neutrino puzzle and give a sufficiently large ν_τ mass to contribute as a hot dark matter component as indicated by the recent COBE data.

1. Introduction

Although the Standard Model of Strong and Electroweak interactions explains all known experimental data, it is now widely accepted that it is unlikely to be the fundamental theory of nature.

In a fundamental theory the various experimentally measurable parameters should be calculable only from few inputs which could be specified in terms of the basic principles of the theory. For example, in String theories[1] one expects that all masses and mixing angles can be determined only from one input, namely the String gauge coupling g_{String} at the String Unification scale $M_{String} \sim O(g_{String} \times M_{Pl})$. However, there is a huge number of String derived models and it is rather impossible to find the unique model among them which might lead to the correct - experimentally tested - predictions at low energies. Although the previously mentioned arbitrariness puts a great obstacle in string model builders, which is not likely to be solved in the near future, we are by now convinced that there is no real rival to string theory. ¿From the phenomenologist's point of view, the alternative way is to attempt to describe the low energy theory, using the experience from string model building and all the possible information that one can extract out of it.

We know already, that there is a number of constraints which should be respected on our way from M_{String} down to m_W. Here there are listed few of them:

• Recent calculations taking into account threshold corrections from String massive states[3, 4, 5, 6, 7], show that the String Unification scale is relatively high, and close to the Plank scale. On the contrary, renormalization group calculations have indicated that minimal supersymmetric Grand Unified Theories (SUSY-GUTs) are in agreement with the precision LEP data when the SUSY-GUT scale M_G is taken close to $M_G \approx 10^{16} GeV$[2]. Thus, the minimal supersymmetric standard model(MSSM) cannot probably be derived directly at the string scale; the above descrepancy between the two scales, would rather suggest that MSSM should be obtained through the spontaneous breaking of some intermediate GUT-like gauge group which breaks at the scale M_G.

• As long as string model building is based on $k = 1$ level of Kac-Moody algebras, it was soon realised that ordinary GUTs (like the SU(5) model), could not be suitable candidates for a string derived model. The reason is that ordinary SU(5) needs necessarily Higgs fields in the adjoint in order to break down to the MSSM, while such representations are not available in $k = 1$ constractions.

The flipped SU(5) model[8, 9, 10], was a first attempt to overcome this difficulty, and has been proposed as a candidate some time ago[9], in the context of the free fermionic formulation of the four-dimensional superstring. It was subsequently discovered[11], that the GUT symmetry $SU(4) \times O(4)$[11], based on the Pati-Salam model[12], can also break down to the standard gauge group without using higgs fields in the adjoint, and therefore could also serve as a candidate Superstring model. We should mention that it is also possible to obtain directly the Standard model gauge group from the string[13], but the proton decay problem[14] as well as the high string unification mentioned above, seem to be naturally solved in the presence of an intermediate gauge group. Moreover, the Grand Unified models inspired from strings[9, 11] offer natural mechanisms of suppressing the neutrino masses[15, 16], explain the baryon asymmetry[17] of the universe etc.

In view of the above theoretical and phenomenological constraints, in my opinon, the GUT model building issue should be reconsidered and worked out systematically. In the present talk, I will propose a specific fermion mass matrix ansatz[18], which was inspired from phenomenological calculations on the flipped string SU(5). The analysis of the charged fermion mass spectrum which is given in section 2, has been done on completely general grounds, and can apply to any SUSY-GUT model. Furthermore, in view of the revived interest[19, 21, 20, 22] of the fermion mass problem in GUTs, this work contributes also to the search[23] for all possible mass matrices with the maximum number of zero entries at the Unification Scale which give the correct predictions at m_W. The discussion on neutrino mass matrix is given in section 3, and applies mainly to the particular string constructions[9, 11, 13].

2. Structure of Fermion Mass Matrices at M_{GUT}

The texture of the quark and lepton mass matrices has the following form at the GUT scale

$$M_U = \begin{pmatrix} 0 & 0 & x \\ 0 & y & z \\ x & z & 1 \end{pmatrix} \lambda_{top}(t_0)\frac{v}{\sqrt{2}}sin\beta, \tag{1}$$

$$M_D = \begin{pmatrix} 0 & ae^{i\phi} & 0 \\ ae^{-i\phi} & b & 0 \\ 0 & 0 & 1 \end{pmatrix} \lambda_b(t_0)\frac{v}{\sqrt{2}}cos\beta \tag{2}$$

$$M_E = \begin{pmatrix} 0 & ae^{i\phi} & 0 \\ ae^{-i\phi} & -3b & 0 \\ 0 & 0 & 1 \end{pmatrix} \lambda_\tau(t_0)\frac{v}{\sqrt{2}}cos\beta \tag{3}$$

$$M_{\nu_{Dirac}} = M_U \tag{4}$$

with $tan\beta \equiv \frac{<\bar{h}>}{<h>}$, $\lambda_b(t_0) = \lambda_\tau(t_0)$, and $v = 246GeV$. We have taken the up-quark matrix to be symmetric, and the down quark matrix to be hermitian. The Dirac neutrino mass matrix, has been taken to be identical to the up-quark mass matrix at the GUT scale, since both arise from the same Yukawa term in these theories[9, 11]. Furthermore, we have chosen to relate the charged lepton mass matrix at the GUT scale with the down quark mass matrix. As a matter of fact, the m_D and m_E matrix elements are not necessarily related in the flipped model. However, our choice minimizes the number of arbitrary parameters and moreover, it leads to definite predictions in the neutrino sector.

Let us discuss first the charged fermion mass matrices. Our ansatz has a total of five zeros in the quark sector (the leptonic sector is directly related to them); two zeros for the up and three for the down quark mass matrix(zeros in symmetric entries are counted only once). These zeros reduce the number of arbitrary parameters at the GUT-scale to eight, namely $x, y, z, a, b, \phi, \lambda_b(t_0)$, and $\lambda_{top}(t_0)$. These non-zero entries should serve to determine the thirteen arbitary parameters of the standard model, i.e., nine quark and lepton masses, three mixing angles and the phase of the Cabbibo-Kobayashi-Maskawa (CKM) matrix. Thus, as far as the charged fermion sector is concerned, we end up with five predictions (we will discuss the predictions in the neutrino sector in the next section.). We may reduce the number of arbitary parameters by one, if we impose more structure in the up-quark mass matrix. We may for example relate the (13), (23) and (22) entries, as follows

$$y = nz^2 \tag{5}$$

$$x = (n-1)z^2 \tag{6}$$

where n can be a number in the range $n \sim (3, 10)$. Although this structure is imposed by hand and not actually necessary[18], the results are more presentable and calculations more easy to handle.

In order to find the structure of the mass matrices at the low energy scale and calculate the mass eigenstates as well as the mixing matrices and compare them with the experimental data, we need to evolve them down to m_W, using the renormalization group equations. The renormalization group equations for the Yukawa couplings at the one-loop level, read

$$16\pi^2 \frac{d}{dt}\lambda_U = \left(I\cdot Tr[3\lambda_U\lambda_U^\dagger] + 3\lambda_U\lambda_U^\dagger + \lambda_D\lambda_D^\dagger - I\cdot G_U\right)\lambda_U, \quad (7)$$

$$16\pi^2 \frac{d}{dt}\lambda_N = \left(I\cdot Tr[\lambda_U\lambda_U^\dagger] + \lambda_E\lambda_E^\dagger - I\cdot G_N\right)\lambda_N, \quad (8)$$

$$16\pi^2 \frac{d}{dt}\lambda_D = \left(I\cdot Tr[3\lambda_D\lambda_D^\dagger + \lambda_E\lambda_E^\dagger] + 3\lambda_D\lambda_D^\dagger + \lambda_U\lambda_U^\dagger - I\cdot G_D\right)\lambda_D, \quad (9)$$

$$16\pi^2 \frac{d}{dt}\lambda_E = \left(I\cdot Tr[\lambda_E\lambda_E^\dagger + 3\lambda_D\lambda_D^\dagger] + 3\lambda_E\lambda_E^\dagger - I\cdot G_E\right)\lambda_E, \quad (10)$$

where λ_α, $\alpha = U, N, D, E$, represent the 3x3 Yukawa matrices which are defined in terms of the mass matrices given in Eq.(1-4), and I is the 3x3 identity matrix. We have neglected one-loop corrections proportional to λ_N^2. $t \equiv \ln(\mu/\mu_0)$, μ is the scale at which the couplings are to be determined and μ_0 is the reference scale, in our case the GUT scale. The gauge contributions are given by

$$G_\alpha = \sum_{i=1}^{3} c_\alpha^i g_i^2(t), \quad (11)$$

$$g_i^2(t) = \frac{g_i^2(t_0)}{1 - \frac{b_i}{8\pi^2} g_i^2(t_0)(t - t_0)}. \quad (12)$$

The g_i are the three gauge coupling constants of the Standard Model and b_i are the corresponding supersymmetric beta functions. The coefficients c_α^i are given by

$$\{c_U^i\}_{i=1,2,3} = \left\{\frac{13}{15}, 3, \frac{16}{3}\right\}, \qquad \{c_D^i\}_{i=1,2,3} = \left\{\frac{7}{15}, 3, \frac{16}{3}\right\}, \quad (13)$$

$$\{c_E^i\}_{i=1,2,3} = \left\{\frac{9}{5}, 3, 0\right\}, \qquad \{c_N^i\}_{i=1,2,3} = \left\{\frac{3}{5}, 3, 0\right\}. \quad (14)$$

In what follows, we will assume that $\mu_0 = M_G \approx 10^{16} GeV$ and $a_i(t_0) = \frac{g_i^2(t_0)}{4\pi} \approx \frac{1}{25}$. In order to evolve the equations(Eqs) (7-10) down to low energies, we also need to do some plausible approximations. First, we find it convenient to diagonalize the up quark mass matrix at the GUT scale and obtain the eigenstates

$$\begin{aligned} m_1(M_{GUT}) &\approx -n(n-1)ptan^2\theta sin^2\theta \\ m_2(M_{GUT}) &\approx (n-1)ptan^2\theta \\ m_3(M_{GUT}) &\approx \frac{p}{cos^2\theta} \end{aligned} \quad (15)$$

with diagonalizing matrix

$$K = \begin{pmatrix} \frac{1}{D_1} & \frac{-sin\theta}{D_2} & \frac{(n-1)sin^2\theta}{D_3} \\ \frac{sin2\theta}{2D_1} & \frac{1}{D_2} & \frac{sin2\theta}{D_3} \\ \frac{nsin\theta}{D_1} & \frac{nsin\theta}{D_2} & \frac{1-nsin^2\theta}{D_3} \end{pmatrix} \quad (16)$$

where $p = \lambda_t(t_0)\frac{v}{\sqrt{2}}sin\beta$, $D_1 \approx \sqrt{1 + sin^2\theta cos^2\theta}$, $D_2 \approx \sqrt{1 + sin^2\theta}$, and $D_3 \approx \sqrt{1 - (2n-1)sin^2\theta}$. In this case all other mass matrices are also rotated by the same similarity transformation, thus

$$\lambda_\alpha \rightarrow \tilde{\lambda}_\alpha = K^\dagger \lambda_\alpha K, \alpha = D, E, N \quad (17)$$

Now assuming that the λ_t coupling is much bigger than all other fermion Yukawa couplings, we may ignore the contributions of the latter in the RHS of the RGEs in Eqs(7-10). In this case all differencial equations reduce to a simple uncoupled form. Thus the top-Yukawa coupling differential equation, for example, may be cast in the form

$$16\pi^2\frac{d}{dt}\tilde{\lambda}_{top} = \tilde{\lambda}_{top}(6\tilde{\lambda}^2_{top} - G_U(t)) \tag{18}$$

with the solution[24, 20]

$$\tilde{\lambda}_{top} = \tilde{\lambda}_{top}(t_0)\xi^6\gamma_U(t) \tag{19}$$

where

$$\gamma_U(t) = \prod_{j=1}^{3}\left(1 - \frac{b_{j,0}\alpha_{j,0}(t-t_0)}{2\pi}\right)^{c^j_\alpha/2b_j}, \tag{20}$$

$$\xi = \left(1 - \frac{6}{8\pi^2}\lambda_{top}(t_0)\int_{t_0}^{t}\gamma^2_U(t)\,dt\right)^{-1/12}. \tag{21}$$

Thus the GUT up quark mass eigenstates given in Eq(15) renormalized down to their mass scale, are given by

$$\begin{aligned} m_u &\approx \gamma_U\xi^3\eta_u n(n-1)ptan^2\theta sin^2\theta \\ m_c &\approx \gamma_U\xi^3\eta_c(n-1)ptan^2\theta \\ m_t &\approx \gamma_U\xi^6\frac{p}{cos^2\theta} \end{aligned} \tag{22}$$

where η_u , n_c are taking into account the renormalization effects from the m_t-scale down to the mass of the corresponding quark. We may combine the equations in Eq(22) and give predictions for the top quark mass and the mixing angle θ in terms of the low energy up and charm quark masses. We obtain

$$m_t \approx \xi^3\frac{n}{n-1}\frac{m_c^2}{m_u}\frac{\eta_u}{\eta_c^2} \tag{23}$$

$$sin\theta = \sqrt{\frac{m_u}{nm_c}} \tag{24}$$

In the basis where the up-quark mass matrix is diagonal, the renormalized down-quark and charged lepton mass matrices are given by

$$m_D^{ren} \approx \gamma_D I_\xi K^\dagger m_D K \tag{25}$$

$$m_E^{ren} \approx \gamma_E K^\dagger m_E K \tag{26}$$

where $I_\xi = Diagonal(1, 1, \xi)$, and $m_{D,E}$, are given in Eqs(2) and (3). In order to make definite predictions in the fermion mass sector, we choose the lepton masses as inputs and express the arbitrary parameters $a, b, \lambda_\tau(\equiv \lambda_b)$ in m_E in terms of m_e, m_μ, and m_τ. Substituting into m_D and diagonalizing m_D^{ren}, we obtain[18]

$$m_d \approx 6.3 \times (\frac{\eta_d}{2})MeV \tag{27}$$

$$m_s \approx 153 \times (\frac{\eta_s}{2}) MeV \tag{28}$$

$$m_b \approx \eta_b \frac{\gamma_D}{\gamma_E} \xi m_\tau \tag{29}$$

Since $\eta_d \approx \eta_s \approx 2$, the predictions for the light quarks $m_{d,s}$ are within the expected range[25]. Now, in order to make a prediction for the bottom quark, we need to know the value of ξ, but the latter depends on the top quark coupling at the GUT scale as well as on the top-mass, through Eq.(21). We can use, however, Eq.(29), to predict the range of the top mass. Thus, using the available limits on the bottom mass, $4.15 \leq m_b \leq 4.35 GeV$, and $\eta_b \approx 1.4$, we can obtain the following range for m_{top}

$$125 GeV \leq m_{top} \leq (165 - 170) GeV \tag{30}$$

The CKM - matrix can be found by diagonalizing m_D^{ren} in Eq(25). For all the above range of the m_{top} mass, it is always possible to find a set of the input parameters, which give CKM-mixing angles within the experimentally acceptable ranges. Its analytic form is rather complicated. As an example, we give the CKM-entries for the specific case where, $m_t = 135 GeV$, $tan\beta = 1.1$,and $\phi = \frac{\pi}{4}$. In this case m_b is predicted to be $4.33 GeV$, while we obtain

$$| V_{CKM} |_{ij} = \begin{pmatrix} 0.9754 & 0.22 & 0.0032 \\ 0.22 & 0.9750 & 0.038 \\ 0.01 & 0.037 & 0.9993 \end{pmatrix} \tag{31}$$

2. Neutrino Masses

There is a lot of experimental evidence today, that the neutrinos have non-zero masses. For example, recent data from solar neutrino experiments[26] show that the deficiency of solar neutrino flux, i.e. the discrepancy between theoretical estimates and the experiment, is naturally explained if the ν_e neutrino oscillates to another species during its flight to the earth. Furthermore, the COBE measurement [27] of the large scale microwave background anisotropy, might be explained [28] if one assumes an admixture of cold ($\sim 75\%$) plus hot ($\sim 25\%$) dark-matter. It is hopefully expected that some neutrino (most likely ν_τ) may be the natural candidate of the hot dark matter component.

Here we would like to address the question of neutrino masses in GUT models which arise [9, 11]in the free fermionic construction of four dimensional strings. Taking into account renormalization effects , it was recently shown[16, 18] that the general see-saw mechanism which occurs naturally in the flipped model, turns out to be consistent with the recent solar neutrino data, while on the other hand suggests that CHOROUS and NOMAD experiments at CERN may have a good chance of observing $\nu_\mu \longleftrightarrow \nu_\tau$ oscillations*

The various tree-level superpotential mass terms which contribute to the neutrino mass matrix in the flipped $SU(5)$ model are the following:

$$\lambda^u_{ij} F^i \bar{f}^j \bar{h} + \lambda^{\phi\nu^c}_{ij} F^i \bar{H} \phi^j + \lambda^{\phi}_{ij} \phi^0 \phi^i \phi^j \tag{32}$$

*for neutrinos in convensional GUTs see[29]

where in the above terms $F^i, \bar{f}^j$ are the $10, \bar{5}$ matter SU(5) fields while $\bar{H}, \bar{h}, h$ are the $\bar{10}, \bar{5}, 5$ Higgs representations and ϕ^i are neutral $SU(5) \times U(1)$ singlets. The Higgs field $\bar{H}$ gets a vacuum expectation value(v.e.v.) of the order of the SU(5) breaking scale ($\sim 10^{16} GeV$), $\bar{h}, h$ contain the standard higgs doublets while ϕ^0 acquires a v.e.v., most preferably at the electoweak scale. The neutrino mass matrix may also receive significant contributions from other sources. Of crusial importance, are the non-renormalizable contributions[30] which may give a direct $M_{\nu^c\nu^c} = M^{rad}$ contribution which is absent in the tree-level potential. Then, the general 9×9 neutrino mass matrix in the basis (ν_i, ν_i^c, ϕ_i), may be written as follows:

$$m_\nu = \begin{pmatrix} 0 & m_U & 0 \\ m_U & M^{rad} & M_{\nu^c,\phi} \\ 0 & M_{\nu^c,\phi} & \mu_\phi \end{pmatrix} \tag{33}$$

where it is understood that all entries in Eq.(33) represent 3×3 matrices. The above neutrino matrix is different from that of standard see-saw matrix-form, since now there are three neutral $SU(5) \times U(1)$ singlets involved, one for each family.

It is clear that the matrix (33) depends on a relatively large number of parameters and a reliable estimate of the light neutrino masses and the mixing angles is a rather complicated task. We are going to use however our knowledge of the rest of the fermion spectrum to reduce sufficiently the number of parameters involved. Firstly, due to the GUT relation $m_U(M_{GUT}) = m_{\nu_D}(M_{GUT})$, we can deduce the form of $m_{\nu_D}(M_{GUT})$, at the GUT scale in terms of the up-quark masses. The heavy majorana 3×3 matrix M^{rad}, depends on the kind of the specific generating mechanism. Here[18] we take it to be proportional to the down quark-matrix at the GUT scale:

$$M^{rad} = \Lambda^{rad} m_D(M_{GUT}) \tag{34}$$

The $M_{\nu^c,\phi}$ and μ_ϕ 3×3 submatrices are also model dependent. In most of the string models however, there is only one entry at the trilinear superpotential in the matrix $M_{\nu^c,\phi}$, which is of the order M_{GUT}. Other terms, if any, usually arise from high order non-renormalizable terms. We will assume in this work only the existence of the trilinear term, since higher order ones will be comparable to M^{rad} and are not going to change our predictions. In particular we will take $M_{\nu^c,\phi} \sim Diagonal(M, 0, 0)$, and $\mu_\phi \sim Diagonal(\mu, 0, 0)$, with $\mu << M \sim M_{GUT}$, thus we will treat Eq.(33) as a 7×7 matrix.

To obtain the neutrino spectrum and lepton mixing, we must introduce values for the two additional papameters M, Λ^{rad} of the neutrino mass matrix in Eq.(33). We assume naturally $M = <\bar{H}> \approx 10^{16} GeV$. The neutrino mass eigenvalues can now be predicted in terms of the scale quantity Λ^{rad} . Thus they can be written as[18]

$$m_{\nu_e} \approx 0, m_{\nu_\mu} = \frac{\Lambda^\mu}{\Lambda^{rad}} \times 10^{-2} eV, m_{\nu_\tau} = \frac{\Lambda^\tau}{\Lambda^{rad}} \times 10 eV \tag{35}$$

For $m_t \approx 130 GeV$ we get $\Lambda^\mu \approx .80 \times 10^{12}$ and $\Lambda^\tau \approx 1.85 \times 10^{12}$.

For the oscillation probabilities, we find[18]

$$P(\nu_e \leftrightarrow \nu_\mu) \approx 3.1 \times 10^{-2} sin^2(\pi \frac{L}{l_{12}}) \tag{36}$$

$$P(\nu_\tau \rightarrow \nu_\mu) \approx 4.0 \times 10^{-3} sin^2(\pi \frac{L}{l_{13}}) \tag{37}$$

$$P(\nu_e \rightarrow \nu_\tau) \approx 4.0 \times 10^{-5} sin^2(\pi \frac{L}{l_{13}}) \tag{38}$$

where L is the source–detector distance and

$$l_{ij} = \frac{4\pi E_\nu}{|m_i^2 - m_j^2|} \tag{39}$$

We can determine the range of Λ^{rad}, using the available solar neutrino data (see for example ref[31] for a systematic analysis of the allowed regions using all available data)

$$5.0 \times 10^{-3} \leq sin^2 2\theta_{ij} \leq 1.6 \times 10^{-2} \tag{40}$$

$$0.32 \times 10^{-5}(eV)^2 \leq \delta m_{ij}^2 \leq 1.2 \times 10^{-5}(eV)^2 \tag{41}$$

Our result in Eq(36) is a bit outside the above range but the mass constraint can be easily satisfied by choosing Λ^{rad} in the range

$$.7 \times 10^{12} \leq \Lambda^{rad} \leq 7 \times 10^{12} \tag{42}$$

Our neutrino masses can also easily be made to fall into the range of the Frejus atmospheric neutrinos

$$10^{-3}(eV)^2 \leq \delta m_{ij}^2 \leq 10^{-2}(eV)^2 \tag{43}$$

but our mixing is much too small. Our results are also consistent with the data on $\nu_\mu \leftrightarrow \nu_\tau$ oscillations[32]

$$sin^2 2\theta_{\mu\tau} \leq 4. \times 10^{-3}, \delta m_{\nu_\mu \nu_\tau}^2 \geq 50(eV)^2 \tag{44}$$

Our results however cannot be made to fall on the $sin^2 2\theta$ vs δm^2 of the BNL $\nu_\mu \leftrightarrow \nu_e$ oscillation results[33].

Moreover, it is always possible to obtain $m_{\nu_\tau} \approx (few \sim 20)eV$, hence one can obtain simultaneously the cosmological HOT-dark matter component, in agreement with the interpretation of the COBE data. Indeed an upper limit on the ν_τ mass can be obtained from the formula

$$7.5 \times 10^{-2} \leq \Omega_\nu h^2 \leq 0.3 \tag{45}$$

Translating this into a constraint on m_{ν_τ}, arising from the relation $m_{\nu_\tau} \approx \Omega_\nu h^2 91.5eV$ where $h = .5 \sim 1.0$ is the Hubble parameter, one gets the range

$$6.8 \leq m_{\nu_\tau} \leq 27eV \tag{46}$$

which can be easily achieved with the above range of Λ^{rad}.

Acknowledgements *I would like to thank the Astroparticle Physics Group, the organizing committee of the workshop, and in particular D.V. Nanopoulos, J. Lopez, I. Giannakis, K. Yuan and G. Kyriazis for kind hospitality.*

3. References

1. M.B. Green, J. Schwarz and E. Witten, *Superstring Theory* Vols.**I,II** (Cambridge U.P.,Cambridge).
2. J. Ellis, S. Kelley and D.V. Nanopoulos, *Phys.Lett.* **B249** (1990) 441; P. Langacker and M. Luo, *Phys.Rev.* **D441** (1991) 817; U. Amaldi, W. de Boer, and H. Fürstenau, *Phys.Lett.* **B260** (1991) 447 ; F. Anselmino,L. Cifarelli and A. Zichichi, CERN-PPE/92-145 and CERN/LAA/MSL/92-011(July 1992); R.G. Roberts and G.G. Ross, *Nucl.Phys.* **B377** (1992) 571; A. Zichichi, *Talk at this Workshop.*
3. V. Kaplunovski, *Nucl.Phys.* **B307** (1988) 145.
4. I. Antoniadis, J. Ellis, R. Lacaze and D.V. Nanopoulos, *Phys.Lett.* **B268**(1991)188.
5. S. Kalara, J.L. Lopez and D.V. Nanopoulos, *Phys.Lett.***B269**(1991)84.
6. L. Ibanẽz, D. Lüst and G.G. Ross, *Phys.Lett.***B272**(1991)251.
7. G.K. Leontaris and N.D. Tracas, *Phys.Lett.***B291**(1991)44.
8. S.M. Barr, *Phys.Lett.***B112**(1982)219;*Phys.Rev.***D40**(1989)2457. J. Derendiger, J. Kim and D.V. Nanopoulos, Phys.Lett.B139(1984)170.
9. I. Antoniadis, J. Ellis, S. Hagelin and D.V. Nanopoulos, *Phys.Lett.***B194** (1987) 231; *Phys.Lett.***B231** (1987)65.
10. J. L. Lopez, D.V. Nanopoulos and A. Zichichi,*The simplest, string derivable, supergravity model and its experimental predictions*, CERN-TH.6667/92.
11. I. Antoniadis and G.K. Leontaris, *Phys.Lett* **B216** (1989) 333;I. Antoniadis G.K. Leontaris and J. Rizos, *Phys.Lett* **B245** (1990)161.
12. J. Pati and A. Salam, *Phys.Rev.***D10** (1974) 275.
13. A. Faraggi, D.V. Nanopoulos and K. Yuan, *Nucl. Phys.***B335** (1990) 437;A. Faraggi, *Nucl. Phys.***B387** (1992) 239.
14. J. Ellis, J.L. Lopez and D.V. Nanopoulos, *Phys.Lett.* **B252**(1990)53; G.K. Leontaris and K. Tamvakis, Phys.Lett.**B260**(1991)333; G.K. Leontaris,*Zeit.Phys.* **C53**(1992)287.
15. G.K. Leontaris *Phys.Lett.* **B 207**(1988)447; G.K. Leontaris and D.V. Nanopoulos *Phys.Lett.* **B 212**(1988)327; G.K. Leontaris and C.E. Vay-

onakis, *Phys.Lett.* **B206**(1988)271; G.K. Leontaris and K. Tamvakis, *Phys.Lett.* **B224**(1989)319; S.Abel,*Phys.Lett.* **B234** (1990)113; A.E. Faraggi,*Phys.Lett.* **B245**(1990)435; G.K. Leontaris and J.D.Vergados *Phys. Lett.* **B258**(1991)111; I. Antoniadis, J.Rizos and K.Tamvakis *Phys.Lett.* **B279**(1992)281; E. Papageorgiu and S. Ranfone *Phys.Lett.* **B282**(1992)89.

16. J. Ellis, J.L. Lopez and D.V. Nanopoulos *Phys.Lett.***B292** (1992)189.
17. J. Ellis, D.V. Nanopoulos and K.A. Olive, *Phys.Lett.***B300** (1992)121;J. Ellis, J. Lopez, D.V. Nanopoulos and K.A. Olive, *Flipped angles and phases; a systematic study*(CERN-TH.6842/93).
18. G.K. Leontaris and J.D. Vergados,IOA-preprint,290/92 *Phys.Lett.*B(To appear).
19. S. Dimopoulos, L.J. Hall and S. Raby, *Phys.Rev.Lett.***68**(1992) 1984; *Phys.Rev* **D45** (1992) 4192; V. Barger et al **Phys.Rev.Lett.** **68**(1992)3394;
20. G. F. Giudice, *MPLA* **7**(1992)2429; G.K. Leontaris and N.D. Tracas, *Phys.Lett.***B303** (1993)50.
21. E.Friere, G. Lazarides and Q. Shafi, *MPLA***7**(1990)2453; K.S. Babu and Q. Shafi,*Bartol preprint* BA-92-70.
22. H.Arason et al., *Phys.Rev.***D46** (1992)3945; P. Ramond, R.G. Roberts and G.G. Ross, *preprint 1993*, (UFIFT-93-05).
23. P. Ramond ,*Talk at this workshop*; S. Raby, *Talk at this workshop*; V. Barger,*Talk at this workshop.*
24. L. Durand and J. Lopez, *Phys.Rev.* **D40** (1989) 207.
25. S. Narison, *Phys.Lett.***B** 197 (1987) 405; *ibid* 216 (1989) 191;J. Gasser and H. Leutwyler, *Phys. Rep.* **87** (1982) 77.
26. K.S. Hirata et al. *Phys. Rev Lett.* **65**(90) 1297; K.S. Hirata et al,*Phys. Rev.***D44**(1991) 2141; SAGE collaboration A.I.Abazov et al, *Phys.Rev.Lett.* **67**(1991) 3332;P. Anselmann et al. *Phys.Lett.***B285**(1992)376.
27. G.F. Smooth et al, Astrophys.J.**396** (1992)L1; E.L.Wright et al. Astrophys.J.**396**(1992)L13.
28. JR.K. Schaefer and Q.Shafi *Nature* **359** (1992)199.
29. S. Dimopoulos, L. Hall, and S. Raby, Lawrence Berkeley Laboratory preprint, LBL-32484; G.K. Leontaris, and N.D. Tracas, Oxford preprint, Sept.92/IOA-281/92; K.S.Babu and Q. Shafi, BA-93-09/1993.
30. J. Ellis, J.L. Lopez and D.V. Nanopoulos *Phys.Lett.* **B292** (1992) 189; S.Kalara, J.L. Lopez and D.V. Nanopoulos, Phys.Lett.B245(1990)421;J. Rizos and K. Tamvakis, *Phys.Lett.***B251**(1990)369.
31. P.I. Krastev and S.T. Petcov, *Phys.Lett.* **B285**(1992)85; S.A. Bladman, D.C. Kennedy and P.G. Langacker,*Phys.Rev.* **D45**(1992)1810.
32. Ch. Berger et al.*Phys. Lett.* **B245**(1990)305.
33. L. Borodovsky et al,*Phys.Rev.Lett.* **68**(1992)247.

TOP MASS PREDICTIONS FROM YUKAWA UNIFICATION

URI SARID, with LAWRENCE J. HALL and RICCARDO RATTAZZI
Theory Group 50A/3115, Lawrence Berkeley Laboratory, 1 Cyclotron Rd.
Berkeley, CA 94720, USA

ABSTRACT

We study in some generality the predictions of third-generation Yukawa unification in SO(10) models containing the minimal supersymmetric standard model. We consider Yukawa coupling evolution and some implications of obtaining the top-bottom mass hierarchy from the Higgs sector.

1. Introduction

The only parameter of the standard model that has been successfully predicted with a high level of significance is the weak mixing angle, which is a prediction of supersymmetric grand unified theories (SUSY GUTs)[1]. All other predictions require additional, model-dependent information about the theory. To obtain quark and lepton mass predictions from grand unified theories, assumptions about the underlying flavor structure of the theory must be made. Of all the flavor predictions of GUTs those pertaining to the heaviest generation are most direct and subject to the fewest assumptions. In this work we pursue a scheme for predicting the top mass which is unique in its simplicity. We attempt as complete and accurate analysis of this prediction as possible: the aim is to predict the top quark mass to within a few GeV. This work was done in collaboration with Lawrence Hall and Riccardo Rattazzi, and is presented in its entirety elsewhere[2].

In the context of grand unified theories, it is sometimes possible to predict the top quark mass from a consideration only of the heaviest generation, with the b and τ masses as inputs from experiment[3,4]. There are three observable masses, m_t, m_b and m_τ, and four parameters, λ_t, λ_b, λ_τ and the ratio $\tan\beta \equiv v_U/v_D$ of the VEVs of the two Higgs doublets. The only possibility for a prediction resulting from consideration of the heaviest generation alone is that the t, b and τ masses originate predominantly from a single Yukawa interaction. This immediately excludes the grand unified gauge group SU(5) from consideration. The simplest supersymmetric grand unified gauge group which allows relations between masses in the up and down sectors is SO(10).

In this work we study the top quark mass prediction which results from the following three assumptions:

(I) The masses of the third generation, m_t, m_b and m_τ, originate from renormalizable Yukawa couplings of the form $\underline{16}_3\,\mathcal{O}\,\underline{16}_3$ in a supersymmetric GUT with a gauge group containing (the conventional) SO(10).

(II) The effective theory beneath the SO(10) breaking scale is the minimal supersymmetric standard model (MSSM).

(III) The two Higgs doublets lie predominantly in a single irreducible multiplet of SO(10).

We find it highly significant that such simple and mild assumptions are sufficient for predicting the top quark mass with an accuracy of a few GeV. We will very rapidly be led to the result that the SO(10) multiplet containing the Higgs doublets is necessarily the $\underline{10}_H$, so that the relevant Yukawa interaction is $\underline{16}_3\,\underline{10}_H\,\underline{16}_3$ (where $\underline{16}_3$ is the third-generation matter supermultiplet). The prediction of the top quark mass from this interaction was first considered by Ananthanarayan, Lazarides and Shafi[3]. In this work, we study: the coupled two-loop renormalization group equations for the three gauge couplings and the three Yukawa couplings, and their solution using an analytic fit; the explanation of the top-bottom mass hierarchy through a large hierarchy in the vacuum expectation values (VEVs) of the light Higgs doublets, namely a large $\tan\beta$; the supersymmetric threshold corrections to the three gauge couplings and the three Yukawa couplings for an arbitrary spectrum of superpartners; threshold corrections at the grand unified mass scale; the extent to which the predicted value of the top quark mass depends on assumption (III); and the extraction of the b quark mass from experiment.

2. Framework

The predicted value of the top mass depends on the renormalization group equations (RGE), the boundary conditions at the GUT and electroweak scales, and the threshold corrections at these two scales. In our analysis we employ two-loop evolution equations for both the Yukawa and the gauge couplings; the relevant details are given in full elsewhere[2].

The tree-level initial conditions at the GUT scale M_G are $g_1 = g_2 = g_3 \equiv g_G$ for the gauge sector, but in the Yukawa sector they depend on the source of the low-energy Yukawa couplings. In SO(10) unification they typically arise from terms of the form $\underline{16}_3\,\mathcal{O}\,\underline{16}_3$, where $\underline{16}_3$ is the chiral supermultiplet for the third generation and the Higgs multiplet $\mathcal{O}$ may be a $\underline{10}_H$ or $\underline{126}_H$ of SO(10). (The $\underline{120}_H$ is antisymmetric and therefore makes no contribution.) To go beyond the SU(5) predictions and exploit the larger SO(10) symmetries, we must *assume* that both light Higgs doublets lie predominantly in a *single* SO(10) multiplet, rather than being arbitrary mixtures of doublets in several representations. Then either $\mathcal{O} \sim \underline{10}_H$, in which case the initial conditions for the Yukawa couplings are

$$\lambda_t = \lambda_b = \lambda_\tau \equiv \lambda_G\,, \qquad (1)$$

or else $\mathcal{O} \sim \underline{126}_H$ from whence $3\lambda_t = 3\lambda_b = \lambda_\tau \equiv \lambda_G$. In the second case, our investigation yields acceptable m_b/m_τ ratios only if $\alpha_3(m_Z) > 0.13$ and $m_b(m_b) < 4.0\,\mathrm{GeV}$ as well as $\lambda_G > 4$. This last requirement, however, when combined with the full SO(10) RGE, implies a Landau pole in the Yukawa coupling less than 20% above the unification scale, which in general allows $\mathcal{O}(1)$ threshold corrections and makes any predictions impossible. Therefore, we will only consider the consequences of $\underline{16}_3\,\underline{10}_H\,\underline{16}_3$. In this case, we must restrict $\lambda_G < 2$ to make sure the Landau pole is at least a factor of 4 above the unification mass; for higher values of λ_G, we simply cannot make any reliable predictions, although these values are by no means ruled out.

3. Large tan β

We discuss in this section some of the implications of the large value of $\tan\beta \equiv v_U/v_D \simeq 50 - 60$ necessitated by the boundary condition $\lambda_t^G = \lambda_b^G = \lambda_\tau^G$. In the standard model and most of its extensions, the hierarchy $m_b \ll m_t$ (and similarly for the τ mass) is imposed through $\lambda_b \ll \lambda_t$; λ_b is the small parameter quantifying the breakdown of the chiral symmetry which protects the bottom quark mass. In multi-Higgs models such as the MSSM, there is also the option of explaining $m_b \ll m_t$ by having the down-type Higgs acquire a much smaller VEV than that of the up-type Higgs, $v_D \ll v_U$. We are *forced* to take this second route by our GUT boundary condition, which implies $\lambda_b \sim \lambda_t \sim 1$. What are the symmetries protecting the VEV of the down-type Higgs and hence the b mass? We will assume as usual that the electroweak symmetry is broken when the scalar potential $m_U^2|H_U|^2+m_D^2|H_D|^2+B\mu(H_UH_D+\text{h.c.})+\frac{1}{8}(g^2+g'^2)(|H_U|^2-|H_D|^2)^2$ develops an instability in the H_U direction through the RG evolution. This breaking is then communicated to H_D by the term μBH_UH_D in the soft SUSY-breaking scalar potential:

$$\frac{-\mu B}{m_U^2+m_D^2} = \frac{1}{2}\sin 2\beta \simeq \frac{1}{\tan\beta} = \frac{v_D}{v_U}. \tag{2}$$

Thus the magnitude of the hierarchy between the up and down sectors is determined by $\mu B/(m_U^2+m_D^2)$, while its direction is set by the direction in which the instability develops in the scalar potential. The denominator is given by $m_U^2+m_D^2 = m_A^2$, the squared mass of the pseudoscalar neutral Higgs boson; evidently, either μ or B or both must be smaller than m_A in order to generate the top-bottom mass hierarchy. And as we argue elsewhere[2], m_A is in turn expected to be smaller than m_0, the common scalar (namely squark and slepton) mass at the electroweak scale.

The symmetry which sets μ to zero is a Peccei-Quinn ($\mathcal{PQ}$) symmetry under which the superfield H_D and the SU(2)-singlet bottom antiquark superfield b^c have equal and opposite charges while all other fields are invariant. It is explicitly broken only by the soft SUSY-breaking term μBH_UH_D and by the term μH_UH_D in the superpotential. We will quantify the degree to which this symmetry is broken by the dimensionless parameter $\epsilon_{\mathcal{PQ}} \equiv \mu/m_0$.

The symmetry which sets B to zero must also set the gaugino mass $m_{1/2}$ and the soft SUSY-breaking trilinear scalar coupling A (and in particular A_t) to zero, as we can see from the fact that they generate B through the RG evolution. The desired symmetry must in fact be a continuous $\mathcal{R}$ symmetry, and we have chosen a particular $\mathcal{R}$ for convenience[2]. Once again, we define a dimensionless symmetry-breaking measure $\epsilon_{\mathcal{R}} \equiv B/m_0 \sim A/m_0 \sim m_{1/2}/m_0$.

We thus have (up to an irrelevant sign)

$$\epsilon_{\mathcal{PQ}}\epsilon_{\mathcal{R}} = \frac{1}{\tan\beta}\left(\frac{m_A^2}{m_0^2}\right). \tag{3}$$

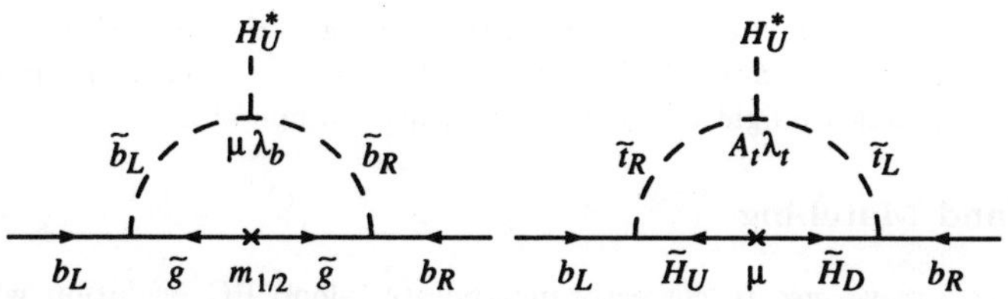

FIG. 1. The leading (finite) 1-loop MSSM contributions to the b quark mass.

We expect both the $\mathcal{PQ}$ symmetry and our $\mathcal{R}$ symmetry to be broken only slightly, for the following reasons. If $\epsilon_{\mathcal{PQ}} \sim 1$, then $\epsilon_{\mathcal{R}} \lesssim 1/50$ which, when combined with the lower bound of order m_Z on the gaugino masses, forces the squarks into the multi-TeV range. Similarly, if $\epsilon_{\mathcal{R}} \sim 1$ then $\epsilon_{\mathcal{PQ}} \lesssim 1/50$ implies very heavy squarks when the LEP lower bound[5] $\mu > m_Z/2$ is satisfied. Since heavy squarks imply considerable fine-tuning to obtain the proper electroweak-breaking scale, it seems likely that both $\epsilon_{\mathcal{PQ}}$ and $\epsilon_{\mathcal{R}}$ are small. Furthermore, the pseudoscalar Higgs mass is expected[2] to be somewhat lighter than the squarks, so $m_A^2 < m_0^2$.

It is not trivial to actually achieve radiative electroweak symmetry-breaking via the RG equations when $\lambda_t^G = \lambda_b^G$, and new approaches may be necessary[6], though significant progress has already been made[7]. For now, we will simply favor the parameter choices $\mu, A, B, m_{1/2} \sim \mathcal{O}(m_Z)$, $m_A \sim 700\,\text{GeV}$, and $m_0 \sim \mathcal{O}(\text{TeV})$.

Both $\mathcal{PQ}$ and $\mathcal{R}$ symmetry breakings enter the scalar potential at tree-level in proportion to $B\mu$, where they generate $v_D \neq 0$ and thereby a tree-level mass for the b. They also contribute to the b mass directly, by allowing radiative couplings of Qb^c to H_U^* which acquires the large VEV. However, any insertion of H_U^* rather than H_D (which indicates an enhancement by a factor $\tan\beta$) must be accompanied by a factor $\epsilon_{\mathcal{PQ}}\epsilon_{\mathcal{R}}$ to make up the $\mathcal{PQ}$ and $\mathcal{R}$ quantum numbers of H_D, so qualitatively we expect no net enhancement. For instance, the dominant radiative couplings of the b quark to the up-type Higgs are through the 1-loop diagrams in Fig. 1. If all masses in the loop were comparable, these diagrams would be enhanced by a factor of $\sim \tan\beta$, so instead of the expected $\sim \pm 1\%$ corrections we would find $\sim \pm 50\%$ corrections. When the $\mathcal{PQ}$ and $\mathcal{R}$ symmetries hold approximately, such large corrections do not arise; the fact that we have a fundamental understanding—in terms of symmetries—of why the b quark is light stabilizes our predictions. With these symmetries, the observed lightness of the b quark is related to a hierarchy in the superpartner spectrum, which has a direct observable consequence: the spectrum must contain a light (electroweak-scale) fermionic superpartner.

As a last comment, we point out that by the same symmetry arguments discussed in this section, other potentially striking effects of a large $\tan\beta$ are tamed by $\epsilon_{\mathcal{PQ}}\epsilon_{\mathcal{R}}$ insertions. For example, attaching a photon in all possible ways to the diagrams of Fig. 1 and inserting a flavor-changing vertex leads to an amplitude for $b \to s\gamma$ which again is proportional to $\epsilon_{\mathcal{PQ}}\epsilon_{\mathcal{R}} \tan\beta \sim 1$, so that no spectacular enhancement is expected over the $\tan\beta \sim 1$ case. Analogous considerations can be extended to other observables such as $B^0\overline{B}^0$, $D^0\overline{D}^0$ and

$K^0\overline{K}^0$ mixing and the electric dipole moment of the neutron. Of course, if $\epsilon_{\mathcal{PQ}}$ and $\epsilon_{\mathcal{R}}$ are not small, so that large $\tan\beta$ is a result of fine-tuning, then $b \to s\gamma$ can become unacceptably large if the superpartners are light (since it is a dimension 5 operator).

4. Running and Matching

The procedure we use to correctly incorporate 2-loop RG evolution with 1-loop matching conditions is described in our paper[2]. In the matching, we encounter functions which we denote by $k_{t,b,\tau}$ and which result from integrating out the superpartners to obtain the 2-Higgs standard model (2HSM). Throughout the paper, we define the quantity $R \equiv m_b/m_\tau$ measured in the 2-Higgs standard model at the Z mass. We also encounter some dependence on the Higgs spectrum, determined essentially by the mass m_A of the physical neutral pseudoscalar Higgs.

The evolution of the Yukawa and gauge couplings in the MSSM, even to 1-loop accuracy, cannot in general be calculated analytically. Numerical solutions are straightforward, and show that, in order to obtain the correct m_b/m_τ ratio, the GUT-scale Yukawa couplings must be $\mathcal{O}(1)$, so the top mass is typically predicted near its fixed point value[3,4,8]. With this in mind, it will be useful and illuminating to obtain simple, explicit approximate solutions by fitting $\lambda_{t,b,\tau}(m_Z)$ numerically to a quadratic polynomial in $1/\lambda_{\mathrm{G}}$. The constant term reflects the independence of the low-energy Yukawa couplings on the initial conditions in the large-λ_{G} limit, a consequence of the fixed point behavior. We define the unperturbed scenario to be the one in which all threshold corrections vanish. When varying λ_{G} between 0.5 and 2, we find in the unperturbed scenario

$$\lambda_{t,b,\tau}(m_Z) \simeq A_{t,b,\tau} + \frac{B_{t,b,\tau}}{\lambda_{\mathrm{G}}} + \frac{C_{t,b,\tau}}{\lambda_{\mathrm{G}}^2} \tag{4}$$

where A, B and C are numerical constants. We have checked that the errors we make in this fit are smaller than 0.2% over the entire range $0.5 \leq \lambda_{\mathrm{G}} \leq 2.0$. We may then solve the quadratic equation $R = (A_b + B_b/\lambda_{\mathrm{G}} + C_b/\lambda_{\mathrm{G}}^2)/(A_\tau + B_\tau/\lambda_{\mathrm{G}} + C_\tau/\lambda_{\mathrm{G}}^2)$ to obtain λ_{G} as a function of R, which in turn gives an explicit solution for $\lambda_{\mathrm{t}}(m_Z)$ as a function $F(R)$ of the experimentally-determined ratio R. The functions (4), as well as the ratio $R = \lambda_{\mathrm{b}}(m_Z)/\lambda_\tau(m_Z)$ and the function $F(R)$, are plotted versus λ_{G} in the paper[2].

5. The Bottom Quark Mass

What is the experimental value of R? Using updated values from the Particle Data Group compilation[9], we have reanalyzed[2] the well-known extraction of the b mass from e^+e^- collisions via QCD sum rules[10]. We estimate

$$m_b(m_b) = \begin{cases} 4.15\,\mathrm{GeV} \pm 0.22\,\mathrm{GeV}, & \alpha_3(m_Z) = 0.11 \\ 4.14\,\mathrm{GeV} \pm 0.28\,\mathrm{GeV}, & \alpha_3(m_Z) = 0.12 \end{cases} \tag{5}$$

and arrive at the final value of R (f_R is described in our paper):

$$R \;=\; \frac{m_b^{\overline{\mathrm{MS}}}(4.1\,\mathrm{GeV})}{m_\tau}\frac{\eta_\tau}{\eta_b}\left[1 + \frac{\alpha_3(m_Z)}{6\pi}\right][1 + f_R(m_A)]$$

$$= \begin{cases} 1.70 \pm 0.09, & \alpha_3(m_Z) = 0.11, \quad m_A = 90\,\text{GeV} \\ 1.61 \pm 0.11, & \alpha_3(m_Z) = 0.12, \quad m_A = 90\,\text{GeV} \\ 1.65 \pm 0.09, & \alpha_3(m_Z) = 0.11, \quad m_A = 1000\,\text{GeV} \\ 1.56 \pm 0.11, & \alpha_3(m_Z) = 0.12, \quad m_A = 1000\,\text{GeV}. \end{cases} \tag{6}$$

6. Threshold Corrections

Next, we investigate the deviations in the top mass prediction induced by threshold corrections at the GUT and SUSY scales. For convenience, we choose to *always* match the full SO(10) theory with the MSSM at the scale M_G of the unperturbed scenario, where g_1 and g_2 met: we *define* from now on the GUT scale to be $\mu_G = 2.7 \times 10^{16}$ GeV. The top mass prediction as a function of R is completely determined once we know the initial conditions $\{\lambda_t^G, \lambda_b^G, \lambda_\tau^G, g_1^G, g_2^G, g_3^G\}$ at the fixed scale μ_G, as well as the functions $k_{t,b,\tau}(\{m_a\})$ of the superpartner masses which match between the MSSM (in which we run with the RGE) and the 2-Higgs standard model (in which we calculate the top pole mass). So we first calculate[2] the perturbations $\epsilon_{t,b,\tau,1,2,3}$ to the initial conditions and the matching functions $k_{t,b,\tau}$ in terms of the various mass thresholds and $\alpha_3(m_Z)$. Then we study the sensitivity of the top mass prediction to these ϵ's and k's. A linear analysis of these perturbations is sufficient for our purposes.

7. Sensitivity Functions

When the initial conditions are perturbed, so are the coefficients of the fit in Eq. (4). We expect the fit parameters to vary linearly with the perturbations as long as these are sufficiently small, and so we write

$$\lambda_\nu(m_Z) = \left[(A_\nu + \sum_n a_{\nu n}\epsilon_n) + \frac{B_\nu + \sum_n b_{\nu n}\epsilon_n}{\lambda_G} + \frac{C_\nu + \sum_n c_{\nu n}\epsilon_n}{\lambda_G^2} \right] (1 + k_\nu) \tag{7}$$

where the sum ranges over $n = t, b, \tau, 1, 2, 3$. The new fit coefficients $\{a_{\nu n}, b_{\nu n}, c_{\nu n}\}$ must be computed numerically. We then solve $R = \lambda_b(m_t)/\lambda_\tau(m_t)$ for λ_G as before, substitute back into Eq. (7) and expand to first order in ϵ_n and k_ν to find

$$\lambda_t(m_Z) = F(R) + \sum_n F_n^{(\epsilon)}(R)\epsilon_n + \sum_\nu F_\nu^{(k)}(R)k_\nu. \tag{8}$$

These functions, and our understanding of the behavior of these functions, are described in the paper[2].

8. Predicting the Top Quark Mass

The final step is the calculation of the position of the pole in the propagator of the top quark within the 2-Higgs standard model. (Note that the pole mass is only a parameter in the calculation of experimental observables; we leave the study of the relation between

$m_t^{\rm pole}$ and the actual "top mass" extracted from collider data to future work.) There are two important radiative corrections to the pole mass: the usual QCD correction from gluon dressing, and Yukawa interaction corrections to the top quark propagator and to the Fermi constant. The result may be written as

$$\begin{aligned} m_t^{\rm pole} &= \lambda_t^{\rm 2HSM}\left[\frac{1}{\sqrt{2^{3/2}G_F}}\left(1+\frac{\Sigma_W(0)}{2m_W^2}\right)\right]\left(1+\frac{\delta m_t}{m_t}+\frac{1}{2}\delta_L^t+\frac{1}{2}\delta_R^t\right)(1+\delta_{\rm QCD}) \\ &= \lambda_t\, 174\,{\rm GeV}\,[1+f_t(m_A)]\,. \end{aligned} \tag{9}$$

In the last line we have substituted $\lambda_t^{\rm 2HSM} \equiv \lambda_t$ and $1+\delta_{\rm QCD} \equiv 1+4\alpha_3/3\pi - \alpha_3/6\pi - (8\alpha_3/4\pi)\ln(m_t/m_Z) \simeq 1.0$, and defined $f_t(m_A) \equiv \Sigma_W(0)/2m_W^2 + \delta m_t/m_t + \frac{1}{2}\delta_L^t + \frac{1}{2}\delta_R^t$. Further details may be found in the paper[2].

To proceed further, we must make some assumptions about the SUSY- and GUT-scale masses. First, consider setting all superpartner masses to be roughly degenerate with the Z mass. As we previously argued, such a spectrum provides no explanation of why $\tan\beta$ is large and the b quark is light; in the absence of $\mathcal{PQ}$ and $\mathcal{R}$ symmetries, we would be forced to tune B to be tiny in order to obtain $v_D \ll v_U$, and the resultant light b mass would be unstable with respect to radiative corrections of the sort in Fig. 1 so the top mass predictions would have huge uncertainties. Furthermore, barring more cancellations between otherwise unrelated diagrams, this scenario leads[11] to a rate for $b \to s\gamma$ many times larger than the standard model, and is therefore excluded by present measurements[12].

Increasing the superpartner masses changes the predictions predominantly in two ways: through k_ν and through ϵ_i. These can be significant, but *only* when they serve to *increase* m_t, assuming the $\mathcal{PQ}$ and $\mathcal{R}$ symmetries are approximately obeyed. In fact, for fixed $\alpha_3(m_Z)$ the top mass prediction depends predominantly (to within a few GeV) on the squark, gluino and higgsino masses rather than the slepton, wino and bino masses. As it turns out, the positive terms proportional to $t_{\tilde{q},\tilde{g}}$ and $t_{\tilde{q},\tilde{h}}$ (from wavefunction renormalization in the k_ν SUSY threshold corrections) are always larger in magnitude than the negative terms proportional to $t_{\tilde{q}}$ and $t_{\sqrt{2}\tilde{g}}$ (from ϵ_3 SUSY threshold corrections to the QCD gauge coupling). Thus heavy superparners inevitably lead to a heavier top.

The top mass prediction is further influenced by $\alpha_3(m_Z)$ and m_A. The resulting dependence of $m_t^{\rm pole}$ on $m_{\tilde{q}}$, $m_{\tilde{g}}$, $m_{\tilde{h}}$, m_A, m_b and $\alpha_3(m_Z)$ is somewhat complicated, so we display it in two complementary ways. First, we list in Table 1 of our paper[2] the predictions of our complete expressions for various choices of the parameters, again assuming the $\mathcal{PQ}$ and $\mathcal{R}$ symmetries hold approximately. Second, an approximate expression can be given in terms of numerically-determined coefficients, as explained in the paper[2].

We now take into account the strong theoretical bias put forward in Sec. 3 towards having μ, B, A_t, and $m_{1/2}$ be significantly smaller than the common scalar mass m_0. Choosing for concreteness $m_{\tilde{h}} = 100\,{\rm GeV}$, $m_{\tilde{g}} = 200\,{\rm GeV}$, $m_{\tilde{w}} = 100\,{\rm GeV}$, $m_{\tilde{q}} = m_{\tilde{\ell}} = 1000\,{\rm GeV}$ and $m_A = 700\,{\rm GeV}$, we obtain the predictions shown in Fig. 2. The solid lines in this figure delimit the predictions we obtain when we let $m_b(m_b)$ vary within our allowed range, according to Eq. (6). The dashed lines correspond to specific guesses for $m_b(m_b)$. Notice that our allowed range for m_b increases with $\alpha_3(m_Z)$ since our estimate for the uncertainty is proportional to $[\alpha_3(m_b)]^2$. It should be remembered that the prediction for

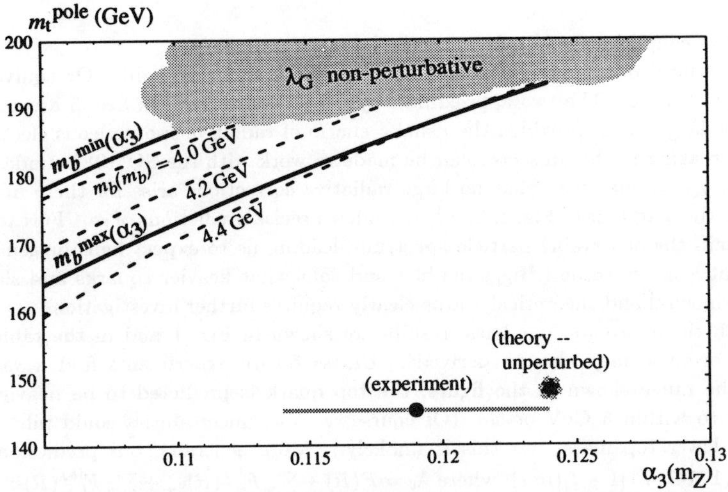

FIG. 2. Our predictions for m_t^{pole} without superheavy corrections using the favored superpartner spectrum, specifically $m_{\tilde{h}} = 100\,\text{GeV}$, $m_{\tilde{g}} = 200\,\text{GeV}$, $m_{\tilde{w}} = 100\,\text{GeV}$, $m_A = 700\,\text{GeV}$ and $m_{\tilde{q}} = m_{\tilde{\ell}} = 1000\,\text{GeV}$. The "cloud" indicates the region where $\lambda_G > 2$. These predictions carry estimated uncertainties of $\sim \pm 5\,\text{GeV}$ from various approximations, from the GUT-scale thresholds discussed in the text, and from the diagrams of Fig. 1. Also shown are the estimated experimental range[14], and the unperturbed MSSM GUT prediction, for $\alpha_3(m_Z)$.

$\alpha_3(m_Z)$ without GUT-scale threshold corrections is around 0.124, so values much lower than this correspond to large GUT-scale corrections to g_3. Furthermore, we find that perturbative Yukawa unification demands $\alpha_3 \lesssim 0.125$. These last two observations are in rough agreement with previous authors[8,13].

We have also studied a variety of threshold corrections arising at the GUT scale due to splittings of various superheavy multiplets. The results are elaborated upon in the paper[2]. We find that, while we cannot eliminate the possibility of large GUT-scale threshold corrections to our predictions, we can show that all those corrections which are generic to SO(10) SUSY GUTs are not expected to change m_t by more than a few GeV.

9. Conclusions

We have computed the top quark mass in supersymmetric SO(10) grand unified theories including all contributions larger than about 5 GeV. The assumption underlying this prediction is that m_t, m_b and m_τ all originate from a single Yukawa interaction with an SO(10) multiplet which completely contains the two low energy Higgs doublets. A crucial feature of the GUT boundary condition $\lambda_b = \lambda_\tau$ is that the resulting value of m_b/m_τ can only be reconciled with experiment if the top quark Yukawa coupling is large. This is why in SO(10), as in SU(5), the top quark is predicted to be heavy. With the GUT boundary condition $\lambda_t = \lambda_b = \lambda_\tau$, the top mass is not just large but actually predictable, and the b

and τ Yukawa couplings are also forced to be large.

The crucial question then arises: why are the b and τ so light? Or equivalently, why is $\tan\beta$ so large? The two approximate symmetries discussed in Sec. 3 are sufficient to guarantee large $\tan\beta$, provided the usual scenario of radiative spontaneous electroweak symmetry breaking in the up sector can be made to work with the SO(10) boundary conditions. They also guarantee that no large radiative corrections arise for the b mass, for example via the diagrams of Fig. 1, and hence allow precise predictions of m_t. Furthermore, they constrain the observable particle spectrum, leading us to expect light higgsinos and gauginos but a heavy second Higgs doublet and somewhat heavier squarks and sleptons. Their experimental and theoretical status clearly requires further investigation.

With these caveats, our basic results are shown in Fig. 1 and in the table[2]; approximate formulae may also be derived[2]. Unless future experiments find a value for α_3 below the range shown in the figure, the top quark is predicted to be heavier than $\sim 170\,\mathrm{GeV}$ to within 5 GeV or so. (Of course, various uncertainties could pile up and result in a lower top mass, but this is unlikely.) More generally, our prediction reads $m_t^{\mathrm{pole}} = \lambda_t\,(174\,\mathrm{GeV})\,[1+f_t(m_A)]$ where $\lambda_t = F(R) + \sum_n F_n^{(\epsilon)}(R)\epsilon_n + \sum_\nu F_\nu^{(k)}(R)k_\nu$, $F(R)$ and the sensitivity functions $F_n^{(\epsilon)}(R)$ and $F_\nu^{(k)}(R)$ as well as the ϵ_n and the k_ν are given elsewhere[2], and $R = \left[m_b^{\overline{\mathrm{MS}}}(4.1\,\mathrm{GeV})/m_\tau\right](\eta_\tau/\eta_b)\,[1+\alpha_3(m_Z)/6\pi]\,[1+f_R(m_A)]$. These yield an analytic prediction of the top mass to within a few GeV—which is more than sufficient for comparison with top mass measurements likely to be made in the near future.

5. Acknowledgements

We would like thank B. Ananthanarayan for useful discussions and G. Anderson for comparisons with his numerical results. This work was supported in part by the Director, Office of Energy Research, Office of High Energy and Nuclear Physics, Division of High Energy Physics of the U.S. Department of Energy under Contract DE-AC03-76SF00098, and in part by NSF grant PHY-90-21139.

6. References

1. H. Georgi, H. Quinn, and S. Weinberg, *Phys. Rev. Lett.* **33**, 451 (1974); S. Dimopoulos, S. Raby, and F. Wilczek, *Phys. Rev.* D **24**, 1681 (1981); S. Dimopoulos and H. Georgi, *Nucl. Phys.* **B193**, 150 (1981); L. Ibañez and G.G. Ross, *Phys. Lett.* B **105**, 439 (1981); M.B. Einhorn and D.R.T. Jones, *Nucl. Phys.* **B196**, 475 (1982); W.J. Marciano and G. Senjanovic, *Phys. Rev.* D **25**, 3092 (1982).
2. L.J. Hall, R. Rattazzi, and U. Sarid, Lawrence Berkeley Laboratory Report No. 33997, hep-ph submission 9306309 (1993).
3. B. Ananthanarayan, G. Lazarides, and Q. Shafi, *Phys. Rev. D* **44**, 1613 (1991); Q. Shafi and B. Ananthanarayan, *Proceedings of the 1991 Trieste Summer School*, edited by E. Gava *et al.* (World Scientific, Singapore, 1992), p. 233 [Bartol Research Institute Report No. BA-91-76 (1991)].
4. H. Arason, D.J. Castaño, B.E. Keszthelyi, S. Mikaelian, E.J. Piard, P. Ramond,

and B.D. Wright, *Phys. Rev. Lett.* **67**, 2933 (1991); S. Kelley, J.L. Lopez, and D.V. Nanopoulos, *Phys. Lett.* B **274**, 387 (1992); V. Barger, M.S. Berger, and P. Ohmann, *Phys. Rev.* D **47**, 1093 (1993).

5. See, for example, F. Zwirner, Report No. CERN-TH.6537/91, and references therein.
6. L.J. Hall, R. Rattazzi, and U. Sarid, in preparation.
7. G.F. Giudice and G. Ridolfi, *Z. Phys.* C **41**, 447 (1988); M. Olechowski and S. Pokorski, *Phys. Lett.* B **214**, 393 (1988); W. Majerotto and B. Mösslacher, *Z. Phys.* C **48**, 273 (1990); P.H. Chankowski, *Phys. Rev.* D **41**, 2877 (1990); M. Drees and M.M. Nojiri, *Nucl. Phys.* **B369**, 54 (1992); B. Ananthanarayan, G. Lazarides, and Q. Shafi, Bartol Research Institute Report No. BA-92-29 (1992).
8. M. Carena, S. Pokorski and C.E.M. Wagner, Max Planck Institute Report No. MPI-Ph/93-10 (1993).
9. K. Hikasa *et al.* (Particle Data Group), *Phys. Rev.* D **45**, part II (1992).
10. E.C. Poggio, H.R. Quinn, and S. Weinberg, *Phys. Rev.* D **13**, 1958 (1976); M.A. Shifman, A.I. Vainshtein, and V.I. Zakharov, *Nucl. Phys.* **B147**, 385 and 448 (1979); B. Guberina, R. Meckbach, R.D. Peccei, and R. Rückl, *Nucl. Phys.* **B184**, 476 (1981).
11. L.J. Hall, A. Rasin, R. Rattazzi and U. Sarid, in preparation.
12. E. Thorndike, CLEO Collab., talk given at the 1993 Meeting of the American Physical Society, Washington, D.C., April 1993.
13. P. Langacker and N. Polonsky, University of Pennsylvania Report No. UPR-0556-T (1993).
14. I. Hinchliffe, invited talk given at XXVIII Rencontres de Moriond, Lawrence Berkeley Laboratory Report No. LBL-33952 (1993).

Supergravity and Model-Building

SUPERGRAVITY BASICS: MODELS WITH PROTON DECAY AND COSMOLOGICAL CONSTRAINTS

R. ARNOWITT[a*†] AND PRAN NATH[b]

[a] *Physics Research Division, Superconducting Super Collider Laboratory Dallas, TX 75237*

[a†] *Center for Theoretical Physics, Department of Physics, Texas A&M University College Station, TX 77843-4242*

[b] *Department of Physics, Northeastern University Boston, MA 02115*

ABSTRACT

Properties of basic supergravity grand unified models are surveyed. Models of this type can be characterized by four additional Gut scale constants beyond those needed to specify the Standard Model. These then determine the properties of the SUSY sector, e.g. 32 particle masses, widths etc. Models with proton decay and R parity invariance are consistent with all present data of LEP, the Tevatron, proton decay and relic density constraints. They make additional predictions which may be testable at the Tevatron e.g. $2m_{\tilde{Z}_1} \cong m_{\tilde{W}_1} \cong m_{\tilde{Z}_2}$ and $m_{\tilde{W}_1} \simeq (\frac{1}{3} - \frac{1}{4})m_{\tilde{g}}$. Future proton decay experiments (e.g. Super Kamiokande and ICARUS) when combined with future LEP200 experiments can explicitly test the validity of these models.

1. Introduction

For some time now, models based on $N = 1$ supergravity grand unification [1, 2] have been the main vehicle for efforts to unify particle interactions. Over the past two years, there has been considerable activity in examining the consequences of such models [3]-[11]. This has been in part due to the observation of a number of people [12] that the LEP and world data determination of the Standard Model coupling constants α_1, α_2 and α_3 at M_Z, combined with the renormalization group equations (RGE) leads to grand unification at a higher mass scale M_G. Grand unification appears to occur if one assumes a minimal supersymmetric particle spectrum with just two Higgs doublets, unification occurring at $M_G \approx 10^{16}$ GeV for SUSY particles of mass $M_S \approx 10^{2-3}$ GeV. The value of the coupling constant at unification is $\alpha_G \simeq 0.04$.

A second reason for the study of these models is that it appears possible to test them experimentally in current and planned experiments. This is due to two rather remarkable, related properties: First, supergravity grand unification allows for the spontaneous breaking of supersymmetry [13, 14] in a fashion leading to realistic models (something that is difficult to do in low energy global SUSY models or in the Planck scale string models). This breaking takes place in the

*Conference speaker
†Permanent address

"hidden" sector and is an explicit consequence of supergravity interactions. While the physics of the hidden sector is at present unknown, its effects on the physical sector can be characterized by just a few, i.e. four, phenomenological parameters for a wide class of theories [1, 15]. Second, the breaking of supersymmetry can trigger the breaking of $SU(2) \times U(1)$ [1] relating these two mass scales. This is done most elegantly via radiative corrections using the RGE to connect the Gut scale to the electroweak scale [16]. As a consequence, one can determine the masses of the 32 new SUSY particles in terms of only four additional Gut scale parameters and the as yet unknown t-quark mass m_t.

In principle, then, one could predict the positions of 28 SUSY particles, once four are discovered (and m_t determined). In practice, of course, what one can do now is determine various allowed mass bands, and predict mass relations for SUSY particles. If the model also possesses proton decay, existing and future bounds on the proton lifetime τ_p can make these predictions more precise. Similarly, the cosmological constraint that the relic density of the lightest supersymmetric particle (LSP) not over close the universe can further restrict the masses. Thus as will be seen these models should become testable in the relatively near future.

The supergravity interactions are themselves Planck scale physics ($M_{P\ell} = (1/8\pi G_N)^{1/2} = 2.4 \times 10^{18}$ GeV) and an important question is how self contained is a supergravity model. Since $M_G/M_{P\ell} \approx 10^{-2}$, one should not expect these models to be accurate to better than a few percent. Thus one is dealing here not yet with a fundamental theory, but rather with an effective theory that is expected to hold below the Planck scale.

2. Supergravity Gut Models

We give here a brief description of supergravity Gut models [17]. We consider here the class of models which obey the following conditions:

(i) There exists a "hidden" sector which is gauge singlet with respect to the physical sector gauge group G which breaks supersymmetry (by a super Higgs [13] or by a gaugino condensate [14]). The hidden sector communicates only gravitationally with the physical sector e.g. for the super Higgs example one has for the superpotential W:

$$W = W_{\text{phys}}(z_a) + W_{\text{hidden}}(z)$$

where $\{z_a\}$ are the physical sector fields and $\{z\}$ the hidden sector fields.

[One finds that it is the supergravity interactions that cause spontaneous breaking of supersymmetry and hence $< z >= O(M_{P\ell})$. The separation of the hidden sector from the physical sector is thus needed to preserve the gauge hierarchy i.e. the large $< z >$ VEVs are tamed by the gravitational couplings which scale by $\kappa = 1/M_{P\ell}$.]

(ii) There exists a Gut sector which breaks G to the Standard Model at scale $Q = M_G$: $G \to SU(3)_C \times SU(2)_L \times U(1)_Y$.

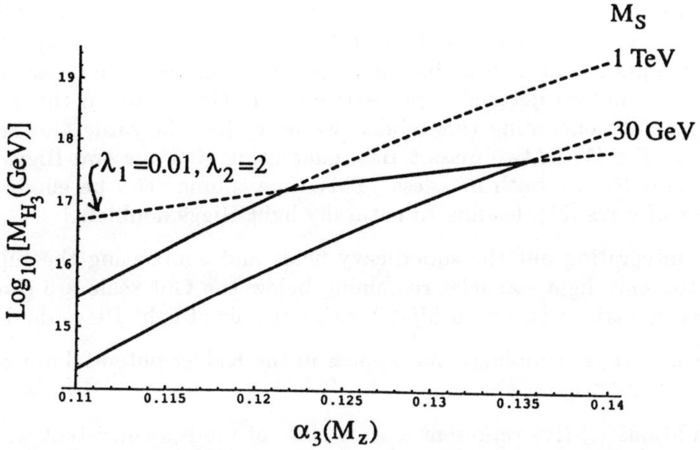

Figure 1: Grand unification constraints for Gut model of Eq. (2.1). For $\lambda_1 > 0.01$, $\lambda_2 < 2$ the horizontal line moves downward and so the solid quadrilateral is the allowed region of grand unification. Thus grand unification occurs provided $\alpha_3(M_Z) < 0.135$ while proton decay ($M_{H_3} > 1 \times 10^{16}$ GeV) implies $\alpha_3(M_Z) \gtrsim 0.114$.

[An example for $SU(5)$ is given by the Gut superpotential[18]

$$W_G = \lambda_1[\frac{1}{3}Tr\Sigma^3 + \frac{1}{2}MTr\Sigma^2] + \lambda_2\bar{H}_X(\Sigma_Y^X + 3M\delta_Y^X)H^Y \tag{1}$$

where Σ_Y^X is a 24 and $\bar{H}_X$ and H^X are $\bar{5}$ and 5 representations of $SU(5)$. Minimizing the effective potential gives diag $<\Sigma_Y^X>= M(2,2,2,-3,-3)$ where $M = O(M_G)$. The superheavy masses are a Higgs color triplet H_3, a color octet, an $SU(2)_L$ triplet and a Standard Model singlet Σ^8, Σ^3 and Σ^0 and the superheavy vector multiplet V. These have masses of $M_{H_3} = 5\lambda_2 M$, $M_{\Sigma^8} = M_{\Sigma^3} = 5\lambda_1 M/2$, $M_{\Sigma^0} = \lambda_1 M/2$ and $M_V = 5gM$ (where $\alpha_G = g^2/4\pi$). If one neglects both SUSY and Gut thresholds, a fit to $\alpha_1(M_Z)$, $\alpha_2(M_Z)$, $\alpha_3(M_Z)$ for grand unification is

$$M_G = 10^{16.19\pm0.34}GeV; \quad M_S = 10^{2.37\pm1.0}GeV \tag{2}$$

with $\alpha_G{}^{-1} = 25.4 \pm 1.7$, where the errors come from $\alpha_3(M_Z) = 0.118 \pm 0.007$ [19]. To stay within the perturbative domain we assume $\lambda_{1,2} \leq 2(\alpha_{\lambda_{1,2}} = \lambda_{1,2}{}^2/4\pi \lesssim 1/3)$ and also $\lambda_{1,2} > 0.01$ (i.e. $\alpha_{\lambda_{1,2}} \gtrsim 10^{-5}$). The result of the grand unification analysis is shown in Fig. 1 [20].

The allowed domain for 30 GeV $< M_S <$ 1 TeV is the interior of the solid quadrilateral. The horizontal line is lower for $\lambda_1 > 0.01$ and $\lambda_2 < 2$. Thus in this model, grand unification occurs provided $\alpha_3(M_Z) < 0.135$. For $\alpha_3(M_Z) < 0.125$, the current 1 σ upperbound on α_3, one has $M_{H_3}/M_G < 10$. As will be seen below, proton decay requires $M_{H_3} \gtrsim 10^{16}$ GeV which implies $\alpha_3(M_Z) \gtrsim 0.114$. The above results are all reasonable, and show that an accurate determination of $\alpha_3(M_Z)$ will indeed determine properties of the Gut sector in this model. As a final point concerning this model, we note that the same Gut scale mass M enters Eq. (1) in two mass terms, guarenting that the two Higgs doublets of H^X and $\bar{H}_X$ are both massless. This "fine tuning" can be eliminated in a number of ways [21], leading to naturally light Higgs doublets.]

(iii) After integrating out the superheavy fields and eliminating the super Higgs field, the only light particles remaining below the Gut scale are those of the supersymmetrized Standard Model with one pair of light Higgs doublets.

(iv) Any super Higgs couplings that appear in the Kahler potential are generation independent.

Conditions (i)-(iv) represent a wide class of models consistent with known data. Thus conditions (ii) and (iii) are implied by the grand unification analysis, i.e. that the Standard Model group holds up to scale M_G with the light particle spectrum of (iii). Conditions (i) and (iv) are needed to suppress flavor changing neutral interactions. Thus it is remarkable that for models where the non-renormalizable parts of the gauge kinetic function $f_{\alpha\beta}$ and Kahler potential are scaled by powers of $\kappa \equiv 1/M_{P\ell}$, conditionss (i)-(iv) lead to a relatively unique class of models [15]. One finds that the renormalizable interactions for models with R parity can be described just below M_G by an effective superpotential with quadratic and cubic parts, $W = W^{(2)} + W^{(3)}$ where

$$W = \mu_0 H_1 H_2 + [\lambda_{ij}^{(u)} q_i H_2 u_j^C + \lambda_{ij}^{(d)} d_i H_1 d_j^C + \lambda_{ij}^{(e)} l_i H_1 d_j^C \tag{3}$$

with an effective potential

$$V = \{\sum_a |\frac{\partial W}{\partial z_a}|^2 + V_D\} + [m_0^2 z_a z_a^+ + (A_0 W^{(3)} + B_0 W^{(2)} + \text{h.c.})] \tag{4}$$

and a universal gaugino mass term $\mathcal{L}_{\text{mass}}^{\lambda} = -m_{1/2}\bar{\lambda}^\alpha \lambda^\alpha$. In Eqs. (2.3) $q_i, l_i, H_{1,2}$ are the quark, lepton and Higgs doublets, $\{z_a\}$ are the set of all light scalar fields, and V_D is the usual D-term. Thus aside from the Yukawa coupling constants of the Standard Model ($\lambda_{ij}^{(u,d,e)}$), the theory at the Gut scale depends on the four soft breaking parameters, m_0, $m_{1/2}$, A_0 and B_0, and the Higgs mixing parameter μ_0. The analysis also shows that these five constants are expected to be of the same order of magnitude, and as will be seen below of electroweak size.

3. Electroweak Breaking

The spontaneous breaking of $SU(2)_L \times U(1)_Y$ arises naturally in supergravity models via radiative corrections to V_H, the Higgs part of the effective potential:

$$V_H = m_1^2(t)|H_1|^2 + m_2^2(t)|H_2|^2 - m_3^2(t)(H_1 H_2 + \text{h.c})$$

$$+\frac{1}{8}[g_2^2(t)+g_Y^2(t)]\,[|H_1|^2-|H_2|^2]^2+\Delta V_1 \tag{5}$$

In Eq. (5), $m_i^2(t)=m_{H_i}^2(t)+\mu^2(t)$, $i=1,2$, and $m_3^2(t)=-B(t)\mu(t)$ are running masses at scale Q: $t=\ln[M_G^2/Q^2]$. From Eqs. (3,4) one sees that they obey boundary conditions at $Q=M_G$ of $m_i^2(o)=m_0^2+\mu_0^2$, $m_3^2(o)=-B_0\mu_0$. Similarly $g_2(t)$ and $g_Y(t)$ are running coupling constants ($g_2(o)=(5/3)g_Y(o)=g=\sqrt{4\pi\alpha_G}$), and ΔV_1 is the one loop correction. Minimizing V_H with respect to the VEVs $v_{1,2}=< H_{1,2} >$ yields

$$\frac{1}{2}M_Z^2(Q)=\frac{\mu_1^2-\mu_2^2\tan^2\beta}{\tan^2\beta-1};\ \sin 2\beta=\frac{2m_3^2}{\mu_1^2+\mu_2^2} \tag{6}$$

where $\mu_i^2=m_i^2+\Sigma_i, \tan\beta=v_2/v_1$ and the loop correction Σ_i is

$$\Sigma_i=\sum_a(-1)^{2j_a}n_aM_a^2\ln[M_a^2/\sqrt{e}Q^2]\partial M_a^2/\partial v_i \tag{7}$$

Here j_a, n_a, $M_a(v_i)$ are the spin, number of helicities and mass of particle a. The results are approximately independent of Q when Q is at the electroweak scale [22], and so it is convenient to set $Q=M_Z$.

The RGE allow one to evaluate μ_i, m_3, g_2, g_Y in terms of the Gut scale parameters. It is convenient to use Eqs. (3.2) to eliminate μ_0^2 and B_0 in terms of the other parameters and $\tan\beta$. Thus one is left with

$$m_0, m_{1/2}, A_0, \tan\beta;\ m_t \tag{8}$$

to describe the SUSY sector of the theory (since M_G and α_G are determined by the grand unification analysis). In addition, the sign of μ_0 is not determined and so there are two branches: $\mu_0<0$ and $\mu_0>0$.

It is of interest to ask under what conditions solutions of Eqs. (6) will exist i.e. spontaneous breaking of $SU(2)\times U(1)$ occurs. One finds the following needs to be satisfied: all of the soft breaking parameters ($m_0, m_{1/2}, A_0, B_0$) cannot be zero, and the top must be heavy ($m_t \gtrsim 90$ GeV). Thus the supergravity induced breaking of supersymmetry does indeed trigger electroweak breaking, and the theory automatically predicts a heavy top quark.

4. Proton Decay

Proton decay is endemic in grand unified models, and one must do something special to inhibit it. It is thus of interest to examine the constraints that the experimental proton decay bounds impose on the parameter space. We consider here the class of models with "$SU(5)$ type" proton decay defined by the following:

(i) The Gut group G contains an $SU(5)$ subgroup [or is $SU(5)$].

(ii) The matter that remains light after G breaks to $SU(3)\times SU(2)\times U(1)$ at $Q=M_G$ is embedded in the usual way in the 10 and $\bar{5}$ representations of the $SU(5)$ subgroup.

(iii) After G breaks, there are only two light Higgs doublets which interact with matter, and these are embedded in 5 and $\bar{5}$ representations of the $SU(5)$ subgroup, the corresponding color triplets becoming superheavy.

(iv) No special symmetry or condition forbids the proton decay interaction.

Under these circumstances there exists a *model independent* SUSY decay mode,

$$p \to \bar{\nu} + K^+ \tag{9}$$

due to the exchange of the superheavy Higgsino color triplets [23, 24].

The above assumptions apply to a number of Gut groups G e.g. $SU(5)$, $O(10)$, E_6, etc. The requirement that only two light Higgs doublets exist below M_G is needed, as discussed, for a satisfactory unification of coupling constants. Suggestions to suppress proton decay generally fall into two catagories. Thus the flipped $SU(5) \times U(1)$ model [25] suppresses the mode of Eq. (9) by violating condition (ii) by "flipping" the quark and lepton embeddings. (However, this is not a fully unified group.) Alternately, one may attempt to violate condition (iv). One may consider a set of color triplet Higgs, $H_i^{(3)}$, $\bar{H}_i^{(3)}$, $i = 1 \cdots n$, (where, by convention, only $H_1^{(3)}$, $\bar{H}_1^{(3)}$ couple to matter) with mass matrix $\bar{H}_i^{(3)}, M_{(3)}^{ij} H_j^{(3)}$. Then proton decay is indeed suppressed if one assumes

$$[M_{(3)}^{-1}]^{11} = 0; \quad \det M_{(3)} \neq 0 \tag{10}$$

While one can generally arrange for Eq. (10) to hold by fine tuning parameters, it appears quite difficult to do this in a natural way (e.g. by a discrete symmetry) and still maintain only one pair of light Higgs doublets [26].

5. Cosmological Constraints

If one considers models with R parity invariance, the lightest supersymmetric particle (LSP) will be stable. For models with proton decay, the LSP turns out to be the lightest neutralino i.e. the $\tilde{Z}_1$. Thus the $\tilde{Z}_1$ is a candidate for cold dark matter. As a minimum condition, one requires that the relic density of the $\tilde{Z}_1$ not overclose the universe. One defines

$$\Omega_{\tilde{Z}_1} = \rho_{\tilde{Z}_1} / \rho_c \tag{11}$$

where $\rho_{\tilde{Z}_1}$ is the mass density of relic $\tilde{Z}_1$ and $\rho_c = 3H_0/8\pi G_N$ is the critical density that closes the universe. One has $\rho_c = 1.88h^2 \times 10^{-29} g/cm^3$ where h is the Hubble constant H_0 in units of 100 km/sec M_{Pc}. Experimentally, current measurements give $0.65 \lesssim h \lesssim 0.75$. Inflationary cosmology requires $\Omega_{\tilde{Z}_1} = 1$ and so a necessary constraint is

$$\Omega_{\tilde{Z}_1} h^2 < 1 \tag{12}$$

While the $\tilde{Z}_1$ is stable and cannot decay, two $\tilde{Z}_1$ can annihilate in the early universe. This can occur via s-channel h and Z poles, or t-channel squarks and sleptons. Following standard procedures [27] one obtains an approximate formula for $\rho_{\tilde{Z}_1}$:

$$\rho_{\tilde{Z}_1} = 4.75 \times 10^{-40} (\frac{T_{\tilde{Z}_1}}{T_\gamma})^3 (\frac{T_\gamma}{2.75°K})^3 N_F{}^{1/2} (\frac{GeV^2}{J(x_p)}) \frac{g}{cm^3} \tag{13}$$

where $(T_{\tilde{Z}_1}/T_\gamma)^3$ is the reheating factor,

$$J(x_f) = \int_0^{x_f} dx < \sigma v >; \quad x = kT/m_{\tilde{Z}_1} \tag{14}$$

v is the relative velocity, σ is the annihilation cross section, $x_f = kT_f/m_{\tilde{Z}_1}$ and T_f is the freeze out temperature. In Eq. (5.4), $< \sigma v >$ is the thermal average of σv. Since $x_f \approx 1/20$, freeze out occurs when the $\tilde{Z}_1$ are non-relativistic and one may use the Boltzman distribution to calculate the thermal average:

$$< \sigma v > = \int_0^\infty dv v^2 (\sigma v) e^{-v^2/4x} / \int_0^\infty dv v^2 e^{-v^2/4x} \tag{15}$$

For annihilation through the h pole, one obtains a Breit-Wigner form for σv,

$$\sigma v = A_h \frac{v^2}{[v^2 - \varepsilon_R)^2 + \gamma_R^2]} \tag{16}$$

where A_h is a known constant, $\varepsilon_R = (m_h^2 - 4m_{\tilde{Z}_1}^2)(m_{\tilde{Z}_1}^2)$ and $\gamma_R = m_h \Gamma_h / m_{\tilde{Z}_1}$. Here Γ_h is the h width, $\Gamma_h \simeq 2 \times 10^{-3}$ GeV. Similar (though more complicated) expressions hold for annihilation through the Z pole, and corresponding expressions for the t-channel sfermion processes.

Previous analyses have assumed the validity of the non-relativistic approximation, $(\sigma v)_{\rm approx} \cong a + bv^2/6$, to carry out the thermal average: $< \sigma v >_{\rm approx} = a + bx$. However, it has been pointed out [28] that such an approximation is bad near a pole. This turns out to be the case here where the approximate evaluation can over estimate $\Omega_{\tilde{Z}_1}$ by a factor of 1000, and under estimate it by a factor of 100 for the h pole annihilation, due to the smallness of the h width [29]. A similar (though less extreme) phenomena occurs for the wider Z pole. Fig. 2, where the total annihilation cross section is used, exhibits this phenomena. The h poles lies at the point where the curve first crosses the axis from positive to negative, and the Z pole is at the second such crossing. Thus in the region prior to the poles, $\Omega_{\rm approx}$ seriously over estimates the true $\Omega_{\tilde{Z}_1}$. Further, it is just in this region where the true $\Omega_{\tilde{Z}_1} h^2 < 1$. This is shown in the example of Fig. 3. One sees there that the region where $\Omega h^2 < 1$ is significantly larger and generally different from that obtained using $\Omega_{\rm approx}$.

6. SUSY Mass Spectrum

We now examine the SUSY mass spectrum arising by letting the five parameters of Eq. (8) range over the full parameter space subject to the following conditions:

(i) Masses do no violate current experimental bounds

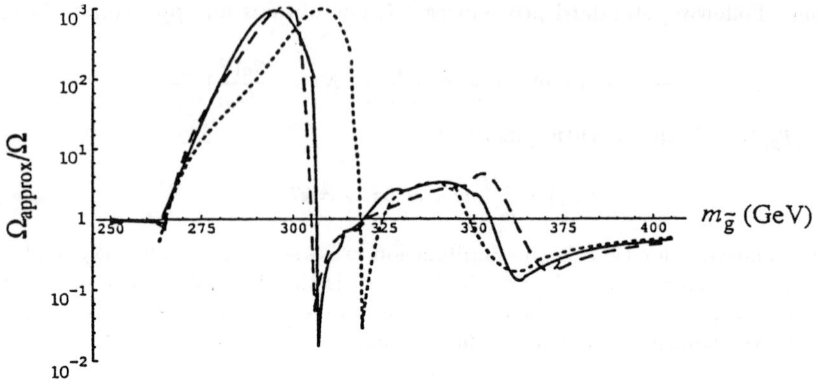

Figure 2: $\Omega_{\text{approx}}/\Omega$ as a function of gluino mass for $m_t = 110$ GeV (dashed curve), 125 GeV (solid curve), and 140 GeV (dotted curve). The h and Z poles occur where $\Omega_{\text{approx}}/\Omega$ decreases rapidly through zero. The curves are for $m_0 = 700$ GeV, $\tan\beta = 2.25$, $\mu > 0$ and $A_t = -0.8(m_t = 110$ GeV), $A_t = -0.4(m_t = 125$ GeV, $A_t = 0.0(m_t = 140$ GeV).

(ii) Radiative breaking of $SU(2) \times U(1)$ occurs

(iii) Experimental bounds on proton decay are obeyed

(iv) No extreme fine tuning occurs: $m_o, m_{\tilde{g}} < 1$ TEV

(v) $M_{H_3}/M_G < 10$

Condition (v) is consistent with the results obtained in the simple Gut model treated in Sec. 2. Also, it is about the largest value one could allow M_{H_3} to take without having to worry about possible Planck scale corrections.

6.1. Current Proton Decay Experiments

We consider first constraints arising from current experimental bounds on proton decay. The current experimental bound on the partial lifetime of $p \to \bar{\nu} + K^+$ is $\tau(p \to \bar{\nu}K^+) > 1 \times 10^{32}$ yr (90% C.L.) [30]. This leads to the following results:

1. $M_{H_3}/M_G < 3$ [3, 4]. For M_{H_3} not too large, the experimental bound limits the parameter space though it is still sizable. One finds $m_0 \gtrsim 500$ GeV, $m_{\tilde{g}} \lesssim 450$ GeV, $|A_t|/m_0 \lesssim 1.5$ and $1.1 \lesssim \tan\beta \lesssim 5$. Thus squarks (with the possible exception of the $\tilde{t}_1$) and probably gluinos require the SSC or LHC to be seen. One also has upper bounds on the t quark and h Higgs of $m_t < 180$ GeV, $m_h < 110$ GeV. Further, if $m_t < 140$ GeV then $m_{\tilde{W}_1} < 100$ GeV when $m_h > 95$ GeV i.e. LEP200 should see either the $\tilde{W}_1$ or the h Higgs (and possibly both). Also, the current data implies a lower bound on M_{H_3} of $M_{H_3} \gtrsim 1 \times 10^{16}$ GeV.

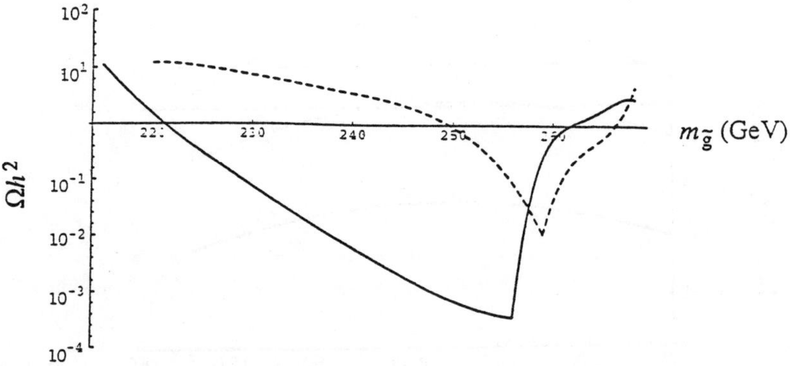

Figure 3: Ωh^2 vs $m_{\tilde{g}}$ (solid curve) and $\Omega_{\text{approx}} h^2$ vs $m_{\tilde{g}}$ (dashed curve) in the h pole region for $m_0 = 700$ GeV, $\tan\beta = 1.88$, $A_t = 0.5 m_0$, $m_t = 125$ GeV, $\mu > 0$. The region where the correct Ωh^2 is less than 1 is considerably larger than for $\Omega_{\text{approx}} h^2$.

2. As M_{H_3}/M_G increases, the bounds on m_0 and $m_{\tilde{g}}$ lessen [31]. Thus for $M_{H_3}/M_G \gtrsim 7$, $m_{\tilde{g}}$ can reach the upper bound of 1 TeV. Also the bands of allowed A_t and $\tan\beta$ widen a little.

3. Over most of the allowed parameter space a remarkable set of scaling laws hold for the charginos, neutralinos and gluino [3, 4, 31]:

$$2m_{\tilde{Z}_1} \cong m_{\tilde{W}_1} \cong m_{\tilde{Z}_2} \tag{17}$$

$$m_{\tilde{W}_1} \simeq \frac{1}{3} m_{\tilde{g}} \text{ for } \mu < 0; \quad m_{\tilde{W}_1} \approx \frac{1}{4} m_{\tilde{g}} \text{ for } \mu > 0 \tag{18}$$

In addition one has near degeneracies for the heavier particles

$$m_{\tilde{W}_2} \cong m_{\tilde{Z}_3} \cong m_{\tilde{Z}_4} >> m_{\tilde{Z}_1} \tag{19}$$

$$m_H \cong m_A \cong m_{H^\pm} >> m_h \tag{20}$$

Eqs. (17-19) arise because $\mu^2 >> M_Z^2, \tilde{m}_2{}^2$, while Eq. (20) is a consequence of fact that generally both m_0 and μ are large.

6.2. Future Proton Decay Experiments

The next generation of proton decay experiments should greatly extend the bounds. Thus Super Kamiokande should be sensitive to the $p \to \bar{\nu} + K^+$ mode up to about 2×10^{33} yr [32], while ICARUS expects to reach 5×10^{33} yr [33]. Figs. (4-6) [31] illustrate what can be expected from these experiments. Fig. 4 shows that when the $\tau(p \to \bar{\nu}K)$ is maximized at fixed value of m_t by letting the other parameters range over the entire parameter space, τ takes on its maximum at about $m_t \simeq 145$ GeV. Fig. 5 shows that ICARUS will be able to test for proton decay for m_0, $m_{\tilde{g}} < 1$ TeV

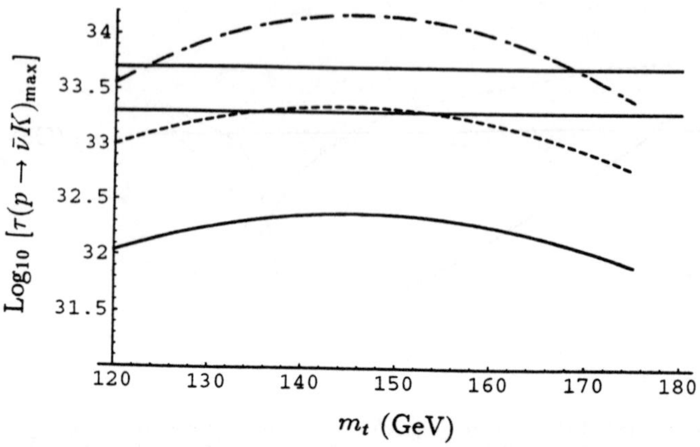

Figure 4: Maximum value of $\tau(p \to \bar{\nu}K^+)$ vs m_t for $M_{H_3}/M_G = 6$, $\mu < 0$ and $m_0 = 1200$ GeV (dot-dashed curve), $m_0 = 800$ GeV (dashed curve), $m_0 = 400$ GeV (solid curve). The lower horizontal line is the upper bound that Super Kamiokande can detect, and the higher horizontal line is the upper bound for ICARUS. $\tau(p \to \bar{\nu}K^+)$ takes it absolute maximum at $m_t \simeq 145$ GeV.

if $M_{H_3}/M_G < 6$ (and Super Kamiokande for $m_0 \lesssim 800$ GeV).

Alternately ICARUS can test for proton decay for $m_0 \lesssim 800$ GeV (and Super Kamiokande for $m_0 \lesssim 600$ GeV) for the entire range $M_{H_3}/M_G < 10$. Thus, for example, should the SSC discover a squark at 500 GeV, ICARUS and Super Kamiokande should see proton decay. Fig. 6 shows the maximum value of $\tau(p \to \bar{\nu}K)$ subject to the constraint $m_{\tilde{W}_1} > 100$ GeV. Thus if ICARUS does not see proton decay, then $m_{\tilde{W}_1} < 100$ GeV and the $\tilde{W}$ should be observed at LEP200. In addition one may show [31] for the entire parameter domain $m_0, m_{\tilde{g}} < 1$ TeV, $M_{H_3}/M_G < 10$, that if $\tau(p \to \bar{\nu}K) > 1.5 \times 10^{33}$ yr (which can be tested at both Super Kamiokande and ICARUS) then either $m_h < 95$ GeV or $m_{\tilde{W}_1} < 100$ GeV i.e. either the h Higgs or the $\tilde{W}_1$ (and possibly both) should be observed at LEP200. If $m_{\tilde{W}_1} < 100$ GeV, then the $\tilde{W}_1$ should also be observable at the Tevatron [34].

6.3. Constraints of Cosmology

The cosmological constraints studied in Sec. 5 that the relic density of the $\tilde{Z}_1$ not overclose the universe, further reduce the allowed parameter space. There still remains, however, a considerable region consistent with both proton decay and cosmology. As seen in Sec. 5, there is considerable $\tilde{Z}_1$ annihilation in the region $2m_{\tilde{Z}_1} \lesssim m_h$ and $2m_{\tilde{Z}_1} \lesssim M_Z$. Since, by Eq. 6.1, the $\tilde{Z}_1$ mass scales with $m_{\tilde{g}}$, the cosmological constraint yields allowed mass bands for $m_{\tilde{g}}$, characteristically about 40 GeV wide for the h pole and 20 GeV wide for the Z pole [29]. This is shown in the example of Fig. 7. Sometimes the two bands overlap, yielding a large region where the relic density constraint is obeyed, as seen in Fig. 8. The allowed region

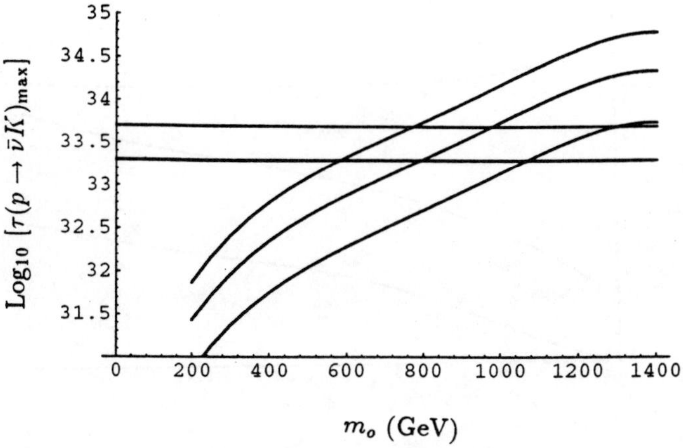

Figure 5: The maximum value of $\tau(p \to \bar{\nu}K^+)$ vs m_0 for $m_t = 150$ GeV, $\mu < 0$ calculated by varying all parameters except m_0 over the entire parameter space. The results are plotted for, $M_{H_3}/M_G = 3$, 6 and 10 (lower, middle and upper curves). The horizontal lines are as in Fig. 4.

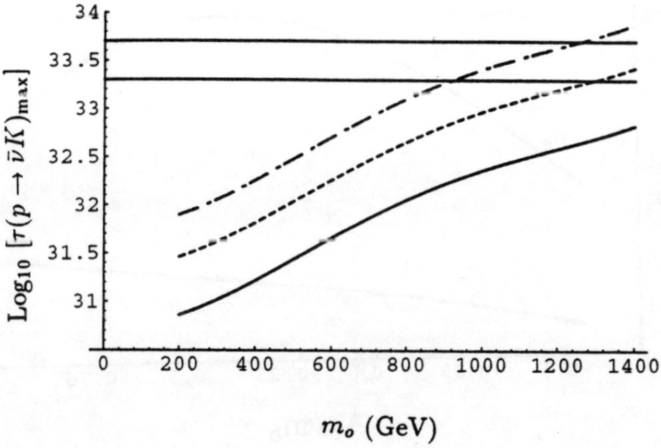

Figure 6: Maximum value of $\tau(p \to \bar{\nu}K^+)$ vs m_0 subject to the constraint $m_{\tilde{W}_1} > 100$ GeV for $M_{H_3}/M_G = 3$, 6, and 10 (solid, dashed, dot-dashed curves). The horizontal lines are as in Figs. 4 and 5.

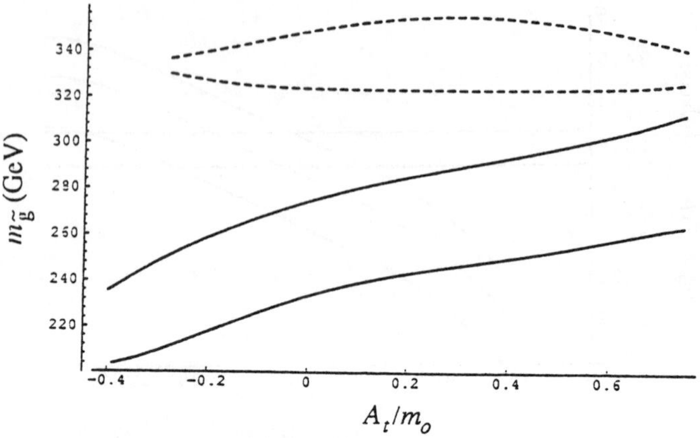

Figure 7: Region in $m_{\tilde{g}}-A_t$ space allowed by the combined proton decay and cosmological constraints for $m_t = 125$ GeV, $m_0 = 600$ GeV, $\tan\beta = 1.73$, $\mu > 0$ and $M_{H_3}/M_G < 6$. The allowed region are those between the solid lines (h pole) and dashed lines (Z pole).

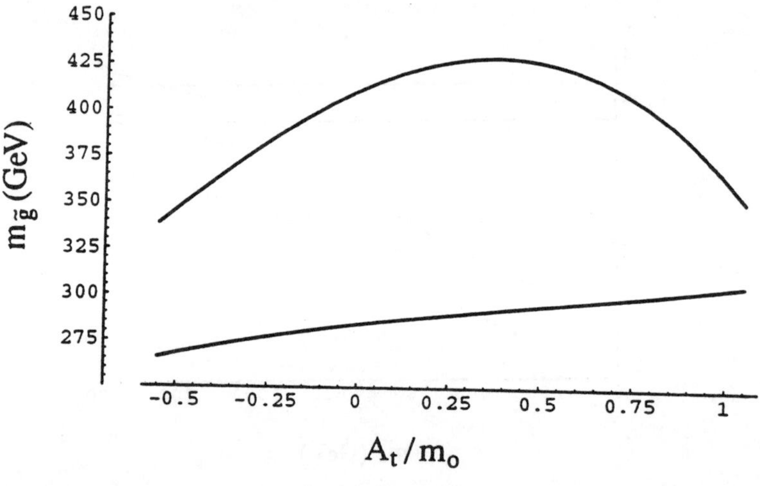

Figure 8: Region in $m_{\tilde{g}} - A_t$ space allowed by combined proton decay and cosmological constraints for $m_t = 140$ GeV, $m_0 = 700$ GeV, $\tan\beta = 2.75\mu > 0$ and $M_{H_3}/M_G < 8$. **The h and Z pole regions coelesce for this case yielding a broad allowed band of gluino masses from 260 GeV to 430 GeV.**

for this example also satisfies the proton decay constraint with $M_{H_3}/M_G < 8$.

In general, the mass spectrum is further restricted by imposing cosmological constraints. Thus the simultaneous requirements of $\Omega_{\tilde{Z}_1}h^2 < 1$ and current proton decay bounds with $M_{H_3}/M_G < 6$ yields $m_h < 105$ GeV, $m_{\tilde{W}_1} < 100$ GeV, $m_{\tilde{Z}_1} < 50$ GeV and $m_t < 165$ GeV. With these constraints the $\tilde{W}_1$ should be accessible at LEP200 and the $\tilde{W}_1$, $\tilde{Z}_2$ and $\tilde{Z}_1$ observable at the Tevatron [34].

7. Conclusions

Basic supergravity models involve four additional parameters beyond those needed in the Standard Model. These extra parameters then specify the SUSY sector of the theory, the masses, production cross sections, widths etc. of the 32 SUSY particles. Current models are consistent with all present data of LEP, the Tevatron, proton decay and with relic density constraints. However, future experiments at LEP200, the Tevatron and future underground experiments will be able to test models with proton decay. While no single experiment can verify whether such models are valid, the combination of these experiments, as seen in Sec. 6 can in fact give clear tests, and this information could become available prior to the comissioning of the SSC or LHC. Of course if any supergravity model is correct, there should be a huge number of SUSY particles produced at the SSC and LHC.

8. Acknowledgements

This research was supported in part by National Science Foundation Grant Nos. PHY-916593 and PHY-9306906.

9. References

1. A. H. Chamseddine, R. Arnowitt and P. Nath, *Phys. Rev. Lett.* **49**, 970 (1982).
2. L. E. Ibañez, *Phys. Lett. Phys. Lett.* **B118**, 73 (1982).
3. R. Arnowitt and P. Nath, *Phys. Rev. Lett.* **69**, 725 (1992).
4. P. Nath and R. Arnowitt, *Phys. Lett.* **B289**, 368 (1992).
5. J. Lopez, H. Pois, D. V. Nanopoulos and K. Yuan, CERN-TH-6628/92-CTP-TAMU-61/92-ACT-19/92.
6. G. G. Ross and R. G. Roberts, *Nucl. Phys.* **B377**, 571 (1992).
7. M. Drees and M. M. Nojiri, *Nucl. Phys.* **B369**, 54 (1992).
8. S. Kelley, J. Lopez, H. Pois, D. V. Nanopoulos and K. Yuan, *Phys. Lett.* **B273**, 43 (1991).
9. K. Inoue, M. Kawasaki, M. Yamaguchi and T. Yanagida, *Phys. Rev.* **D45**, 387 (1992).
10. P. Ramond, talk at Recent Advances in the Superworld, The Woodlands, April,

1993.

11. B. Ananthanarayan, G. Lazarides and Q. Shafi, *Phys. Rev.* **D44**, 1613 (1991).
12. P. Langacker, Proc. PASCOS 90-Symposium, Eds. P. Nath and S. Reucroft (World Scientific, Singapore 1990); J. Ellis, S. Kelley and D. V. Nanopoulos, *Phys. Lett.* **B249**, 441 (1990); **B260**, 131 (1991); U. Amaldi, W. de Boer and H. Furstenau, *Phys. Lett.* **B260**, 447 (1991); F. Anselmo, L. Cifareli, A. Peterman and A. Zichichi, *Nuov. Cim.* **104A**, 1817 (1991); **115A**, 581 (1992).
13. J. Polonyi, Univ. Budapest Rep. No. KFKI-1977-93, 1977 (unpublished).
14. H. P. Nilles, *Phys. Lett.* **B115**, 193 (1981); S. Ferrara, L. Girardello and H. P. Nilles, *Phys. Lett.* **B125**, 457 (1983).
15. R. Barbieri, S. Ferrara and C. A. Savoy, *Phys. Lett.* **B119**, 343 (1982); L. Hall, J. Lykken and S. Weinberg, *Phys. Rev.* **D27**, 2359 (1983); P. Nath, R. Arnowitt and A. H. Chamseddine, *Nucl. Phys.* **B227**, 121 (1983).
16. K. Inoue et. al., *Prog. Theor. Phys.* **68**, 9927 (1982); L. Ibañez and G. G. Ross, *Phys. Lett.* **B110**, 227 (1982); L. Alvarez-Gaumé, J. Polchinski and M. B. Wise, *Nucl. Phys.* **B250**, 495 (1983); J. Ellis, J. Hagelin, D. V. Nanopoulos and K. Tamakis, *Phys. Lett.* **B125**, 2275 (1983); L. E. Ibañez and C. Lopez,. *Phys. Lett.* **B128**, 54 (1983); *Nucl. Phys.* **B233**, 545 (1984); L. E. Ibañez, C. Lopez and C. Muños, *Nucl. Phys.* **B256**, 218 (1985).
17. For a review of supergravity Gut models see P. Nath, R. Arnowitt and A. H. Chamseddine, *Applied $N = 1$ Supergravity* (World Scientific, Singapore, 1984); H. P. Nilles, *Phys. Rep.* **110**, 1 (1984).
18. E. Witten, *Nucl. Phys.* **B177**, 477 (1981); S. Dimopoulos, H. Georgi, *Nucl. Phys.* **B193**, 1501 (1981); N. Sakai, *Zeit. f. Phys.* **C11**, 153 (1981).
19. S. Bethke, XXVI Conference on High Energy Physics, Dallas, 1992; G. Alterelli, Les Rencontres de Physique de la Valee D'Aoste, LaThuile, 1993.
20. D. Ring and S. Urano, in preparation.
21. B. Grinstein, *Nucl. Phys.* **B206**, 387 (1982); H. Georgi, *Phys. Lett.* **115B**, 380 (1982); K. Inoue, A. Kakuto and T. Tankano, *Prog. Theor. Phys.* **75**, 664 (1986); A. Anselm and A. Johansen, *Phys. Lett.* **B200**, 331 (1988); A. Anselm, *Sov. Phys. JETP* **67**, 663 (1988); K. S. Babu and S. M. Barr, Bartol preprint (1993).
22. G. Gamberini, G. Ridolfi and F. Zwirner, *Nucl. Phys.* **B331**, 331 (1990); R. Arnowitt and P. Nath, *Phys. Rev.* **D46**, 1981 (1992).
23. S. Weinberg, *Phys. Rev.* **D26**, 187 (1982); N. Sakai and T. Yanagida, *Nucl. Phys.* **B197**, 533 (1982); S. Dimopoulos, S. Raby and F. Wilczek, *Phys. Lett.* **B112**, 133 (1982); J. Ellis, D. V. Nanopoulos and S. Rudaz, *Nucl. Phys.* **B202**, 43 (1982); S. Chadha and M. Daniels, *Nucl. Phys.* **B229**, 105 (1983); B. A. Campbell, J. Ellis and D. V. Nanopoulos, *Phys. Lett.* **B141**, 229 (1984).
24. R. Arnowitt, A. H. Chamseddine and P. Nath, *Phys. Lett.* **B156**, 215 (1985); P. Nath, R. Arnowitt and A. H. Chamseddine, *Phys. Rev.* **D32**, 2348 (1985).

25. I. Antoniadis, J. Ellis, J. Hagelin and D. V. Nanopoulos, *Phys. Lett.* **B208**, 209 (1988).
26. C. D. Coughlan et. al., *Phys. Lett.* **B158**, 401 (1985); K. S. Babu and S. M. Barr, Bartol preprint (1993).
27. B. W. Lee and S. Weinberg, *Phys. Rev. Lett* **39**, 165 (1977); D. A. Dicus, E. Kolb and V. Teplitz, *Phys. Rev. Lett.* **39**, 168 (1977); H. Goldberg, *Phys. Rev. Lett.* **50**, 1419 (1983); J. Ellis, J. S. Hagelin, D. V. Nanopoulos, K. Olive and M. Srednicki, *Nucl. Phys.* **B238**, 453 (1984).
28. K. Griest and D. Seckel, *Phys. Rev.* **D43**, 3191 (1991); P. Gondolo and G. Gelmini, *Nucl. Phys.* **B360**, 145 (1991).
29. R. Arnowitt and P. Nath, *Phys. Lett.* **B299**, 58 (1993) and Erratum, P. Nath and R. Arnowitt, *Phys. Rev. Lett.* **70**, 3696 (1993).
30. Particle Data Group, *Phys. Rev.* **D45**, Part 2, (1992).
31. R. Arnowitt and P. Nath, CTP-TAMU-32/93-NUB-TH-3066/93-SSCL-Preprint-440 (1993).
32. Y. Totsuka, Proc. XXIV Conf. on High Energy Physics, Munich, 1988, Eds. R. Kotthaus and J. H. Kuhn (Springer Verlag, Berlin, Heidelberg, 1989).
33. ICARUS Detector Group, Int. Symposium on Neutrino Astrophysics, Takayama, 1992.
34. P. Nath and R. Arnowitt, *Mod. Phys. Lett.* **A2**, 331 (1987); R. Barbieri, M. Frigeni and M. Mangano, *Nucl. Phys.* **B367**, 28 (1991); H. Baer and X. Tata, *Phys. Rev.* **D47**, 2739 (1993).

TESTING UNIFIED SUPERSYMMETRIC MODELS

S. KELLEY

Department of Physics
Maharishi International University
Fairfield, IA 52557-1069, USA

Abstract

I define the Standard Supersymmetric Model (SSM) as the minimal supersymmetric extension of the Standard Model with gauge coupling unification and universal soft supersymmetry breaking at the unification scale. This well-defined model has a five-dimensional space of unknown parameters $(m_t, tan_\beta, m_{1/2}, m_0, A)$. I outline the top-down and bottom-up methods of solving this model. Thresholds may be treated either by dropping heavy particles from the RGE's or by considering loop corrections to the vacuum energy. Substantial regions of the parameter space are consistent with all experimental constraints. I consider the relation of the SSM to more realistic models such as supersymmetric $SU(5)$, Flipped $SU(5) \times U(1)$, the String-Inspired Standard Model, and string-derived models. I briefly discuss sparticle spectroscopy and flavor changing neutral currents (FCNC's) as sample methods for determining the unknown parameters of the SSM, and discriminating between the SSM and more realistic models.

1. Introduction

Unified supersymmetric gauge theories are very likely the framework to describe physics below the Planck scale. In addition to the theoretical motivation that supersymmetry solves the gauge hierarchy problem, the excellent agreement of LEP data with the predictions of gauge coupling constant unification in supersymmetric models[1] provides a strong basis for belief in supersymmetry. Recent precision LEP measurements have sparked an explosion of work on unified supersymmetric models. In sorting through the many different models available, the elucidation of a Standard Supersymmetric Model (SSM), comparable to the Standard Model in definiteness and number of parameters, proves useful. The SSM is defined as the minimal extension of the Standard Model with gauge coupling unification and universal soft supersymmetry breaking at the unification scale. The SSM has an unknown parameter space of five dimensions, $(m_t, tan_\beta, m_{1/2}, m_0, A)$.

In this talk, I present the basic properties of the SSM, the techniques used to extract predictions over the complete five-dimensional parameter space, the relation of the SSM to more realistic models, and discuss sparticle spectroscopy and flavor changing neutral currents as sample tests of unified supersymmetric models. Section 2 defines the SSM, and Section 3 outlines the top-down and bottom-up methods of solving the SSM. Section 4 considers how thresholds may be treated by dropping the heavy particles from the RGE's or considering loop corrections to the vacuum energy. Section 5 considers the relation of the SSM to more realistic models. Two sample experimental tests of unified supersymmetric models, sparticle spectroscopy and flavor changing neutral currents (FCNC's), are discussed in Sections 6 and 7. Conclusions are presented in Section 8.

Unification and supersymmetry go hand in hand. Supersymmetry ensures the cancellation of quadratic divergences necessary to reliably connect unification-scale physics with accessible low-energy predictions. Unification ensures that the model has only a few parameters, introducing correlations between the observables of the low-energy model which can be experimentally tested. Because of quantum loop-effects, every prediction of a unified supersymmetric theory is sensitive to all the fundamental parameters of the theory. In principle, any five exact experimental measurements determine the parameters of the SSM, and each additional measurement tests the theory. In practice, uncertainty in experimental measurements gives an allowed region in parameter space which narrows, and possibly disappears, with increasing experimental precision.

A strict definition of a valid scientific theory requires that the theory be falsifiable. The SSM could be falsified in two ways, observing processes like proton decay or lepton flavor violation which are forbidden in the SSM, or showing that every point in the parameter space is inconsistent with some experimental measurement. Thus it is important to develop general methods for accurately extracting the predictions of a theory over its entire parameter space.

However, theoretical considerations clearly indicate that the SSM is only an approximation to the complete theory of nature. Moreover, quantum gravitational effects probably introduce an uncertainty into every prediction of the SSM.[2] However, the ability to test the SSM without non-renormalizable corrections from quantum gravity is the foundation for investigating these corrections. Because of this, precision calculations and experimental tests of the SSM are crucial. The definiteness and simplicity of the SSM make it the ideal example for developing tools to test unified supersymmetric models. These tools are generally applicable to more realistic models, but easier to develop in the context of a simple, well-defined model.

String theory provides a framework to quantify the effects of quantum gravity, and realistic string models reduce to unified supersymmetric models below the Planck scale. Therefore, the techniques to test the SSM are a necessary subset of those needed to test string theories. Also, because of the huge diversity of possible string models, the field-theoretic parameterization afforded by string-inspired models is a necessary complement to the "needle in the haystack" approach of specific string-derived models.

The SSM should be taken seriously as a well-defined, testable model, which incorporates most of the features the complete theory of nature is likely to contain. The methods of extracting the predictions of the SSM and the process of experimentally testing it are thus extremely important. However, the main goal of testing the SSM is to discriminate the fine details of what lies beyond.

2. The SSM

The SSM is defined as the minimal supersymmetric extension of the Standard Model with gauge coupling unification and universal soft supersymmetry breaking at the unification scale. The SSM superpotential

$$W = D_I^C \lambda_1^{IJ} Q_j H + U_I^C \lambda_2^{IJ} Q_J \overline{H} + E_I^C \lambda_3^{IJ} L_J H + \mu H \overline{H} \tag{2.1}$$

involves a dimensionful Higgs mixing term, μ. By adding a gauge singlet to the model and replacing the last term in the superpotential Eq. (2.1) with

$$W_{singlet} = \lambda_7 \phi H \overline{H} + \lambda_8 \phi^3 \tag{2.2}$$

the Higgs mixing can be dynamically generated without a dimensionful coupling.[3] Although the final, unconstrained, five-dimensional parameter space of the model with a dimensionful Higgs mixing and the model with an extra singlet are identical, the structure, solution, and constraints differ between the two models. The analysis of the model defined by Eq. (2.2) is computationally more complex. In what follows, only the superpotential Eq. (2.1) will be considered.

The Yukawa couplings $\lambda_1, \lambda_2, \lambda_3$ in Eq. (2.1) are general 3×3 matrices in generation space. A superfield basis in flavor space which keeps the gauge couplings diagonal may be chosen so that

$$\lambda_1 = \lambda_d \quad \lambda_2 = \lambda_u K \quad \lambda_3 = \lambda_e \tag{2.3}$$

Where λ_d, λ_u, λ_e are diagonal matrices with real entries and K is the KM matrix. This generational structure of the Yukawa couplings underlies attempts to understand the observed fermion masses and mixings,[4] and calculation of supersymmetric contributions to FCNC's. However, for most other applications of the SSM, taking a unit KM matrix and neglecting all but the top, bottom, and tau Yukawas is an excellent approximation which will be used throughout this paper, except for the section on FCNC's.

The boundary conditions for the model at the unification scale are particularly simple, involving only ten parameters: the unification scale M_x, the unified gauge coupling, α_x, the third generation Yukawas λ_b, λ_t, λ_τ, the universal soft supersymmetry-breaking parameters, $m_{1/2}$, m_0, A, the supersymmetric Higgs mixing parameter μ, and the soft supersymmetry-breaking Higgs mixing parameter B. From these parameters renormalized at M_x, the coefficients in the SSM Lagrangian at any lower renormalization point may be determined through the RGE's.[5] Compared to a theory with general, unrelated soft supersymmetry-breaking gaugino masses, scalar masses, and trilinear couplings, this is an immense reduction in the number of parameters.[6] The known experimental values of $m_b, m_\tau, m_Z, \alpha_3, \alpha_{em}$ and $\sin^2\theta$ give boundary conditions at low energies which constrain the ten-dimensional parameter space at the unification scale. Various strategies for doing this are considered in the next section.

3. Solving the SSM

In this section, I examine different strategies to solve the constraints on the parameter space of the SSM neglecting thresholds. There are two basic strategies for doing this: top-down and bottom-up. Threshold effects greatly complicate the problem and will be considered in the next section.

The top-down method chooses the ten parameters at the unification scale $(M_x, \alpha_x, \lambda_b, \lambda_t, \lambda_\tau, m_{1/2}, m_0, A, \mu, B)$ and then computes all the parameters in the Lagrangian at m_Z using the renormalization group equations. Assuming only real vev's of the neutral Higgs, the scalar potential at tree level is

$$V = (m_{\bar{H}}^2 + \mu^2)|\bar{v}|^2 + (m_H^2 + \mu^2)|v|^2 + 2B\mu\bar{v}v + \frac{1}{8}(g_2^2 + \frac{3}{5}g_1^2)(|\bar{v}|^2 - |v|^2)^2 \quad (3.1)$$

Extremizing this potential with respect to $\bar{v}$ and v determines the Higgs vev's, which in turn determine M_Z and $\tan\beta = \bar{v}/v$.

$$\frac{2v\bar{v}}{v^2 + \bar{v}^2} = \sin 2\beta = \frac{2B\mu}{(m_{\bar{H}}^2 + m_H^2 + 2\mu^2)}$$
$$\frac{g_2^2}{2}(v^2 + \bar{v}^2) = m_Z^2 = \frac{2(m_H^2 - m_{\bar{H}}^2 \tan^2\beta)}{\tan^2\beta - 1} - 2\mu^2 \quad (3.2)$$

An acceptable point in parameter space must also yield a potential bounded from below with the global minimum having non-zero vev's for only the two neutral Higgs.

For the points in the 10-d parameter space with acceptable minima, this process determines the entire SSM Lagrangian, including known experimental parameters like α_3, $\sin^2\theta$, α_{em}, m_b, m_τ, m_Z. These experimental values constrain the 10-d parameter space at the unification scale. The top-down method may be computationally simplified by expressing all the dimensionful parameters in units of $m_{1/2}$. Once the potential has been minimized to determine $v/m_{1/2}$ and $\bar{v}/m_{1/2}$, the constraint $m_Z^2 = g^2(v^2+\bar{v}^2)/2$ determines $m_{1/2}$. However when thresholds are included, the dimensionful values of the parameters are required to construct the potential, and this simplification is not possible.

The bottom-up method attempts to simplify and speed up this process by directly determining some of the parameters at the unification scale from known low-energy parameters. The gauge sector provides the classic example whereby, at one loop and neglecting thresholds, α_{em} and α_3 predict M_x, α_x, and $\sin^2\theta$.

$$\sin^2\theta = \frac{1}{5} + \frac{7\alpha_{em}}{15\alpha_3}$$
$$\frac{1}{\alpha_X} = \frac{3}{20\alpha_{em}} + \frac{3}{5\alpha_3}$$

$$\ln\left(\frac{M_U}{m_Z}\right) = \frac{\pi}{30}\left(\frac{3}{\alpha_{em}} - \frac{8}{\alpha_3}\right) \tag{3.3}$$

The prediction of $\sin^2\theta$ represents a spectacular success of the SSM. More careful treatment of this prediction, including higher-loops and thresholds, has been used to further constrain SSM and SSM-like models.[7]

The one-loop RGE's for the Yukawa couplings involve only the gauge couplings and the Yukawa couplings. The fermion masses m_b, m_t, m_τ and $\tan\beta$ allow $\lambda_b, \lambda_t, \lambda_\tau$ to be determined at low energies and evolved to calculate the three Yukawas at the unification scale, independent of the soft supersymmetric-breaking parameters. Since it is convenient to use m_Z as the low-energy renormalization point, the physical fermion masses must be converted to running $\overline{MS}$ masses renormalized at m_Z.[8]

Thus, from two unknowns, m_t and $\tan\beta$, and the known fermion masses, m_b and m_τ, the Yukawas can be determined at m_Z and evolved to the unification scale. Specifying the soft supersymmetry-breaking parameters $m_{1/2}, m_0$, and A at the unification scale allows the entire set of RGE's (except the two for μ and B which do not enter any of the other RGE's) to be evolved to determine all the parameters at m_Z. Having chosen $\tan\beta$ and knowing the experimental value of m_Z determines the two Higgs vev's through Eq. (3.2). The minimization conditions $\partial V/\partial v = \partial V/\partial \bar{v} = 0$ can now be solved for the remaining unknowns, μ and B, renormalized at m_Z. In this method, the constraints from m_b, m_τ, m_Z (and m_t when it has been determined) are implemented at the beginning of procedure rather than at the end. This eliminates the computational problem encountered in the top-down method of searching a large dimensional space for a smaller dimensional subspace satisfying constraints.

The parameter space of the SSM is bounded in the $(m_t, \tan\beta)$ plane by a combination of perturbative unitarity, the experimental lower bound on the top mass, and electroweak breaking. The lower bound on the gluino mass bounds $m_{1/2}$ from below, and naturalness bounds $m_{1/2}$ from above, as well as bounding the magnitude of ξ_0 and ξ_A. However the naturalness bounds require a precise definition of how much fine tuning is natural, and are somewhat a matter of taste.[6,9] Large five-dimensional areas of parameter space are consistent with all known experimental constraints.

The lightest neutralino provides a natural dark matter candidate. The SSM has an exact R-parity and the lightest supersymmetric particle (LSP) is stable. Areas of parameter space where the lightest supersymmetric particle is a chargino or areas where the relic neutralino density is too great are excluded in the SSM, although R-parity breaking extensions of the SSM could possibly rescue these areas. However, cosmologically, the most interesting areas of parameter space are those where the LSP is a neutralino with relic density near the critical density.[10]

4. Thresholds

The most straightforward method of including threshold effects is to drop heavy particles from the RGE's at scales below the mass of the heavy particles. This method has the advantage of maintaining the leading log accuracy of the RGE's. However, the method has two computational drawbacks. First, the RGE's and scalar potential become substantially more complicated once a non-supersymmetric threshold has been integrated out.[11] This is because the dimensionless parameters in the scalar potential are no longer simply related to gauge and Yukawa couplings by supersymmetry, leading to a proliferation of parameters. Second, the threshold structure depends on the final solution, which in general must be determined by iteratively approaching a self-consistent solution. Despite these two drawbacks, significant progress has been made in determining and solving the non-supersymmetric RGE's resulting from integrating out the various thresholds in the SSM.[12]

A computationally simpler approach uses the one-loop corrections to the vacuum energy[13] to construct a scalar potential independent of the renormalization point to one-loop order[14]

$$V = V_{tree} + \Delta V$$
$$\Delta V = \frac{1}{64\pi^2} STr\mathcal{M}^4 \left(ln\frac{\mathcal{M}^2}{Q^2} - \frac{3}{2} \right) \tag{4.1}$$

where V_{tree} is given by Eq. (3.1), $STr\ f(\mathcal{M}^2) = \Sigma_j (-1)^{2j}(2j+1) Tr\ f(\mathcal{M}_j^2)$, and $\mathcal{M}_j^2$ are the field-dependent spin-j mass matrices. The term ΔV incorporates threshold effects to the scalar potential and quantities derived from it (e.g. Higgs masses). The computation using the bottom-up scheme is particularly simple.[6]

A) Use $m_b(m_Z), m_t(m_Z), m_\tau(m_Z)$ and $\tan\beta(m_Z)$ to compute λ_1, λ_2, λ_3 at m_Z and evolve up to M_U.

B) Choose $\xi_0 = m_0/m_{1/2}$, $\xi_A = A/m_{1/2}$ at M_U and evolve the RGE's down to m_Z.

C) For each choice of $m_{1/2}$, numerically solve the equations $\partial V/\partial v = \partial V/\partial \bar{v} = 0$ for μ and B. This method requires only one iteration of the RGE's for each point in the $(m_t, \tan\beta, \xi_0, \xi_A)$ parameter space and numeric solution of the minimization equations for each value of $m_{1/2}$. The procedure is sufficiently efficient to search the whole five-dimensional parameter space. In addition, the property that the scalar potential is independent of the renormalization point to one-loop order provides a highly non-trivial numerical check on the procedure.[6]

The potential Eq. (4.1) requires a slight adjustment to be truly independent of the renormalization point: as it is, only derivatives of Eq. (4.1) have this property. Using ϕ to denote the collection of field vev's and $t = ln(Q/m_Z)$, the corrected potential is

$$V = V_{tree}(\phi, t) + \Delta V(\phi, t) - \Delta V(0, t) \tag{4.2}$$

This prescription can be justified in several ways.[15] Adding a field-independent piece to V does not change the derivatives of V. Furthermore, the renormalized potential should resemble the tree level potential with suitably modified coefficients and wave-function renormalizations, which clearly vanishes at the origin of field space as Eq. (4.2) does. Explicit computation for a representative point in the SSM parameter space confirms the one-loop renormalization independence of Eq. (4.2).

$$\frac{\partial V}{\partial t}|_{t=0} = -0.0007 \tag{4.3}$$

The departure of Eq. (4.3) from zero is of the size expected from round-off error in the single-precision computer calculation used. For comparison, a similar calculation including only the stop and sbottom contributions in ΔV gives

$$\frac{\partial V}{\partial t}|_{t=0} = -0.0224 \tag{4.4}$$

The cancellation in Eq. (4.3) is only observed when all particles contributing to ΔV are included.

In numerically computing the derivative in Eqs. (4.3) and (4.4), the potential is computed at the renormalization points m_Z ($t = 0$) and a renormalization point slightly higher than m_Z ($t = \delta$). Several subtle points necessary to numerically observe the vanishing of $\partial V/\partial t$ are worth mentioning. In running the RGE's between m_Z and M_U, and back, no thresholds should be integrated out. This would, in effect, double count the influence of the threshold in V: once in ΔV, and once by integrating out the threshold in the RGE's. The masses in ΔV should be renormalized at $t = 0$. Note that the potential is formally t-independent to one-loop no matter where the masses in the supertrace are renormalized. However, only when these masses are renormalized at $t = 0$ do the two-loop subleading-log terms and the linear t-term in the potential vanish. In computing V_{tree} at $t = \delta$, all the parameters in V_{tree} must be renormalized at $t = \delta$, including the wave-function renormalization of the vev's h and $\bar{h}$. Two-loop and higher-loop effects (higher derivatives of V with respect to t) increase logarithmically with Q/m_Z, and signal a progressive deterioration of the one-loop approximation as the mass of the thresholds increase. Although it is difficult to formulate precise criteria for where the approximation breaks down, higher-order terms certainly become important for sparticle masses above about a TeV.

Although the inclusion of threshold effects shifts the detailed quantitative predictions of the SSM, the qualitative picture of the presently allowed regions in parameter space remains much the same. However, the tree-level result that $m_h < m_Z$ is substantially modified, and the lightest Higgs could be much heavier.[16]

5. More Realistic Models

Fundamental considerations clearly indicate that the SSM is not a complete theory. The gauge couplings in the SSM diverge above the unification scale. Furthermore, quantum gravitational effects require a fundamental change in the very framework of quantum field theory, which the SSM is formulated in, at a scale not much larger than the gauge unification scale. GUT's provide a natural structure to preserve the unification of gauge couplings at scales larger than the gauge unification scale. String theory provides a natural unified framework for quantum field theory and gravity.

Supersymmetric $SU(5)$[17] unifies the SSM gauge group within $SU(5)$. The fields added to the SSM are superheavy X and Y gauge superfields, an extra adjoint chiral superfield whose vev breaks $SU(5) \to SU(3) \times SU(2) \times U(1)$, and $D^C, \bar{D}^C$ chiral superfields filling out the $SU(5)$ representations containing the SSM Higgs doublets. Since all of these fields have masses near the gauge unification scale, the renormalizable Lagrangian at scales much below M_U is that of the SSM with couplings slightly modified by GUT threshold corrections. The fermionic components of the superheavy $D^C, \bar{D}^C$ superfields mediate dimension-five, non-renormalizable, proton-decay operators. Proton decay is absent in the SSM and would provide definite evidence for physics beyond the SSM. Consideration of experimental limits on the proton lifetime rule out large areas of the supersymmetric $SU(5)$ parameter space.[18] The results can be interpreted as showing that the model is either highly predictive or highly constrained, depending on one's point of view. In addition, the group $SU(5)$ gives the generally successful prediction $\lambda_\tau(M_U) = \lambda_b(M_U)$.[19] However, contour plots of m_b in the plane of the most relevant parameters, $(\tan\beta, m_t)$,[20] eliminate all but a narrow strip near the edge of the area of perturbative Yukawas. The effect of less relevant variables, especially α_3, must also be considered in a detailed analysis.[21]

Essentially, testing the parameter space of supersymmetric $SU(5)$ reduces to testing the parameter space of the SSM with the addition of the proton decay constraint and the constraint $\lambda_b(M_U) = \lambda_\tau(M_U)$. Increasingly precise measurements of $\sin^2\theta, \alpha_3, m_b$, and τ_P will either restrict the supersymmetric $SU(5)$ parameter space to unnaturally high supersymmetry breaking masses or, if τ_P is observed, impose two constraints on the five-dimensional parameter space of the SSM: one from τ_P, and one from $\lambda_b(M_U) = \lambda_\tau(M_U)$. A theoretical problem with supersymmetric $SU(5)$ is the fine-tuning required to split light Higgs doublets from the superheavy Higgs triplets. It is possible to remedy this problem in extensions of the model[22] although these extensions usually introduce additional problems. Another problem, in the context of the string, is that the adjoint chiral Higgs superfield needed to break $SU(5)$ cannot be obtained in the standard $k = 1$ construction.[23]

Supersymmetric Flipped $SU(5) \times U(1)$[24] gives an alternative GUT comparable in simplicity to supersymmetric $SU(5)$. Since a **10** and $\overline{\mathbf{10}}$ representation breaks $SU(5) \times U(1) \to SU(3) \times SU(2) \times U(1)$, Flipped $SU(5) \times U(1)$ naturally arises in string constructions.[25] In addition, a missing partner mechanism splits the masses of the Higgs doublets and triplets without fine tuning. Flipped $SU(5) \times U(1)$ reduces to the SSM at scales much below M_U[26] in much the same way as supersymmetric $SU(5)$. In Flipped $SU(5) \times U(1)$, it can be shown that the onset of supersymmetry breaking, the gauge unification scale, and the $SU(5) \times U(1) \rightarrow SU(3) \times SU(2) \times U(1)$ scales must be within about an order of magnitude of each other.[26] The departure from universal soft supersymmetry breaking in the resulting SSM because of the splitting between these three scales is therefore small. Dimension-six operators give the dominant contribution to proton decay in Flipped $SU(5) \times U(1)$. Experimental limits on τ_P are just now reaching the sensitivity needed to begin to constrain the model.[27] Because of the flipped embeddings, the relation $\lambda_b(M_U) = \lambda_\tau(M_U)$ need not hold in the flipped model, though this relation may be preserved in string-derived models due to an underlying $SO(10)$ symmetry.

The most promising ultimate explanation of the SSM is the string. In addition to $SU(5) \times U(1)$ models, $SU(3) \times SU(2) \times U(1)$-like models naturally emerge in many string constructions. Moreover, gauge coupling unification is a prediction of string models and the gauge unification scale is calculable in any particular model.[28] A string unification scale consistent with the SSM and low-energy measurements is possible in orbifold models with somewhat awkward choices of the moduli.[29] However, for generic values of the moduli, and for large classes of models constructed in the free fermionic formulation of the string, the string unification scale is too high, about 10^{18} GeV.[30] A simple remedy to this problem is to add extra vector-like representations of chiral superfields with masses chosen to give the desired string unification scale while preserving the successful SSM prediction of $\sin^2\theta$. Using only representations with Standard Model quantum numbers, the unique minimal choice is an extra $Q, \bar{Q}$ pair with mass $\mathcal{O}(10^{13}\,\text{GeV})$ and an extra $D^C, \bar{D}^C$ pair with mass $\mathcal{O}(10^5\,\text{GeV})$. This minimal model is called the String Inspired Standard Model (SISM)[31] and can be easily recast as a Flipped $SU(5) \times U(1)$ SISM[32] by adding an extra $\mathbf{10}, \overline{\mathbf{10}}$ pair of chiral superfields which contain $Q, \bar{Q}$ and $D^C, \bar{D}^C$ representations. Many non-minimal choices exist with larger numbers of representations, though allowing only Standard Model quantum numbers, at least one $Q, \bar{Q}$ pair is always needed. This requirement can be relaxed by allowing representations with non-standard hypercharge.[33]

Ultimately, string theory should predict the fermion masses and mixings. Experimental bounds have been shown to be consistent with extremely simple patterns for the Yukawa matrices at the unification scale.[4] These textures provide a reachable target for string models. One approach attempts to fix the orbifold moduli to give the observed fermion masses and mixings.[34] String derivations of the soft supersymmetry parameters validate the assumption of predominantly universal soft supersymmetry and add small non-universal corrections.[35] In the top-down approach, each particular string model potentially fixes the high-energy boundary conditions. In the bottom-up approach, experiment can be used to constrain the high-energy parameters and give an indication of the classes of string models required. Considering the huge number of plausible string models and the computational effort required to derive predictions, an artful combination of these complementary approaches seems to be necessary.

6. Sparticle Spectroscopy

A remarkable consequence of the self-interacting dynamics inherent in the framework of quantum field theories is that every physical process depends on all the parameters of the theory. In principle, this means that for a theory with n parameters, any n physical measurements completely determine the theory, if these measurements can be made with arbitrary precision. However, in practice, experimental measurements have a finite precision, and different parameters in the theory enter each process at different orders in the perturbative loop expansion. This makes the strategy of which processes are used to determine which parameters important.

The most direct strategy is to actually produce the particle of interest. Definitive proof for supersymmetry would be the production of sparticles. If the masses of several sparticles could be measured, the resulting sparticle spectroscopy could be used to determine the parameters of the SSM and search for departures from the SSM. The sparticles corresponding to the two light generations have a very simple dependence on only three of the SSM parameters.

$$m_{\tilde{p}}^2 = m_0^2 + c_{\tilde{p}}(m_{\tilde{p}})m_{1/2}^2 + 2\left[T_3^{\tilde{p}} - \frac{3}{5}Y^{\tilde{p}}\tan^2\theta\right]m_W^2 \tag{6.1}$$

Measurements of three sparticle masses can be converted into a determination of $m_{1/2}$ and m_0 with fractional uncertainties comparable to that of the sparticle mass measurements.[36]

Sufficiently accurate determination of more sparticle masses could be used to discriminate between different extensions of the SSM such as extensions of the gauge group, additional Yukawas, generational-dependent extra heavy gauge bosons, and non-universal sypersymmetry breaking, which all leave distinct imprints on the sparticle spectrum. For example, for large enough sparticle masses and $m_{\tilde{g}} = c_{\tilde{g}}m_{1/2}$ (so that the D-terms in Eq. (6.1) can be neglected) the quantity

$$\Delta_{ij} = \frac{m_i^2 - m_j^2}{m_{\tilde{g}}^2} = \frac{c_i - c_j}{c_{\tilde{g}}^2} \tag{6.2}$$

gives, e.g. $\Delta_{\tilde{e}_L\tilde{e}_R} = 0.062(0.048)$ and $\Delta_{\tilde{u}_L\tilde{d}_R} = 0.088(0.061)$ in the SISM (SSM).[31] These measurements would discriminate between the SISM and the SSM independent of the soft supersymmetry-breaking parameters used.

7. Flavor Changing Neutral Currents

It is also possible to deduce information about particles inaccessible directly at presently available energies through their virtual influence in loop corrections to precision experiments. This is the strategy which has been used to bound the top and Higgs masses by requiring consistency of the different radiatively corrected electroweak processes measured at LEP.[37] Flavor changing neutral currents offer a particularly sensitive probe of virtual loop effects. Because flavor changing neutral currents vanish at tree level, the influence of the new superpartners can potentially be comparable to that of the Standard Model fields.

In calculating supersymmetric contributions to flavor changing neutral currents by summing over mass eigenstates, the final result is buried within cancellations between the contributions of six different squark mass eigenstates involving 6×6 matrices of quark-squark mixing angles. An alternative, mass-insertion result[38] may be derived from the mass-eigenstate result. In the mass-eigenstate approach, the flavor changing resides in flavor non-diagonal gauge couplings while in the mass-insertion approach, the gauge couplings are flavor diagonal and the flavor changing resides in flavor off-diagonal elements of the sparticle mass matrices. The difference between the two approaches is the choice of flavor basis and the result is independent of this choice.

For example, to lowest order in mass insertions, the gluino box contributions to $\Delta S = 2$ processes gives rise to an effective interaction Lagrangian[39]

$$\begin{aligned}
\mathcal{L}^{eff}_{\Delta S=2} = \frac{\alpha_s^2}{216 M_{\tilde{q}}^2} \Bigg\{ & \left(\frac{\delta\tilde{m}^2_{d_L s_L}}{M_{\tilde{q}}^2}\right)^2 [66\tilde{f}(x) + 24x f(x)](\bar{d}_i \gamma_\mu P_L s_i)(\bar{d}_j \gamma^\mu P_L s_j) \\
& + \left(\frac{\delta\tilde{m}^2_{d_R s_R}}{M_{\tilde{q}}^2}\right)^2 [66\tilde{f}(x) + 24x f(x)](\bar{d}_i \gamma_\mu P_R s_i)(\bar{d}_j \gamma^\mu P_R s_j) \\
& + \left(\frac{\delta\tilde{m}^2_{d_L s_L}}{M_{\tilde{q}}^2}\right)\left(\frac{\delta\tilde{m}^2_{d_R s_R}}{M_{\tilde{q}}^2}\right)\Big([-72\tilde{f}(x) + 504x f(x)](\bar{d}_i P_L s_i)(\bar{d}_j P_R s_j) \\
& \qquad + [120\tilde{f}(x) + 24x f(x)](\bar{d}_i P_L s_j)(\bar{d}_j P_R s_i)\Big) \\
& + \left(\frac{\delta\tilde{m}^2_{d_L s_R}}{M_{\tilde{q}}^2}\right)^2 x f(x)\big[324(\bar{d}_i P_R s_i)(\bar{d}_j P_R s_j) - 108(\bar{d}_i P_R s_j)(\bar{d}_j P_R s_i)\big] \\
& + \left(\frac{\delta\tilde{m}^2_{d_R s_L}}{M_{\tilde{q}}^2}\right)^2 x f(x)\big[324(\bar{d}_i P_L s_i)(\bar{d}_j P_L s_j) - 108(\bar{d}_i P_L s_j)(\bar{d}_j P_L s_i)\big] \\
& + \left(\frac{\delta\tilde{m}^2_{d_L s_R}}{M_{\tilde{q}}^2}\right)\left(\frac{\delta\tilde{m}^2_{d_R s_L}}{M_{\tilde{q}}^2}\right)\tilde{f}(x)\big[108(\bar{d}_i P_L s_i)(\bar{d}_j P_R s_j) \\
& \qquad - 324(\bar{d}_i P_L s_j)(\bar{d}_j P_R s_i)\big]\Bigg\} \qquad (7.1)
\end{aligned}$$

with

$$f(x) = \frac{1}{6(1-x)^5}(-6\ln x - 18x\ln x - x^3 + 9x^2 + 9x - 17) \tag{7.2a}$$

$$\tilde{f}(x) = \frac{1}{3(1-x)^5}(-6x^2\ln x - 6x\ln x + 1x^3 + 9x^2 - 9x - 1) \tag{7.2b}$$

where $M_{\tilde{q}}$ is the universal (or average) down-squark mass and $x = M_{\tilde{g}}^2/M_{\tilde{q}}^2$. The $\Delta B = 2$ effective interaction can be read directly from Eq. (7.1) by the substitution $s \to b$. Requiring that each chiral contribution not exceed the experimental value of ΔM_K, ΔM_B and ϵ_K gives upper bounds on the various mass insertions summarized in Table 1:

Phenomenological Upper Bounds on $(\delta\tilde{m}^2_{\bar{d}_A s_B})$ and $(\delta\tilde{m}^2_{\bar{d}_A b_B})$			
	$\left(\frac{\delta\tilde{m}^2_{\bar{d}_A s_B}}{M^2_{\tilde{q}}}\right)\left(\frac{\delta\tilde{m}^2_{\bar{d}_{A'} s_{B'}}}{M^2_{\tilde{q}}}\right)$		$\left(\frac{\delta\tilde{m}^2_{\bar{d}_A b_B}}{M^2_{\tilde{q}}}\right)\left(\frac{\delta\tilde{m}^2_{\bar{d}_{A'} b_{B'}}}{M^2_{\tilde{q}}}\right)$
$(\alpha\beta)(\alpha'\beta')$	ΔM_K	ϵ_K	ΔM_B
$(LL)^2$ or $(RR)^2$	$(0.10)^2$	$(0.0080)^2$	$(0.27)^2$
$(LL)(RR)$	$(0.0060)^2$	$(0.00049)^2$	$(0.073)^2$
$(LR)^2$ or $(RL)^2$	$(0.0082)^2$	$(0.00066)^2$	$(0.082)^2$
$(LR)(RL)$	$(0.044)^2$	$(0.0035)^2$	$(0.14)^2$

TABLE 1. All numbers in the table must be multiplied by the factor $(M_{\tilde{q}}/\mathrm{TeV})^2$. Numerical values assume $\sqrt{x} \sim M_{\tilde{g}}/M_{\tilde{q}} = 1$. Stricter (weaker) results generally apply to $x < 1$ ($x > 1$). Bounds derived from ϵ_K assume maximal CP violation. Bounds from ΔM_B must be scaled by $(160\ \mathrm{MeV}/f_B)$.

All the constraints are useful (i.e. $\delta\tilde{m}^2_{AB}/M^2_{\tilde{q}} < 1$), even for heavy squark masses $M_{\tilde{q}} \gg 1$ TeV. This severely restricts the flavor structure of a general supersymmetric theory.

In the SSM, non-diagonal squark mass matrices are generated by renormalization effects between the unification scale and M_Z. The resulting low-energy mass insertions are predominantly left-left, and exhibit the following approximate flavor dependence:[39]

$$\frac{[\delta\tilde{m}^2_{\bar{d}_L d_L}]_{ij}}{M^2_{\tilde{q}}} = c_{LL}(\xi_0, \xi_A)[K^\dagger \lambda_u^2 K]_{ij} \tag{7.3}$$

where λ_u is the diagonalized charge 2/3 quark Yukawa matrix and K is the KM matrix. The function $c_{LL}(\xi_0, \xi_A)$ may be calculated by running the one-loop matrix RGE's and one finds that $|c_{LL}(\xi_0, \xi_A)| < 1$ for $|\xi_0|$, $|\xi_A| < 5$. Combining this with bounds on the various KM matrix elements reveals that the SSM safely satisfies all the bounds in Table 1.

The recent CLEO observation of $B \to K^*\gamma$[40] suggests that $b \to s\gamma$ may severely constrain the SSM. In the Standard Model with two Higgs doublets, calculations of $b \to s\gamma$ bound the mass of the charged Higgs.[41] However, in the limit of exact supersymmetry, the chargino contributions cancel the two Higgs SM contributions and $BR(b \to s\gamma) = 0$.[42] Despite the complexities of a full calculation in the SSM including two-loop QCD corrections, the resulting operator mixing, and all the supersymmetric contributions, preliminary calculations reveal that $b \to s\gamma$ is a very promising constraint on the parameters of the SSM.[43]

Most nontrivial extensions of the SSM contain extra Yukawa couplings, and these generically lead to large FCNC's.[44] In supersymmetric Flipped $SU(5)\times U(1)$,[24] for example, there can be potentially large FCNC's generated above the GUT scale:[39,45]

$$[\delta\tilde{m}^2_{\tilde{d}_L d_L}]_{ij} = -\frac{1}{8\pi^2}[\lambda_6^* \lambda_6^T]_{ij} \ln\left(\frac{M_{Pl}}{M_{GUT}}\right)[3m_0^2 + A^2] \tag{7.4a}$$

$$[\delta\tilde{m}^2_{\tilde{d}_R d_R}]_{ij} = [\delta\tilde{m}^2_{\tilde{d}_L d_L}]^*_{ij} \tag{7.4b}$$

$$[\delta\tilde{m}^2_{\tilde{d}_R d_L}]_{ij} = -\eta^1_{ij} v = \frac{vA}{8\pi^2}[\lambda_6 \lambda_6^\dagger \lambda_d + \lambda_d \lambda_6^* \lambda_6^T]_{ij} \ln\left(\frac{M_{Pl}}{M_{GUT}}\right) \tag{7.4c}$$

$$[\delta\tilde{m}^2_{\tilde{d}_L d_R}]_{ij} = [\delta\tilde{m}^2_{\tilde{d}_R d_L}]^*_{ij} \tag{7.4d}$$

where λ_6 is an *a-priori* unknown Yukawa coupling associated with a see-saw neutrino mass mechanism.

¿From the phenomenological constraints in Table I, useful bounds on the unknown GUT Yukawa λ_6 result. The strongest bounds come from the $(\delta\tilde{m}^2_{LL})(\delta\tilde{m}^2_{RR})$ contributions to ΔM_K, ΔM_B:

$$\left|[\lambda_6^* \lambda_6^T]_{12}\right| \ln\left(\frac{M_{Pl}}{M_{GUT}}\right) < 0.47\frac{\xi_0^2 + 6}{3\xi_0^2 + \xi_A^2}\left(\frac{M_{\tilde{q}}}{1\,\mathrm{TeV}}\right) \tag{7.5a}$$

$$\left|[\lambda_6^* \lambda_6^T]_{13}\right| \ln\left(\frac{M_{Pl}}{M_{GUT}}\right) < 1.28\frac{\xi_0^2 + 6}{3\xi_0^2 + \xi_A^2}\left(\frac{M_{\tilde{q}}}{1\,\mathrm{TeV}}\right) \tag{7.5b}$$

using the approximate low-energy relation $M^2_{\tilde{q}} \simeq (\xi_0^2 + 6)m^2_{1/2}$. Equation (7.5) provides important knowledge about the unknown matrix λ_6 and the pattern of soft supersymmetry breaking. For instance, if supersymmetry breaking takes the form of either m_0 or A, then Eq. (7.5) requires certain elements of λ_6 to be quite small—smaller than expected from superstring theories, which relate Yukawa couplings to gauge couplings $\lambda \approx g \approx 0.7$. One may be forced to conclude that these couplings vanish at the tree level in such theories—or that soft supersymmetry breaking takes the form of $m_{1/2}$. Note that because of Eq. (7.4b,d), the Flipped contribution to CP violation ϵ_K vanishes. Flipped $SU(5) \times U(1)$ can, however, contribute significantly to "direct" CP violation ϵ', $K_L \to \pi^0 ee$ and $K_L \to \pi^0 \nu\bar{\nu}$, as shown in ref. 39.

Perhaps the most interesting process in Flipped $SU(5) \times U(1)$ is $\mu \to e\gamma$. A non-trivial constraint on the elements of the arbitrary unitary matrix U, which embeds the lepton doublets along with the up-quark triplets in the GUT representations, may be derived from the limits on $BR(\mu \to e\gamma)$:[39]

$$|U_{13}^* U_{23}| \ln\left(\frac{M_{Pl}}{M_{GUT}}\right) \lesssim \frac{3.6 \times 10^{-3}}{\lambda_t^2} \left(\frac{0.276}{E(x_{\tilde{B}})}\right) \left(\frac{M_{\tilde{l}}}{100\,\mathrm{GeV}}\right)^2 \sqrt{\frac{BR(\mu \to e\gamma)}{4.9 \times 10^{-11}}} \qquad (7.6)$$

where $x_{\tilde{B}} = m_{\tilde{B}}^2 / M_{\tilde{l}}^2$ and $E(1) = .276$. The constraints from $BR(\tau \to \mu\gamma)$ and $BR(\tau \to e\gamma)$ are not interesting at present experimental limits. A distinctive signature of Flipped $SU(5) \times U(1)$ is that the final-state electrons and photons are always left-handed. In conventional supersymmetric $SU(5)$ the final-state electrons and photons are always right-handed, but since the unknown matrix U appearing in Eq. (7.6) gets replaced by the KM matrix, the branching ratio for $\mu \to e\gamma$ is suppressed well below presently accessible levels.

8. Conclusions

The SSM constitutes a simple, well-defined basis for testing unified supersymmetric models. The experimental predictions of the SSM are unfolded over a five-dimensional parameter space $(m_t, \tan\beta, m_{1/2}, m_0, A)$. Large regions of this parameter space are consistent with all known experimental results. Two complementary methods of solving the SSM are the top-down and bottom-up methods. Thresholds effects may be included in evolving the RGE's or in one-loop corrections to the vacuum energy. More realistic models like supersymmetric $SU(5)$, Flipped $SU(5) \times U(1)$, the SISM, and string-derived models are well approximated by the SSM. In addition, more realistic models can give rise to processes like proton decay or lepton flavor violation which are forbidden in the SSM, as well as constraints on the parameter space of the SSM like $\lambda_b(M_U) = \lambda_\tau(M_U)$. String-derived model are beginning to give explicit predictions for the Yukawa couplings and soft supersymmetry-breaking parameters. Sparticle spectroscopy and flavor changing neutral currents are examples of tests which might determine the parameters of the SSM and provide a means to discriminate between more realistic models.

Despite the simplicity of the SSM, it is still an active area of research. Only in the last few years has its constrained parameter space been identified and searched. Combining the merits of the two methods of treating thresholds, complete two-loop calculations, and estimation of the accuracy of one-loop and two-loop treatments all seem possible and worthwhile projects.

The interplay between the SSM, string-inspired, and string-derived models is very lively. Perhaps the most exciting prospect arises from the glimpse that string theory gives to physics beyond the Planck scale, physics beyond space-time. It seems that the ultimate explanation and source of the laws of nature transcends space and time. However, the scientific method has developed and operates within space and time. Perhaps in the search for ultimate truth, the traditional method of testing theories by proving them false must be supplemented by more direct methods of cognizing the structure of natural law.

9. References

1. J. Ellis, S. Kelley and D. V. Nanopoulos, *Phys. Rev. Lett.* **249** (1990) 441;
   P. Langacker and M. Luo, *Phys. Rev.* **D44** (1991) 817;
   U. Amaldi, W. de Boer and H. Fürstenau, *Phys. Lett.* **260B** (1991) 447.
2. L. J. Hall and U. Sarid, LBL preprint LBL-32905 (1992).
3. J. P. Derendinger and C.A. Savoy, *Nucl. Phys.* **B237** (1984) 307;
   J. Ellis, J. S. Hagelin, S. Kelley and D. V. Nanopoulos, *Nucl. Phys.* **B311** 1988 1.
4. H. Georgi and C. Jarlskog, *Phys. Lett.* **86B** (1979) 297;
   S. Dimopoulos, L. J. Hall and S. Raby, *Phys. Rev. Lett.* **68** (1992) 1984.
5. S. Bertolini, F. Borzumati, A. Masiero, and G. Ridolfi, *Nucl. Phys.* **B353** (1991) 591.
6. S. Kelley, J. L. Lopez, D. V. Nanopoulos, H. Pois and K. Yuan, *Nucl. Phys.* **B398** (1993) 3.
7. J. Ellis, S. Kelley and D. V. Nanopoulos, *Phys. Lett.* **260B** (1991) 131;
   F. Anselmo, L. Cifarelli, A. Peterman and A. Zichichi, *Nuovo Cimento* **105A** (1992) 581;
   P. Langacker and N. Polonsky, *Phys. Rev.* **D47** (1993) 4028.
8. S. Kelley, J. L. Lopez and D. V. Nanopoulos, *Phys. Lett.* **274B** (1992) 387.
9. R. Barbieri and G. Giudice, *Nucl. Phys.* **B306** (1988) 63.
10. J. Ellis, D. V. Nanopoulos, K. A. Olive, and M. Srednicki, *Nucl. Phys.* **B238** (1984) 453;
    S. Kelley, J. L. Lopez, D. V. Nanoupolos, K. Yuan, and H. Pois, *Phys. Rev.* **D47** (1993) 2461.
11. P. H. Chankowski, *Phys. Rev.* **D41** (1990) 2877.
12. H. E. Haber and R. Hempfling, SCIPP 91/33 (1992).
13. S. Coleman and E. Weinberg, *Phys. Rev.* **D7** (1973) 1888.
14. G. Gamberini, G. Ridolfi, and F. Zwirner, *Nucl. Phys.* **B331** (1990) 331.
15. B. Kastening, *Phys. Lett.* **283B** (1992) 287.
16. J. Ellis, G. Ridolfi, and F. Zwirner, *Phys. Lett.* **257B** (1991) 83.

17. S. Dimopoulos and H. Georgi, *Nucl. Phys.* **B193** (1981) 150;
N. Sakai, *Z. Phys.* **C11** (1982) 153;
A. Chamseddine, R. Arnowitt and P. Nath, *Phys. Rev. Lett.* **105** (1982) 970.

18. B. Campbell, J. Ellis and D. V. Nanopoulos, *Phys. Lett.* **141B** (1984) 229;
R. Arnowitt and P. Nath, *Phys. Rev.* **D38** (1988) 1479.

19. M. S. Chanowitz, J. Ellis, and M. K. Gaillard, *Nucl. Phys.* **B128** (1977) 506;
A. J. Buras, J. Ellis, M. K. Gaillard, and D. V. Nanopoulos, *Nucl. Phys.* **B315** (1978) 66;
D. V. Nanopoulos, and D. A. Ross, *Nucl. Phys.* **B157** (1979) 273; *Phys. Lett.* **108B** (1982) 351; *Phys. Lett.* **118B** (1982) 99.

20. J. Ellis, S. Kelley, and D. V. Nanopoulos, *Nucl. Phys.* **B373** (1992) 55.

21. P. Langacker and N. Polonsky, University of Pennsylvania preprint UPR-0556T (1993).

22. A. Masiero, D. V. Nanopoulos, K. Tamvakis, and T. Yanagida, *Phys. Lett.* **115B** (1982) 380;
B. Grinstein, *Nucl. Phys.* **B206** (1982) 387;
K. Inoue, A. Kakuto and T. Tankano, *Prog. Theor. Phys.* **75** (1986) 664;
R. Barbieri, G. Dvali and A. Strumia, *Nucl. Phys.* **B391** (1993) 487.

23. H. Dreiner, J. L. Lopez, D. V. Nanopoulos, and D. Reiss, *Phys. Lett.* **216B** (1989) 289.

24. I. Antoniadis, J. Ellis, J. S. Hagelin, and D. V. Nanopoulos, *Phys. Lett.* **194B** (1987) 231.

25. I. Antoniadis, J. Ellis, J. S. Hagelin, and D. V. Nanopoulos, *Phys. Lett.* **231B** (1989) 65.

26. J. Ellis, J. S. Hagelin, S. Kelley, and D. V. Nanopoulos, *Nucl. Phys.* **B311** (1988) 1.

27. J. S. Hagelin and S. Kelley, MIU preprint MIU-THP-61/92 (1993).

28. J. Minahan, *Nucl. Phys.* **B298** (1988) 36;
V. S. Kaplunovsky, *Nucl. Phys.* **B307** (1988) 145.

29. L. E. Ibanez, D. Lust, and G. G. Ross, *Phys. Lett.* **272B** (1991) 251.

30. I. Antoniadis, J. Ellis, S. Kelley, and D. V. Nanopoulos, *Phys. Lett.* **272B** (1991) 31.

31. S. Kelley, J. L. Lopez, D. V. Nanopoulos, *Phys. Lett.* **278B** (1992) 140.

32. J. L. Lopez, D. V. Nanopoulos, and A. Zichichi, CERN-TH.6903/93 (1993).

33. A. E. Faraggi, *Phys. Lett.* **302B** (1993) 202.

34. B. Carlos, F. Gomez, and C. Muñoz, *Phys. Lett.* **292B** (1992) 42.

35. B. Carlos, J. A. Casas, and C. Muñoz, CERN-TH.6681/92 (1992);
V. S. Kaplunovsky and J. Louis, *Phys. Lett.* **306B** (1993) 269.

36. A. E. Faraggi, J. S. Hagelin, S. Kelley, and D. V. Nanopoulos, *Phys. Rev.* **D45** (1991) 3272.

37. J. Ellis and F. L. Fogli, *Phys. Lett.* **232B** (1989) 139;
P. Langacker, *Phys. Rev. Lett.* **63** (1989) 1920.

38. M. Dugan, B. Grinstein, and L. J. Hall, *Nucl. Phys.* **B255** (1985) 413.

39. J. S. Hagelin, S. Kelley, and T. Tanaka, MIU preprint MIU-THP-59/92.

40. E. Thorndike, CLEO Collab., talk given at the 1993 Meeting of the American Physical Society, Washington D.C., April 1993.

41. V. Barger, M. Berger, and R. J. Phillips, *Phys. Rev. Lett.* **70** (1993) 1368;
J. Hewett, *Phys. Rev. Lett.* **70** (1993) 1045.

42. R. Barbieri and G. Giudice, CERN-TH.6830/93.

43. J. L. Lopez, D. V. Nanopoulos, and G. T. Park, Texas A & M University preprint CTP-TAMU-16/93 (1993).

44. L. J. Hall, V .A. Kostelecky, and S. Raby, *Nucl. Phys.* **B267** (1986) 415.

45. F. Gabbiani and A. Masiero, *Phys. Lett.* **209B** (1988) 289;
A. Faraggi, J. L. Lopez, D. V. Nanopoulos, and K. Yuan, *Phys. Lett.* **221B** (1989) 337.

THE STRONG CP PROBLEM, STRING THEORY AND THE NELSON-BARR MECHANISM

ROBERT G. LEIGH†

Santa Cruz Institute for Particle Physics
University of California, Santa Cruz, CA 95064

ABSTRACT

We review recent work on the strong CP problem in the context of realistic string-inspired models. We discuss the various solutions, review the conjecture that CP is generally a gauged discrete symmetry in string theory and then consider models of the Nelson-Barr type. We note that squark non-degeneracy spoils the Nelson-Barr structure at the one loop level. We stress that string theory expectations, as well as naturalness arguments, make it very difficult to avoid the constraints on non-degeneracy.

The strong CP problem is well known. In a CP non-invariant theory there is an additional operator, proportional to a parameter θ, which leads to a large contribution to the neutron electric dipole moment. By comparison to the experimental data, it is found that θ must be less than $\sim 10^{-9}$. There are two contributions to θ: a bare value, present in theories with explicit CP violation, and a contribution proportional to the overall phase of the mass matrix of the coloured fields, $\theta_{QFD} \sim arg\ Det\ m$. That there is no reason for these two contributions to combine to leave such a small remainder constitutes the strong CP problem.

There are three well known solutions to the strong CP problem. The first scenario simply has no observable θ-parameter. This arises, for example, when $m_u = 0$. This solution however is manifestly inconsistent with lowest order chiral perturbation theory,[1] which gives $m_u/m_d \simeq 0.5$. Recently however it has been suggested that an ambiguity at next-to-leading order may allow a zero up-quark mass. This subject is at best controversial, and we will refrain from commenting further on it here, except to note that such a solution to the strong CP problem would be relatively easy to implement; the required symmetry structure is very simple.

The second solution to the strong CP problem has been discussed extensively. Given a spontaneously broken anomalous symmetry (the Peccei-Quinn symmetry), θ may be dynamically relaxed to zero through the vacuum expectation value of the

†Address after September 1, 1993: Department of Physics and Astronomy, Rutgers University, Piscataway, NJ 08855-0849.

axion.[2, 3] The axion solution is tightly constrained by astrophysical and cosmological considerations.[4] In the context of string theory, there are many potential sources of axions.[5] It is well known that the gravity supermultiplet gives rise to a model-independent axion, which couples universally to all of the gauge groups of a given four-dimensional model:

$$\begin{aligned} *da = H &= dB - \omega_Y + \omega_L \\ \Delta a = *dH &= *(R \wedge R - F \wedge F). \end{aligned} \tag{1}$$

The problem with this axion is that its decay constant is of order the string scale, much larger than that required by the usual cosmological arguments.

Other axions may arise from other sources, depending upon the particular compactification studied, such as from internal components of the antisymmetric tensor field. However, the associated PQ symmetries are generally violated[6] by worldsheet non-perturbative effects and thus the associated axions are inappropriate for the strong CP problem.

In the presence of an 'anomalous' $U(1)$ in the spectrum of a string model, there is an interesting effect on the model independent axion. The anomaly is cancelled by the transformation of a Green-Schwarz term:

$$\delta\Gamma_{GS} \propto \int B \wedge F \rightarrow \int A \wedge *da. \tag{2}$$

Thus the associated gauge boson, A, gains mass by eating the model independent axion. The local $U(1)$ symmetry is therefore broken and the model-independent axion is no longer present in the spectrum. However, there is a remaining *global* $U(1)$ symmetry, which is spontaneously broken. There is always an associated Fayet-Iliopoulos D-term[7] present; in general there exists a supersymmetric vacuum in which the D-term is cancelled by vev's of some charged scalar fields. This breaking of the global symmetry leads to a new axion with decay constant of order the scalar vev; perhaps, it may thus be possible to attain a decay constant at much smaller scales.

Even if an appropriate axion can be found, there are additional problems. First, if other gauge groups become strong at a high scale, the axion potential will be dominated by instanton effects of this gauge group and the QCD θ angle will not generically be cancelled. This will happen for example, in models in which gaugino condensation occurs. Essentially, one axion is necessary for each strong gauge group.

In addition to this problem, it is widely believed that quantum gravitational effects, such as from black holes or wormholes, lead to violations[8] of global symmetries (such as PQ symmetries) through higher dimensional operators whose scale is set by the Planck mass. Such operators give direct contributions to the axion potential and lead to a minimum away from $\theta = 0$ unless they can be suppressed through some very high dimension. It is known[9] that this may occur accidentally in some models: an array of gauge and discrete symmetries can be constructed such that an approximate PQ symmetry exists which ensures a sufficiently small θ. This mechanism is perhaps the most likely axion scenario within the context of string theory, though we note the rather delicate symmetry structure involved.

Given the problems outlined above, it is natural to consider the third solution to the strong CP problem, the Nelson-Barr mechanism.[10] In this scenario, one supposes that CP is a good symmetry of the underlying theory, but is spontaneously violated. In this case, the bare value of θ is zero and one arranges things so that the tree level mass matrix is complex but possesses a real determinant. The game is then to race against perturbative corrections to keep θ small enough. The scale of CP violation should be low enough that higher dimensional operators not spoil the structure of the theory; we will take this to be roughly of order 10^{11} GeV, a scale which can be naturally generated in string models (given $m_{susy} \sim$ TeV).

It is this scenario that we will explore in the context of string-inspired models in this paper. To begin however, we will make a few remarks concerning CP as a spontaneously violated symmetry.

1. CP as a Spontaneously Violated Symmetry

It is not at all obvious *a priori* that the Nelson-Barr mechanism can be implemented in string theory. We must first argue that CP *can be* a spontaneously violated symmetry. Until recently it was quite possible that CP was simply not a good symmetry (it could for example have been violated by non-perturbative effects). However, it turns out that CP is an example of a *gauged* discrete symmetry. Because of this it is protected from all manner of violation, whether non-perturbative stringy phenomena or otherwise (such as quantum gravitational effects). The fact that it is seen to be violated in our low energy world must then be a consequence of spontaneous breakdown, perhaps through complex vev's of heavy scalar fields.

In fact it is possible to demonstrate[11] the gauge symmetry that gives rise to the four-dimensional CP symmetry in a wide range of simple string compactifications. In ten-dimensions, the heterotic string theory possesses no parity symmetry as the theory is chiral. There is however a charge conjugation symmetry which can easily be seen to be equivalent to an $SO(32)$ (or $E_8 \times E_8$) gauge transformation. In the simplest toroidal compactifications, the four dimensional theory is non-chiral; a parity symmetry can be constructed which has just the right transformation properties on the states. It is a proper (ten-dimensional) Lorentz transformation. Four-dimensional charge conjugation is a combination of the aforementioned gauge rotation and a proper Lorentz transformation.

Thus in these simplest of theories, both C and P can be thought of as gauge symmetries: they arise as combinations of ordinary gauge and general cooordinate transformations. For suitable values of moduli, many other compactifications possess a gauged CP symmetry. It is reasonable to conjecture, as a general property of string theory, that CP is indeed a gauge symmetry.

If this is true, CP can be violated by complex expectation values of fields. In the remainder of this paper we will suppose that CP is spontaneously violated by complex expectation values of observable sector matter fields at a relatively low scale ($\sim 10^{11}$ GeV). In many models, CP might be violated by complex vev's of hidden sector fields or moduli. There is little hope that this could give rise to a successful implementation of the Nelson-Barr mechanism, and we will simply assume that this is not realized in the models of interest. We will see that Nelson-Barr models quite generically have severe problems; the origin of these problems is in the soft supersymmetry breaking (supergravity) sector.

2. One Loop contributions to θ

In supersymmetric models of spontaneous CP violation where one has arranged the vanishing of θ at tree-level, there are a variety of possible contributions to θ at one-loop order. These are given by

$$\delta\theta = Im \text{ Tr } \left[m_u^{-1}\delta m_u + m_d^{-1}\delta m_d\right] - 3\text{Arg } \tilde{m}_3 \, . \tag{3}$$

$\delta m_{u,d}$ represent the one-loop corrections to the tree level quark mass matrices, $m_{u,d}$, and $\tilde{m}_3$ is the gluino mass including one-loop contributions. The dominant contributions at one-loop will come from gluino-squark or quark-squark exchange, respectively. In order to analyze these quantities, we first need to make a few more stipulations about the underlying model. To obtain vanishing θ at tree level[12, 13] the tree level gluino mass must be real (a non-zero phase represents a contribution to θ). All terms in the Higgs potential must also be real. To accomplish this we assume, as discussed above, that supersymmetry breaking dynamics do not spontaneously break CP, so that at some large scale the theory is completely CP-invariant and, in particular, all soft supersymmetry breaking terms are real.

The supersymmetric Nelson-Barr model that we will consider is defined as follows: in addition to the usual quark and lepton families, we have an additional pair of isosinglet down quark fields, q and $\bar{q}$, as well as some singlet fields, $\mathcal{N}_i$ and $\bar{\mathcal{N}}_i$. It is straightforward to consider models with several q and $\bar{q}$ fields, and with additional types of singlets. The terms in the superpotential which give rise to the quark mass matrix are*

$$W = \mu q\bar{q} + \gamma^{ij}\mathcal{N}_i q\bar{d}_j + H_1\lambda_{ij}Q_i\bar{d}_j \tag{4}$$

(the terms in the superpotential involving u quarks and leptons will not be important for our considerations). I should mention here that one framework for obtaining such a mass matrix is suggested by E_6 models, such as those which often appear in superstring theories.[14] In these models, generations of quarks and leptons arise from the **27** representation, as could the q and $\bar{q}$ fields (as members of the **10** of $SO(10)$).

Given the couplings above, the fermion mass matrix has the Nelson-Barr structure:

$$m_F = \begin{pmatrix} m_d & M_D\vec{a} \\ 0 & \mu \end{pmatrix}, \tag{5}$$

*$Q\bar{q}$ terms can be forbidden via either gauged $U(1)$ or discrete symmetries.

where $m_d = \lambda_d H_1$ and the vector $\vec{a}$ is defined as:

$$a^i = \frac{1}{M_D}\gamma^{ji}\mathcal{N}_j \quad ; \quad \vec{a}^\dagger\vec{a} = 1. \tag{6}$$

If the $\mathcal{N}$ fields have complex vev's, CP is spontaneously broken, but nevertheless the determinant of m_F is real. In these expressions, μ and M_D are of order the intermediate scale.† We note that in the limit in which we integrate out the heavy modes, the phases in $\vec{a}$ lead to an unsuppressed CKM phase (as long as $\mu \sim M_D$). Since the dominant one-loop graphs involve squark exchange, the form of the scalar mass matrices is of particular interest. Consider first the $\phi\phi^*$ type terms. For the squarks in the **3** representation of $SU(3)$, these take the form, on the full 4×4 set of states:

$$\mathcal{M}^2_{LL} = \begin{pmatrix} \tilde{m}_d^2 + m_d^T m_d & M_D m_d^T \vec{a} \\ \vec{a}^\dagger m_d M_D & \tilde{m}_q^2 + M_D^2 + \mu^2 \end{pmatrix}. \tag{7}$$

Similarly, for the $\bar{\mathbf{3}}$ squarks, we have:

$$\mathcal{M}^2_{RR} = \begin{pmatrix} \tilde{m}_{\bar{d}}^2 + m_d m_d^T + M_D^2 \vec{a}\vec{a}^\dagger & \mu M_D \vec{a} \\ \mu M_D \vec{a}^\dagger & \tilde{m}_{\bar{q}}^2 + \mu^2 \end{pmatrix}. \tag{8}$$

Finally, for the $\phi\phi$-type matrix, which connects the 3 and $\bar{3}$ squarks, we have

$$\mathcal{M}^2_{RL} = \begin{pmatrix} A_d < H_1 > + \mu_H \frac{H_2}{H_1} m_d & M_5^2 \vec{b} \\ 0 & A_\mu \mu \end{pmatrix} \tag{9}$$

where μ_H is the real coefficient of the $H_1 H_2$ term in the superpotential, and M_5 and $\vec{b}$ are defined by

$$M_5^2\, b^i = A_\gamma^{ji}\mathcal{N}_j + \left(\frac{\partial W}{\partial \mathcal{N}_j}\right)^* \gamma^{ji} \quad ; \quad \vec{b}^\dagger\vec{b} = 1 \tag{10}$$

Note that $\vec{b}$ receives contributions from both soft terms as well as F-terms and in general $\vec{b}$ is not proportional to $\vec{a}$. In fact $\vec{b}$ is of the utmost importance to our discussion. From the form of the mass matrices given above, it can be seen that there is mixing of light with heavy states that will lead to effects unsuppressed by inverse powers of the heavy mass scale

$$\sim (m_{susy}\mu)\ \frac{1}{\mu^2}\ (m_{susy}\mu) \tag{11}$$

unless a detailed constraint is satisfied. This constraint is

$$M_5^2\, \vec{b} = A_\mu\, M_D\, \vec{a}. \tag{12}$$

† μ may itself be proportional to the *real* vev of some other scalar field. Plausible mechanisms have been suggested for obtaining such vev's. In particular, in "intermediate scale scenarios," it has been noted that the $\mathcal{N}$ fields can readily obtain vev's of order $m_I = \sqrt{m_{3/2} m_{Pl}}$.

We will see the disastrous effects of this potential non-decoupling in the following.

In a general supergravity setting, the matrices $\tilde{m}_d^2$, $\tilde{m}_{\bar{d}}^2$ and A_d are completely arbitrary. It is well-known that flavor changing neutral current processes (predominantly $K^o - \bar{K}^o$ mixing) constrain these to be approximately degenerate, *i.e.*

$$\tilde{m}_{\bar{d}}^2 = \tilde{m}_{\bar{d}}^2 \times \mathbf{1} + \delta m_{\bar{d}}^2 \; ; \;\; \tilde{m}_Q^2 = \tilde{m}_d^2 \times \mathbf{1} + \delta\tilde{m}_d^2 .$$
$$\tilde{m}_{\bar{q}}^2 = \tilde{m}_{\bar{d}}^2 + \delta\tilde{m}_{\bar{q}}^2 \; ; \;\; \tilde{m}_q^2 = \tilde{m}_d^2 + \delta\tilde{m}_q^2 \qquad (13)$$

with δm^2 small[‡] compared to m_{susy}^2. We will find in the following discussion that the constraints on these quantities from θ are *much* stronger than from FCNC's.

Let us now ask how severe are the constraints arising from the smallness of θ. First consider the gluino mass diagram shown in Fig. 1.

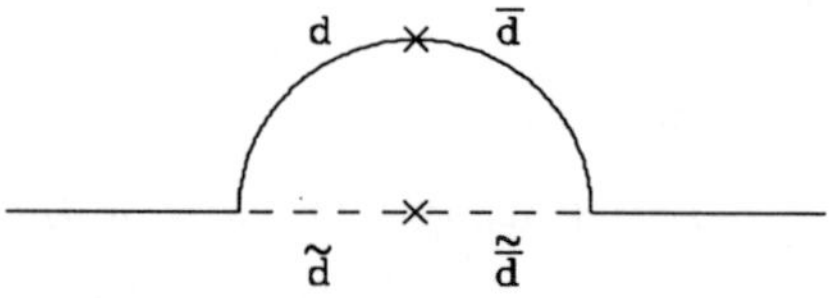

Fig. 1

If the heavy eigenstates are not decoupled (see discussion above) then θ gets a potentially large contribution[13] from:

$$\text{Im } \delta m_\lambda \sim \frac{\alpha_s}{4\pi}\frac{M_D^2}{(M_D^2+\mu^2)}\frac{M_5^2}{M_D}\text{Im } \vec{a}^\dagger\vec{b}. \qquad (14)$$

Indeed, this diagram leads to the requirement that the phases of the vectors $\vec{a}$ and $\vec{b}$ line up to about one part in 10^{-7}. Certainly the simplest way to satisfy this is $\vec{b} = \vec{a}$; we will assume this to be the case in the remainder of this section. Later we will investigate this condition and argue that it is not natural.

The light fermion contributions to the gluino mass lead to a weaker limit on proportionality. If A_d is not proportional to the unit matrix, one will obtain a complex result, in general. This will give a limit suppressed by powers of the b-quark mass over the susy-breaking scale:

$$\frac{\text{Im } \vec{a}^\dagger \delta A_d m_d^T \vec{a}}{m_{susy}^3}\frac{< H_1 > M_D^2}{M_D^2+\mu^2} < 10^{-7}. \qquad (15)$$

More significant limits arise from the graph of Fig. 2. From one proportionality violating insertion and one degeneracy violating insertion we obtain:

$$\frac{\text{Im } \vec{a}^\dagger \delta A_d \lambda_{\bar{d}}^{-1}\left(\delta\tilde{m}_{\bar{d}}^2 - \delta\tilde{m}_{\bar{q}}^2 \times \mathbf{1}\right)\vec{a}}{m_{susy}^3}\frac{M_D^2}{M_D^2+\mu^2} < 10^{-6}, \qquad (16)$$

‡To be more precise, the sd components are bounded at the 10^{-2} or 10^{-3} level by $K^o - \bar{K}^o$.

and

$$\frac{\mathrm{Im}\,\vec{a}^{\dagger}\delta A_d \lambda_d^{-1}\vec{a}}{m_{susy}^3}\,\frac{(M_5^2 - A_\mu M_D)^2\mu^2}{(M_D^2+\mu^2)^2} < 10^{-6}. \tag{17}$$

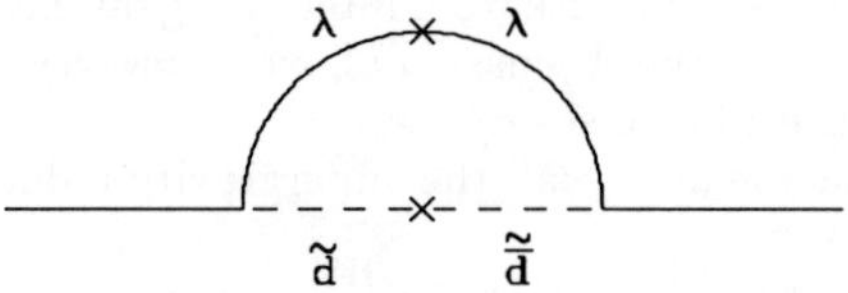

Fig. 2

From Fig. 2, with two degeneracy-violating insertions, we find contributions to θ of order:

$$10^{-1}\frac{\alpha_s}{4\pi}\,\frac{A_d \mathrm{Im}\,\vec{a}^{\dagger} m_d \delta\tilde{m}_Q^2 m_{\bar{d}}^{-1}\left(\delta\tilde{m}_{\bar{d}}^2 - \delta\tilde{m}_{\bar{q}}^2 \times \mathbf{1}\right)\vec{a}}{m_{susy}^5}\,\frac{M_D^2}{M_D^2+\mu^2}, \tag{18}$$

which lead to the constraints

$$\frac{\left(\delta\tilde{m}_{\bar{q}}^2, \delta\tilde{m}_{\bar{d}}^2\right)}{m_{susy}^2}\,\frac{(\delta\tilde{m}_Q^2)}{m_{susy}^2} < 10^{-9}. \tag{19}$$

Non-degeneracies also lead to a contribution of order

$$10^{-1}\frac{\alpha_s}{4\pi}\,\frac{\mathrm{Im}\,\vec{a}^{\dagger} m_d \delta\tilde{m}_d^2 m_{\bar{d}}^{-1}\vec{a}}{m_{susy}^3}\,\frac{(M_5^2 - A_\mu M_D)M_D}{M_D^2+\mu^2}. \tag{20}$$

We note that Eqs. (17) and (20) vanish in the limit in which Eq. (12) is satisfied but Eqs. (16) and (18) represent large contributions independent of this.

3. Discussion

Given these bounds on parameters, we must ask how natural it is to expect that they are satisfied. The strongest constraint comes from $\vec{b} \neq \vec{a}$. One notes from the definitions of these vectors that they are apparently unrelated. There is one possibility: that the minimum of the vacuum energy occurs at such values of the fields so that (see Eq. 10)

$$\frac{\partial W}{\partial \mathcal{N}_i} \propto m_{susy}\mathcal{N}_i^*. \tag{21}$$

From a careful study of this minimization problem, one can show that such relations can only hold within a supergravity theory with minimal Kähler potential: any non-degeneracy at greater than a part in 10^7, such as in the mass matrix of the $\mathcal{N}$ fields, will show up as $\theta > 10^{-9}$. The degeneracy and proportionality constraints require satisfying limits on $\delta\tilde{m}_d^2$, $\delta\tilde{m}_{\bar{d}}^2$, $\delta\tilde{m}_{\bar{q}}^2$ and δA_d, which are considerably more stringent than those obtained from K^o-$\bar{K}^o$ mixing. It is hard to comprehend how they could be satisfied in the absence of a detailed theory of flavor. These constraints require as well a condition on M_5, or perhaps some other condition on parameters. For example, the equality $M_5^2 = M_D A_\mu$ would eliminate light-heavy squark couplings, so that contributions to θ which arise from integrating out heavy fields would vanish; see Eqs. (17) and (20) and the discussion above. However, this condition requires exact degeneracy (at the Planck scale).

To see what this means, recall the supergravity induced potential:

$$V = e^K[(\frac{\partial W}{\partial \phi^i} + d_i W) g^{i\bar{j}} (\frac{\partial W}{\partial \phi^j} + d_j W)^* - 3|W|^2]. \tag{22}$$

with

$$d_i = \frac{\partial K}{\partial \phi^i} \quad ; \quad g_{i\bar{j}} = \frac{\partial^2 K}{\partial \phi^i \partial \phi^{j*}}. \tag{23}$$

If we assume that supersymmetry is broken in a hidden sector, there are two sets of fields: z_i, responsible for supersymmetry breaking, and the "visible sector fields," y_i, with

$$W = g(y) + h(z). \tag{24}$$

Universality[15] is the assumption that there is an approximate $U(n)$ symmetry of the Kähler potential, K, where n is the number of chiral multiplets in the theory. Frequently one takes simply

$$K = \sum \phi_i^* \phi_i. \tag{25}$$

Clearly, this is an extremely strong assumption. The Yukawa couplings of the theory exhibit no such symmetry. It does not hold, for example, for a generic superstring compactification, where the symmetry violations are simply $\mathcal{O}(1)$. It is these violations of universality that lead to non-degeneracy and non-proportionality in the scalar mass matrices. We can characterize the violations of universality quite precisely. For small y, we can expand K in powers of y. Rescaling the fields, we can write

$$K = k(z, z^*) + y_i y_i^* + \ell_{ij}(z, z^*) y_i y_j^* + h_{ij}(z, z^*) y_i y_j + \ldots. \tag{26}$$

There is no reason, in general, why ℓ_{ij} should be proportional to the unit matrix, so the zz^* components of the metric will contain terms involving $y_i y_j^*$ which are non-universal. Plugging into Eq. (22) yields non-universal mass terms for the visible sector fields. In general, there is no symmetry which can forbid these couplings; for example, $\ell = z^* z$ cannot be eliminated by symmetries. Violations of proportionality arise in a similar manner.

We can discuss the likelihood of enjoying the required universality on two fronts. Recently the soft supersymmetry breaking sector that might arise in string theory has been analyzed.[16] It is found that if the dilaton F-term dominates supersymmetry breaking, then the supergravity is minimal, with a Kähler potential given by Eq. (25). However, the analysis is valid at string tree level only; non-universal corrections are expected at order α_{str}/π. Furthermore, in models of gaugino condensation, this does not happen anyway; supersymmetry breaking is instead dominated by F-terms of moduli. Since, unlike the dilaton, moduli are non-universally coupled the Kähler potential is in principle the most general real functional allowed by gauge invariance, as in Eq. (26). In either case, it is certainly true that the high degrees of degeneracy and proportionality necessary for the Nelson-Barr mechanism are very difficult to understand (at least without some detailed theory of flavor).

Apart from string theory, we can give general arguments based on naturalness (in the sense of 't Hooft[17]) for the expected size of non-degeneracy. In Nelson-Barr theories, as in those studied above, there are typically coloured fields that live in different representations of the gauge group. Hence we can really not expect that universality holds (at the Planck scale) to better than the order of gauge couplings. For fields in the same representation, we expect universality to be violated by powers of Yukawa couplings. To avoid this, it would be necessary to have a complicated arrangement of flavor symmetries. As yet, we have been unable to find an example.

Let us now put these naturalness and stringy arguments aside and simply *insist* upon minimal supergravity at the Planck scale; we can then ask whether or not renormalization group effects lead to a large θ. The answer to this question is that quite generically, θ is in fact too large. There is however one case where this can be avoided: if there exist no gauge symmetries that distinguish $\bar{d}$ from $\bar{q}$, then θ can be small enough if we require a rather mild condition on the couplings γ_{ij}.

Alternatively, one can ask whether or not there is some (small) range of parameters for which θ is small enough. One could, for example, try to exploit the fact that $M_D \ll \mu$ would suppress all of the above contributions to θ. Unfortunately, this strategy is limited by the fact that the induced KM phase or the phase entering the SUSY box graph would be of order $\frac{M_D^2}{\mu^2}$ and so this ratio must be $>10^{-2}$ in order to generate large enough ϵ.

Thus we are left with the strong feeling that the Nelson-Barr mechanism is very hard to implement in supersymmetric string-inspired models. If it is to succeed at all, a great deal of knowledge of flavor symmetries would be necessary. Certainly such symmetry would be very complicated, and one wonders if such a solution is really satisfactory at all. The $m_u = 0$ solution or even the automatic axion scenario are likely to be much easier to attain.

I wish to thank Michael Dine and Alex Kagan for a most enjoyable collaboration and the organizers of this workshop for their hospitality.

References

1. H. Georgi and I. McArthur, Harvard University Report No. HUTP-81/A011 (unpublished); D.B. Kaplan and A.V. Manohar, *Phys. Rev. Lett.* **56**, 2004 (1986); K. Choi, C.W. Kim and W.K. Sze, *Phys. Rev. Lett.* **61**, 794 (1988); J. Donoghue and D. Wyler, *Phys. Rev.* **D45** (1992) 892; K. Choi, *Nucl. Phys.* **B383** (1992) 58.
2. R.D. Peccei and H.R. Quinn, *Phys. Rev. Lett.* **38** (1977) 1440; *Phys. Rev.* **D16** (1977) 1791.
3. S. Weinberg, *Phys. Rev. Lett.* **40** (1978) 223; F. Wilczek, *Phys. Rev. Lett.* **40** (1978) 279.
4. J. Preskill, M. Wise and F. Wilczek, *Phys. Lett.* **120B** (1983) 127; L. Abbott and P. Sikivie, *Phys. Lett.* **120B** (1983) 133; M. Dine and W. Fischler, *Phys. Lett.* **120B** (1983) 137.
5. J. Lopez and D. Nanopoulos, *Phys. Lett.* **241B** (1990) 392.
6. M. Dine, N. Seiberg and E. Witten, *Nucl. Phys.* **B278** (1986) 769; *Nucl. Phys.* **B289** (1987) 319.
7. M. Dine, N. Seiberg and E. Witten, *Nucl. Phys.* **B289** (1987) 589.
8. M. Kamionkowski and J. March-Russell, *Phys. Lett.* **282B** (1992) 137; R. Holman *et al.*, *Phys. Lett.* **282B** (1992) 132; S.M. Barr and D. Seckel, *Phys. Rev.* **D46** (1992) 539.
9. G. Lazarides, C. Panagiotakopoulos and Q. Shafi, *Phys. Rev. Lett.* **56** (1986) 432; J. Casas and G. Ross, *Phys. Lett.* **192B** (1987) 119; M. Dine, SCIPP-92-27.
10. A. Nelson, *Phys. Lett.* **136B** (1984) 387; S.M. Barr, *Phys. Rev. Lett.* **53** (1984) 329.
11. M. Dine, R.G. Leigh and D.A. MacIntire, *Phys. Rev. Lett.* **69** (1992) 2030; K. Choi, D.B. Kaplan and A.E. Nelson, *Nucl. Phys.* **B391** (1993) 515.
12. S.M. Barr and A. Masiero, *Phys. Rev.* **D38** (1988) 366; S. Barr and G. Segrè, preprint BA-93-01 (1993).
13. M. Dine, A. Kagan and R.G. Leigh, SCIPP 93/05, SLAC-PUB-6090, to appear in *Phys. Rev.* **D**.
14. P.H. Frampton and T.W. Kephart, *Phys. Rev. Lett.* **65** (1990) 1549.
15. L. Hall, J. Lykken and S. Weinberg, *Phys. Rev.* **D27** (1983) 2359.
16. L. Ibáñez and D. Lüst, *Nucl. Phys.* **B382** (1992) 305; B. de Carlos, J.A. Casas and C. Muñoz, *Phys. Lett.* **299B** (1993) 234; V.S. Kaplunovsky and J. Louis, CERN-TH-6809/93, UTTG-05-93.
17. G. 't Hooft, in *Recent Developments in Gauge Theories*, G. 't Hooft *et. al.*, Eds., Plenum (New York) 1980.

FLAVOR SYMMETRIES AND THE PROBLEM OF SQUARK DEGENERACY IN SUPERGRAVITY THEORIES

ALEXANDER L. KAGAN

Stanford Linear Accelerator Center
Stanford University, Stanford, CA 94309

ABSTRACT

In the supersymmetric standard model there are no symmetries which insure squark degeneracy and, hence, adequate suppression of flavor changing neutral currents. We review this situation in the context of supergravity theories. We then show that additional gauged horizontal symmetries can yield realistic quark mass matrices while at the same time constraining the Kähler potential of supergravity theories so that sufficient squark degeneracy is obtained. This approach suggests likely patterns for squark masses, and indicates that there could be significant supersymmetric contributions to $B^0 - \bar{B}^0$ and $D^0 - \bar{D}^0$ mixing and CP violation in the K and B systems.

Low energy supersymmetric models require an extremely high degree of alignment among squark and quark mass eigenstates. This suggests that squark masses of different families should be highly degenerate. For example, at one loop there are new contributions to the real part of $K^o - \bar{K}^o$ mixing which are too large unless [1, 2, 3]

$$\frac{\delta\tilde{m}_q^2}{m_{susy}^2}\frac{\delta\tilde{m}_{\bar{q}}^2}{m_{susy}^2}\frac{1}{m_{susy}^2} < 10^{-10}\ \mathrm{GeV}^{-2} \tag{1}$$

while for the imaginary part, the limit is about two orders of magnitude stronger. So for squark masses below a TeV a degeneracy of better than one part in 100 is necessary.* There are also limits on degeneracy from other processes: $B^o - \bar{B}^o$ mixing, $b \to s\gamma$, $\mu \to e\gamma$, *etc.* There are additional constraints on the size of certain CP-violating angles coming from $K^o - \bar{K}^o$, and the neutron and electron electric dipole moments.

The question is, can one naturally satisfy all of the above constraints? In some early models of supersymmetry breaking, these conditions were automatically satisfied because the breaking of supersymmetry was fed to squarks by gauge interactions.[5] In hidden sector supergravity theories, however, which provide the

*In an alternative approach, Nir and Seiberg[4] have recently argued that flavor symmetries can be used to obtain sufficient alignment of down squark and quark mass eigenstates without squark degeneracy.

basis for much of our thinking about low energy supersymmetry, the situation is far less clear. It is often said that this degeneracy is perhaps reasonable, since, after all, gravity is "flavor blind." On closer examination, however, this argument is seen to be without substance. In models of this type there are operators which one can add to the theory, corresponding to so-called non-minimal supergravity, not suppressed by any (even approximate) symmetry, which give rise to an $\mathcal{O}(1)$ breaking of the degeneracy[6]. This problem has been discussed in numerous places. In the context of supergravity theories, for example, it is considered in ref. [7]. In ref. [8] this situation was anticipated for string theory. Strategies for naturally raising the supersymmetry breaking scale into the multi-TeV region, *e.g.*, via addition of technicolor interactions, to alleviate this problem were proposed. Explicit departures from universality in simple orbifold models have been computed in ref. [9]. Kaplunovsky and Louis[10] recently have reviewed this problem in the framework of string theory. They note that if supersymmetry breaking is associated principally with the dilaton, one will obtain some degree of degeneracy. However, they have also pointed out serious theoretical difficulties with such a scenario. A recent numerical analysis of the low energy spectrum for this case predicts an uncomfortably light Higgs mass[11]. The more likely scenario, in which supersymmetry breaking is associated with string moduli, will, in general, lead to non-degeneracy of squark masses.

In early work on hidden sector supergravity models, it was suggested that one should simply postulate a large, approximate, flavor symmetry among squarks.[12] Indeed, while various other solutions to this problem might be contemplated, flavor symmetries seem a most natural framework. There is an immediate issue which one must face in this approach. Whatever horizontal symmetry there may be is clearly very badly broken by the ordinary quark mass matrices. So one must show that an adequate degree of squark degeneracy is compatible with a realistic quark mass spectrum. One might further hope that models of this kind would explain the many puzzling features of the fermion mass spectrum. We do not attempt this here. In particular, our model requires a rather large range of quark Yukawa couplings, not unlike the situation in the supersymmetric standard model.

We will begin with a review of the scalar potential in standard model supergravity theories with general Kähler potential, paying special attention to the origin of non-universality of scalar masses. We then discuss results recently obtained [13] for a simple model containing a gauged $SU(2)_H$ horizontal symmetry, in which there is sufficient mass degeneracy to satisfy the limits coming from the real part of $K^o - \bar{K}^o$ mixing. The limit on the imaginary part of K^o-$\bar{K}^o$ mixing, two orders of magnitude smaller, is more problematic. Additional, although plausible, assumptions about the parameters of the model are required. The interplay between the most general Kähler potential consistent with $SU(2)_H$, and the resulting soft terms in the canonical basis for quarks and squarks, is discussed in some detail.

1. Non-Universality of Scalar Masses in Supergravity Theories

In this section we review some aspects of supergravity theories, and explain the origin of the non-degenerate and non-proportional terms in the scalar potential. For definiteness, we consider the case of $N = 1$ supergravity theories, with supersymmetry broken in a hidden sector. Such a theory is described, in general, by three functions, a real function K, which is the Kähler potential, the superpotential W, and a function f which describes the gauge couplings. Here we focus on the form of the scalar potential. Defining a metric on the space of fields by

$$g_{ij} = \frac{\partial^2 K}{\partial \phi_i^* \partial \phi_j} \tag{2}$$

and defining also

$$d_i = \frac{\partial K}{\partial \phi_i} \tag{3}$$

the general potential is

$$V = e^{\frac{K}{M^2}}[(\frac{\partial W}{\partial \phi_i} + d_i \frac{W}{M^2}) g_{ij}^{-1} (\frac{\partial W}{\partial \phi_j} + d_j \frac{W}{M^2})^* - 3\frac{|W|^2}{M^2}]. \tag{4}$$

g_{ij}^{-1} is the inverse of the metric, $M = M_p/\sqrt{8\pi}$ and the sum is over all scalar fields ϕ_i. The kinetic energy is given by

$$g_{ij} D^\mu \phi_i^* D_\mu \phi_j. \tag{5}$$

Our assumption that supersymmetry is broken in a hidden sector means that there are two sets of fields: z_i, responsible for supersymmetry breaking, and the "visible sector fields," y_i;

$$W = g(y) + h(z). \tag{6}$$

The scales are such that

$$\langle \frac{\partial h}{\partial z_i} + \frac{1}{M^2}\frac{\partial K}{\partial z_i} h \rangle \sim m_{susy} M. \tag{7}$$

As discussed long ago by Hall, Lykken and Weinberg[12], universality is the assumption that there is an approximate $U(n)$ symmetry of the Kähler potential, where n is the number of chiral multiplets in the theory. In minimal supergravity one takes simply

$$K = \sum \phi_i^* \phi_i, \tag{8}$$

which leads to universal masses for the scalars, as well as proportionality between Yukawa and scalar trilinear couplings. Clearly, this is an extremely strong assumption. In the supersymmetric standard model the Yukawa couplings of the theory exhibit no such symmetry. As already mentioned, it does not hold for generic superstring compactifications[9, 10] where the symmetry violations are simply $\mathcal{O}(1)$.

We can characterize the violations of universality quite precisely. For small y, we can expand K in powers of y,

$$K = k(z, z^*) + y_i^* y_i + \ell_{ij}(z, z^*) y_i^* y_j + p_{ij}(z, z^*) y_i y_j + \ldots \tag{9}$$

We must rescale the y_i fields so that they are canonically normalized. From eq. (5) the rescaled fields, y_i', are related to the y_i by a matrix A, *i.e.*, $y' = Ay$, whose inverse satisfies

$$A^{-1\dagger}(I + \langle \ell \rangle)A^{-1} = I, \tag{10}$$

where the quantity in parenthesis is just $\langle g_{ij} \rangle$. We note, for later reference, that A^{-1} can be expressed as a product of matrices

$$A^{-1} = \tilde{A}^{-1} U. \tag{11}$$

$\tilde{A}^{-1}$ is the product of a unitary matrix which diagonalizes $\langle g_{ij} \rangle$, and a diagonal matrix whose entries are the inverse square roots of the eigenvalues of $\langle g_{ij} \rangle$, up to phases. U is an arbitrary unitary matrix which simply reflects the fact that the canonically normalized basis is fixed up to unitary transformations. In general, there is no reason why ℓ_{ij} should be proportional to the unit matrix, so the $z^* z$ components of the inverse metric will contain terms involving $y_i'^* y_j'$ which are non-universal. Plugging into eq. (4) yields non-universal mass terms for the visible sector fields.

Let us focus on the squark sector of the supersymmetric standard model. Among the ℓ_{ij} in eq. (9) will be the following terms

$$\ell_{ij}^Q Q_i^* Q_j + \ell_{ij}^{\bar{d}} \bar{d}_i^* \bar{d}_j + \ell_{ij}^{\bar{u}} \bar{u}_i^* \bar{u}_j, \tag{12}$$

where Q and $\bar{u}$, $\bar{d}$ are $SU(2)_L$ doublets and singlets, respectively. Assuming, for simplicity, only one hidden sector field z and $k(z, z^*) = zz^*$, non-universal masses for the Q squarks will be given by

$$e^{\frac{K}{M^2}} |\langle \frac{\partial h}{\partial z} + \frac{1}{M^2} \frac{\partial K}{\partial z} h \rangle|^2 \left(\frac{-\partial^2 \ell_{ij}'^Q}{\partial z^* \partial z} + \frac{\partial \ell_{ik}'^Q}{\partial z^*} \frac{\partial \ell_{kj}'^Q}{\partial z} \right) Q_i'^* Q_j', \tag{13}$$

where ℓ'^Q is just ℓ^Q transformed to a canonically normalized basis, *i.e.*, $\ell'^Q = (A_Q^{-1})^\dagger \ell A_Q^{-1}$. The matrix A_Q is just the canonical transformation, defined as in eq. (10), appropriate for the left-handed squark doublets. Analogous expressions hold for the $\bar{u}$ and $\bar{d}$ squark masses. Note that there is no symmetry one can add which will forbid the troublesome couplings in K. For example, $\ell_{ij}^Q = l_{ij} z^* z$ cannot be eliminated by symmetries. Thus, applying 't Hooft's naturalness criterion[14], we expect these mass terms to be of order m_{susy}^2, given hidden sector vev's of order M, implying non-degenerate squark masses.

We now move to discussion of Yukawa and scalar trilinear couplings. We will always choose the canonically normalized basis for fermions to be identical to that of their scalar superpartners, given in eq. (10).† In a canonically normalized basis the down quark mass matrix $m^d_{ij} d_i \bar{d}_j$ becomes

$$m'^d = (A_Q^{-1})^T m_d A_{\bar{d}}^{-1}. \tag{14}$$

One might worry that such a transformation would spoil the quark mass and mixing hierarchies, given arbitrary ℓ_{ij}. However, this is not the case. For example, suppose the various Yukawa couplings are chosen so that the original mass matrix entries satisfy

$$m^d = \begin{pmatrix} \sim m_d & \sim m_d & \sim m_d \\ \sim m_d & \sim m_s & \sim m_s \\ \sim m_d & \sim m_s & \sim m_b \end{pmatrix}, \tag{15}$$

with similar assumptions for the up-quark matrix. Since the eigenvalues of $I + \langle \ell^Q_{ij} \rangle$, $I + \langle \ell^{\bar{d}}_{ij} \rangle$, or of $\tilde{A}^{-1}$ in eq. (11), are of order unity, one can always find a canonically normalized quark basis in which, again, the mass matrix entries will exhibit the same hierarchy as above, modulo factors of 2 or 3.‡ This corresponds to an appropriate choice of U in eq. (11). Since the canonical normalization for left-handed up and down quarks is essentially the same, by $SU(2)_L$ invariance, the mixing angle hierarchy is also preserved.

We can now discuss non-proportionality between Yukawa and scalar trilinear couplings. The ℓ_{ij} terms in the Kähler potential lead to non-universal corrections of cubic, and higher dimensional couplings, if present, via the $z^* y_i$ components of $g^{-1}_{\bar{i}j}$. From eq. (4) these are given by

$$-e^{\frac{K}{M^2}} \langle \frac{\partial h}{\partial z} + \frac{1}{M^2} \frac{\partial K}{\partial z} h \rangle \left(\frac{\partial \ell'_{ij}}{\partial z^*} y'^*_i + 2 \frac{\partial p'_{ij}}{\partial z^*} y'_i \right) \frac{\partial g'^*(y')}{\partial y'^*_j} + h.c., \tag{16}$$

where again we've taken $k(z^*, z) = z^* z$, and primes denote transformation to a canonically normalized basis.§

Again we focus on the squark sector. In particular, we choose a canonically normalized basis in which the quark Yukawa couplings lead to mass matrices of the form given in eq. (15), as discussed above. It follows from eq. (16) that the resulting scalar trilinear couplings would lead to a left-right down squark mass matrix satisfying the hierarchy

$$\tilde{m}^2_{LR} \approx \begin{pmatrix} A_{11} m_d & A_{12} m_s & A_{13} m_b \\ A_{21} m_s & A_{22} m_s & A_{23} m_b \\ A_{31} m_b & A_{32} m_b & A_{33} m_b \end{pmatrix}, \tag{17}$$

†The kinetic energy for fermions ψ_i, to be compared with eq. (5), is given by $g_{ij} \bar{\psi}_i \not{D} \psi_j$.

‡One way to see this is to express eq. (15) as the sum of three unit-rank mass matrices, with a hierarchy of eigenvalues. Performing the transformation in eq. (14), it is clear that one obtains a matrix which is again the sum of three hierarchichal unit-rank matrices whose eigenvalues are of same order as the original ones.

§Additional deviations from proportionality arise, in general, if 'hidden sector' fields couple to the observable sector in the superpotential via non-renormalizable terms.[6]

where the A_{ij} are of order m_{susy}. Of course an analogous expression follows for the corresponding up squark matrix. As we'll see in the next section, such violations of proportionality are small enough for the $K^o - \bar{K}^o$, $B^o - \bar{B}^o$ and $D^o - \bar{D}^o$ systems. So it is non-degeneracy of scalar masses but not non-proportionality which, in general, leads to problematic flavor violations in hidden sector supergravity theories.

In the next section we discuss a simple model with gauged $SU(2)_H$ horizontal symmetry. We'll see that constraints on the Kähler potential due to this symmetry lead to large squark degeneracy.

2. Flavor Symmetries and Squark Degeneracy

In this section we discuss the simplest supersymmetric model incorporating gauged non-abelian horizontal symmetries. The model does not address the problems of fermion mass and mixing hierarchy; these are put in by hand. However it does lead to large squark mass degeneracy and suppression of flavor violation. The gauge group is that of the standard model times an additional $SU(2)_H$. For purposes of enumerating the different particles and couplings, we will label the states by the quantum numbers they might have in an $SU(5) \times SU(2)_H$ unification. Note that we are not assuming an underlying $SU(5)$ symmetry, but simply using $SU(5)$ to classify the states. The three generations are then assumed to form doublets and singlets of the $SU(2)_H$. The states are

$$\bar{5}_a = (\bar{5}, 2) \quad 10_a = (10, 2) \quad \bar{5}_s = (\bar{5}, 1) \quad 10_s = (10, 1). \tag{18}$$

The Higgs particles are taken to be two singlets of $SU(2)_H$; this will avoid the problem of flavor changing currents mediated by Higgs particles. We will denote these by H_1 and H_2. To break $SU(2)_H$, one adds three fields Φ_i^a, $i = 1, 2, 3$, transforming as $(1, 2)$. The model is then free of both perturbative and non-perturbative anomalies. Alternatively, one can add an $SU(2)_H$ doublet of right-handed neutrinos, in which case only two Φ_i singlets are added.

Again we assume that supersymmetry is broken in the hidden sector of an underlying supergravity theory. We also assume that, after supersymmetry breaking, the potential for the fields Φ_i is such that these fields obtain large vev's. This assumption may seem unnatural, but it is often satisfied in string theories. First, there are "D-flat" directions (*i.e.*, directions in which the auxiliary D-fields in the $SU(2)_H$ gauge supermultiplet vanish) where some of these fields have vev's. F-flatness is known to arise in string theory in at least two ways. Moduli of string compactifications with $(2, 2)$ world sheet supersymmetry are F-flat.[15] At points of enhanced symmetry, the moduli are typically charged (*e.g.*, at orbifold points); the enhanced symmetry could be our horizontal symmetry. Generically, however, the moduli *do* appear in the superpotential of the matter fields (some of these couplings may be exponentially suppressed at large radius).[15] F-flatness is also known to arise in the presence of discrete R-symmetries.[16] In either case, if some of the Φ fields acquire negative masses upon supersymmetry breaking, they can acquire large vev's. We will require that these be smaller than M by a factor of $\mathcal{O}(10)$. We will not attempt here to explain how this factor might arise but simply argue that in a theory with small couplings it is not unnatural. ¶

To keep the discussion simple, we will assume that two singlets, Φ_1 and Φ_2, obtain vev's: $|\langle\Phi_1\rangle| = (0, \phi)^T$ and $|\langle\Phi_2\rangle| = (\phi, 0)^T$. ‖ In order to understand how this breaking of the $SU(2)_H$ symmetry feeds down to other fields, we need to examine the lagrangian more carefully. Let us focus first on the quark fields. Denoting quark doublets by Q and singlets by $\bar{d}$ and $\bar{u}$, the superpotential just below M_p contains dimension-four terms:

$$W_q = \lambda_1 \epsilon_{ab} Q_a \bar{d}_b H_1 + \lambda_2 \epsilon_{ab} Q_a \bar{u}_b H_2 + \lambda_3 Q_s \bar{d}_s H_1 + \lambda_4 Q_s \bar{u}_s H_2 \tag{19}$$

These give rise to $SU(2)_H$ symmetric terms in the mass matrix. Clearly we need to assume that λ_1 and λ_2 are small (this might be arranged by means of a discrete symmetry).

$SU(2)_H$-violating terms arise from higher dimension couplings, of which there are a wide variety. For example, in the d-quark sector, we have:

$$\frac{1}{M}(\lambda_5^i \epsilon_{ab} \Phi_a^i Q_b \bar{d}_s H_1 + \lambda_6^i \epsilon_{ab} \Phi_a^i Q_s \bar{d}_b H_1) + \frac{1}{M^2}(\lambda_7^{ij} \epsilon_{ab} \epsilon_{cd} \Phi_a^i \Phi_c^j Q_b \bar{d}_d H_1 + ...) \tag{20}$$

Similar terms are present in the up-quark sector. Some points should be noted immediately. First, ϕ/M (times coupling constants) cannot be too small; in the limit that $\phi \to 0$, there is no mixing of the third generation with the first two. As we remarked above, the $SU(2)$ symmetric terms in the light quark matrices must be small, so m_c and m_s must go as ϕ^2/M^2; this quantity thus cannot be much smaller than 10^{-2}. With this restriction on ϕ, there is no difficulty in obtaining reasonable fermion masses and KM angles, provided one is willing to take several Yukawa couplings to be small and comparable, as in the standard model. Hopefully, other horizontal schemes could be more predictive. We leave the exploration of this question to future work.

We choose the various Yukawa couplings so that the down quark mass matrix entries in the original interaction basis satisfy the hierarchy of eq. (15), with similar assumptions for the up-quark matrix.** We then pick a canonically normalized basis in which the mass matrices are again of the same form, as discussed in the previous section. We return to this point shortly. The quark masses and eigenstates then take on a particularly simple form. For example, the down masses are given by

$$m_d = m_{11} - \frac{m_{12} m_{21}}{m_s}, \quad m_s = m_{22} - \frac{m_{23} m_{32}}{m_b}, \quad m_b = m_{33}. \tag{21}$$

The down mass eigenstates are given by

$$|d_i\rangle = x_{ij}^d |j\rangle, \qquad |\bar{d}_i\rangle = \bar{x}_{ij}^d |\bar{j}\rangle, \tag{22}$$

¶One possible origin of this scale is through the appearance of a Fayet-Iliopoulos D-term.[17]

‖The precise alignment of Φ_1 and Φ_2 will not be important to us, as long as it is D-flat.

**If $m_{31}^d \sim m_{32}^d \sim m_b$, then the mixings of the first and second generation $\bar{d}$ quarks with the third generation quarks are large, and the $SU(2)_H$ symmetry will not lead to sufficient degeneracy for the $K - \bar{K}$ system.

where i=1,2,3 correspond to d, s, b, respectively. The canonical basis element $|1\rangle$ corresponds to the vector $(1,0,0)^T$, *etc.* With up and down mass matrices of the form of eq. (15), the KM angles are essentially generated in the down sector, and the x^d_{ij} are given by

$$x^d_{ii} \sim 1, \quad x^d_{21} \sim \frac{m^d_{12}}{m_s} \sim V_{us}, \quad x^d_{32} \sim \frac{m^d_{23}}{m_b} \sim V_{cb}, \quad x^d_{31} \sim \frac{m^d_{13}}{m_b} \sim V_{ub}. \tag{23}$$

The remaining x_{ij} follow from orthonormality of the eigenstates. The $\bar{x}_{ij}$ are obtained by complex conjugating the above and interchanging indices on the m_{ij}. Expressions for the up masses and eigenstates are completely analogous. Knowledge of the x_{ij} and $\bar{x}_{ij}$ will be required to estimate the various off-diagonal squark mass matrix entries of relevance to FCNC.

We begin our discussion of the squark sector in the original interaction basis. Canonical normalization of the quark and squark fields depends on the form of the Kähler potential and will be discussed below. The $SU(2)_H$ symmetry significantly restricts the form of K. It can be expanded in powers of the squark fields, Φ_i, *etc.*

$$\begin{aligned} K = k(z^*, z, \Phi_i^*, \Phi_i) + Q_s^* Q_s + Q^* Q + \ell^{Q_s} Q_s^* Q_s + \ell^Q Q^* Q \\ +(\ell_i^1 \Phi_i Q^* Q_s + ...) + (\ell_{ij}'^1 \Phi_i Q \Phi_j Q^* + ...) + ... \end{aligned} \tag{24}$$

The ℓ coefficients are functions of hidden sector fields, z, z^*, and we've omitted $SU(2)_H$ indices on Q, Φ_i but terms with all possible contractions should be understood. For illustrative purposes we've only shown terms of relevance for the $SU(2)_L$ squark doublets explicitly, but of course there will be analogous terms for the remaining scalars.

Leading terms for the squark fields in K lead to $SU(2)_H$-symmetric squark mass terms of the form

$$\begin{aligned} V_{soft} = m_1^2|Q_a|^2 + m_2^2|Q_s|^2 + m_3^2|\bar{u}_a|^2 + m_4^2|\bar{u}_s|^2 + ... \\ +A_1\lambda_1 Q\bar{d}H_1 + A_2\lambda_2 Q\bar{u}H_1 + ... + h.c., \end{aligned} \tag{25}$$

Here, m_i and A_i are of order m_{susy}. Terms linear and quadratic in Φ give rise to symmetry-breaking terms of the type:

$$\delta V^2_{soft} = \frac{m^2_{susy}}{M_p}(\gamma_1 \Phi_1 Q Q_s^* + ...) + \frac{m^2_{susy}}{M_p^2}(\gamma_1' \Phi_1 Q \Phi_2 Q^* + ...) \tag{26}$$

and

$$\begin{aligned} \delta V^3_{soft} = \tfrac{m_{susy}}{M_p} \lambda_5^1 Q \bar{d}_s H_1 (\eta_1 \Phi_1 + \eta_2 \Phi_2 + \eta_3 \Phi_2^*) \\ + \tfrac{m_{susy}}{M_p^2} \lambda_7^{11} Q \bar{d} H_1 (\eta_1' \Phi_1 \Phi_1 + \eta_2' \Phi_1 \Phi_2 + \eta_3' \Phi_1 \Phi_2^*) + ... \end{aligned} \tag{27}$$

Again, terms with all possible contractions of $SU(2)_H$ indices should be understood. Here γ, γ', η and η' are dimensionless numbers. The couplings η_3 and η_3' may seem surprising since they are not among the usual allowed soft breaking terms. These terms, however, are *supersymmetric* terms, arising because the superpotential of the effective theory will in general contain terms like $m_{susy}\Phi_i\Phi_j$.

Of course, we've to transform the resulting squark mass matrices from the interaction basis, (q_a, q_s), to canonically normalized basis. For the Kähler potential of our model it is not difficult to show that the inverse of the normalization matrix for left-handed squarks, $\tilde{A}_Q^{-1}$, (corresponding to $\tilde{A}^{-1}$ in eq. (11)), is of the form

$$\tilde{A}_Q^{-1} = \begin{pmatrix} \frac{1}{(1+\langle\ell^Q\rangle)^{\frac{1}{2}}} + O(\frac{\phi^2}{M^2}) & O(\frac{\phi^2}{M^2}) & O(\frac{\phi}{M}) \\ O(\frac{\phi}{M}) & \frac{1}{(1+\langle\ell^Q\rangle)^{\frac{1}{2}}} + O(\frac{\phi^2}{M^2}) & O(\frac{\phi}{M}) \\ O(\frac{\phi}{M}) & O(\frac{\phi}{M}) & \frac{1}{(1+\langle\ell^{Q_s}\rangle)^{\frac{1}{2}}} + O(\frac{\phi^2}{M^2}) \end{pmatrix}. \quad (28)$$

As discussed above, it is convenient to choose canonically normalized basis in which the quark mass matrices are of the form of eq. (15), as in the original interaction basis. The associated unitary transformation for left-handed squarks, corresponding to choice of U in eq. (11), in fact leads to an overall matrix A_Q^{-1} of the same form as $\tilde{A}_Q^{-1}$ above, (where A_Q transforms from the original interaction basis to the desired canonically normalized basis). Rescaling of the right-handed squarks is completely analogous.

The down squark mass matrices are of three types; $\tilde{m}^2_{LL}$, $\tilde{m}^2_{LR}$, and $\tilde{m}^2_{RR}$, where L and R refer to left-handed and right-handed squarks, respectively. For example, transforming to the desired canonically normalized basis, as above, one obtains

$$\tilde{m}^2_{LL} = diag(\tilde{m}_1'^2, \tilde{m}_1'^2, \tilde{m}_2'^2) + \delta\tilde{m}^2, \quad (29)$$

where

$$\tilde{m}_1'^2 = \frac{\tilde{m}_1^2}{1+\langle\ell^Q\rangle} \quad \tilde{m}_2'^2 = \frac{\tilde{m}_2^2}{1+\langle\ell^{Q_s}\rangle}. \quad (30)$$

The first and second terms originate in V_{soft} and δV^2_{soft}, respectively. Note that the (13), (23), (31) and (32) entries of $\delta\tilde{m}^2$ are proportional to $\frac{\phi}{M}$ and the remaining entries are proportional to $\frac{\phi^2}{M^2}$, leading to large suppression of flavor violation. The $\tilde{m}^2_{LR}$ matrix corresponding to a down quark matrix of the form of eq. (15) will be of the same form as obtained in the case of no horizontal symmetry, *i.e.*, eq. (17), with the exception of the (13), (23), (31) and (32) entries[††] which will be of order $\frac{\phi}{M}$ smaller. The up-sector is, of course, completely analogous.

To evaluate the size of FCNC effects we need to know the squark mass matrix elements in the quark mass eigenstate basis. For example, for $\tilde{m}^2_{ds}$ we find, using eqs. (22), (23),

$$\tilde{m}^2_{ds} = (\tilde{m}_2'^2 - \tilde{m}_1'^2)x^*_{13}x_{23} + \delta\tilde{m}^2_{12} + (\delta\tilde{m}^2_{11} - \delta\tilde{m}^2_{22})x_{21} + \delta\tilde{m}^2_{13}x_{23} + \delta\tilde{m}^2_{32}x^*_{13} + \dots \quad (31)$$

[††]In eq. 14 of ref. [13] these entries were incorrectly given. The corresponding $d\bar{b}$ entries in the last column of Table 2b are an order of magnitude larger then stated there.

In fact, the matrix elements satisfy the promising hierarchy $\tilde{m}^2_{ds} << \tilde{m}^2_{db}, \tilde{m}^2_{sb}$, where the l.h.s. is approximately proportional to $\frac{\phi^2}{M^2}$, and the r.h.s. to $\frac{\phi}{M}$. Similar results hold for $\tilde{m}^2_{RR}$ and for the up-sector.

We now discuss implications for FCNC's. Bounds on off-diagonal down squark masses [2, 3] from the $K^o - \bar{K}^o$ and $B^o - \bar{B}^o$ mass differences are summarized in Table 1. Estimates of these quantities in the $SU(2)_H$ model are collected in Table 2. We've taken $\frac{\phi}{M} \approx \frac{1}{10}$, and we've also indicated the dependence on the dimensionless couplings in δV^2_{soft}.[‡‡] Similar estimates are obtained in the up-sector for $\tilde{m}^2_{LL}$ and $\tilde{m}^2_{RR}$, but $\tilde{m}^2_{LR}$ entries will be related to the up quark mass matrix and will be somewhat larger. This will turn out to be of significance for $D^o - \bar{D}^o$ mixing.

Table 1a. Bounds on various components of the squark mass matrix from $K^o - \bar{K}^o$ mixing. Here, m_λ is the gluino mass and $\tilde{m}$ is a typical first-second generation squark mass. All quantities in GeV^{-2}. We have taken $F_K = 170$ MeV and $B_K = 1$.

$\frac{m_\lambda}{\tilde{m}}$	$\frac{\tilde{m}^4_{ds}}{\tilde{m}^6}$	$\frac{\tilde{m}^2_{ds}\tilde{m}^2_{\bar{d}\bar{s}}}{\tilde{m}^6}$	$\frac{\tilde{m}^4_{\bar{d}\bar{s}}}{\tilde{m}^6}$
1	2×10^{-9}	3×10^{-11}	10^{-10}
.1	4×10^{-10}	10^{-10}	2×10^{-11}

Table 1b. Bounds on various components of the squark mass matrix from $B^o - \bar{B}^o$ mixing. We have taken $F_B = 230$ MeV and $B_B = 1$.

$\frac{m_\lambda}{\tilde{m}}$	$\frac{\tilde{m}^4_{db}}{\tilde{m}^6}$	$\frac{\tilde{m}^2_{db}\tilde{m}^2_{\bar{d}\bar{b}}}{\tilde{m}^6}$	$\frac{\tilde{m}^4_{\bar{d}\bar{b}}}{\tilde{m}^6}$
1	10^{-8}	10^{-9}	2×10^{-9}
.1	2×10^{-9}	2×10^{-9}	7×10^{-10}

Table 2a. Predictions of the $SU(2)_H$ model for $K^o - \bar{K}^o$ mixing for representative values of squark masses. The dimensionless parameters γ, η, *etc.*, in this and the following table are as defined in the text. All quantities in GeV^{-2}.

$\tilde{m}$	$\frac{\tilde{m}^4_{ds}}{\tilde{m}^6}$	$\frac{\tilde{m}^2_{ds}\tilde{m}^2_{\bar{d}\bar{s}}}{\tilde{m}^6}$	$\frac{\tilde{m}^4_{\bar{d}\bar{s}}}{\tilde{m}^6}$
300 *GeV*	$10^{-9}\gamma'^2$	$10^{-9}\gamma'^2$	$3 \times 10^{-12}\eta'^2$
1 *TeV*	$10^{-10}\gamma'^2$	$10^{-10}\gamma'^2$	$3 \times 10^{-14}\eta'^2$

‡‡Rescaling via A_Q^{-1} introduces dependence of the entries in Table 2 on additional dimensionless parameters originating in the Kähler potential, at same orders in $\frac{\phi}{M}$ as those given. These parameters are again nominally of $\mathcal{O}(1)$. For simplicity this dependence is ommitted since we are only interested in order of magnitude estimates.

Table 2b. Predictions of the $SU(2)_H$ model for $B^o - \bar{B}^o$ mixing for representative values of squark masses.

$\tilde{m}$	$\frac{\tilde{m}^4_{db}}{\tilde{m}^6}$	$\frac{\tilde{m}^2_{db}\tilde{m}^2_{\bar{d}\bar{b}}}{\tilde{m}^6}$	$\frac{\tilde{m}^4_{d\bar{b}}}{\tilde{m}^6}$
$300\ GeV$	$10^{-7}\gamma^2$	$10^{-7}\gamma^2$	$3\times10^{-11}\eta^2$
$1\ TeV$	$10^{-8}\gamma^2$	$10^{-8}\gamma^2$	$3\times10^{-13}\eta^2$

We see that for ~ 300 GeV squarks and gluinos, some of the dimensionless couplings in δV^2_{soft} will have to be $\sim \frac{1}{10} - \frac{1}{3}$ to obtain satisfactory Δm_K and Δm_B. If the squarks of the first two generations have masses of order 1 TeV then all dimensionless couplings in the scalar potential can be $\mathcal{O}(1)$. Provided the third generation squarks are comparatively light[18] (perfectly possible in this sort of model), this does not imply any fine tuning of Higgs parameters since the first two generations have small Yukawa couplings.

As remarked above, the $\tilde{m}^2_{LR}$ entries have interesting consequences for $D^o - D^o$ mixing. Given the current experimental bound of $\frac{\Delta m_D}{m_D} < 6.97\times10^{-14}$ one obtains the following constraints, taking $F_D = 200\ MeV$ and $B_D = 1$:

$$\left(\frac{\tilde{m}^2_{u\bar{c}}}{\tilde{m}^2}\right)^2 \frac{1}{\tilde{m}^2} < 10^{-9}\ \mathrm{GeV}^{-2} \tag{32}$$

for $\tilde{m}_g \sim \tilde{m}$; the bound is 7×10^{-10} for $\tilde{m}_g \sim 0.1\ \tilde{m}$ By way of comparison, in the $SU(2)_H$ model we expect for the above quantities $2\times10^{-10}\eta'^2$ for 300 GeV squark masses, and $2\times10^{-12}\eta'^2$ for 1 TeV squark masses. Recent HQET calculations[19] for the standard model lead to $\frac{\Delta m_D}{m_D} \sim 10^{-17} - 10^{-16}$. So it is clear that Δm_D can be one to two orders of magnitude larger in the $SU(2)_H$ model.

What about constraints from CP violation? The bounds from ϵ_K on the imaginary parts of the various quantities constrained by Δm_K are about two orders of magnitude stronger than indicated in Table 1. Constraints on non-degeneracy will depend on the size of the phases entering these quantities. For example, if the phases are of order unity, as would be expected in models with explicit CP violation, then for 1 TeV first and second generation squarks some of the γ' couplings in δV_{soft} would have to be of order $\frac{1}{10}$. It is interesting that for this choice of parameters, large CP-violating supersymmetric contributions in the B-system are possible[20], This is because the bound from Δm_B can be saturated (corresponding to $\gamma \sim \frac{1}{3}$), while the phases of relevance can be of order unity.

In scenarios of spontaneous CP violation the relevant phases might naturally be smaller than unity[21], in which case the ϵ_K constraint is less restrictive. An alternative strategy for accomodating CP violation bounds is to increase the amount of squark degeneracy by adding additional abelian discrete or continuous horizontal symmetries. Such symmetries might also play a role in more elaborate models which attempt to explain the fermion mass hierarchies. One notices that all terms in eqs.(24) or (26) which contribute to off-diagonal entries in $\tilde{m}^2_{LL}$, and $\tilde{m}^2_{RR}$ (in the canonically normalized or original interaction basis), can, in principle, be eliminated by additional symmetries. The smallness of off-diagonal squark mass matrix entries would be limited by the x_{ij}, $\bar{x}_{ij}$, or KM angles. This can lead to further suppression of order θ_c^2 for $K^o - \bar{K}^o$ (corresponding to the third term in eq. (31)) and of order $\left(\frac{V_{cb}}{\gamma}\right)^2$ for $B^o - \bar{B}^o$. Moreover, in many cases, the lowest dimension operators would be CP-conserving, providing further suppression of Im $K^o - \bar{K}^o$.

3. Conclusion

In general, squarks will not be degenerate in supergravity theories for standard model gauge group. However, we've seen that additional non-abelian horizontal symmetries can overcome this problem by restricting the form of the Kähler potential. In fact, a very simple gauged horizontal $SU(2)$ model exists which leads to adequate suppression of flavor changing neutral currents.[13] The more stringent CP violation constraints, however, require some dimensionless parameters in the scalar potential to be of $\mathcal{O}(\frac{1}{10})$. The choice of $SU(2)$ is essentially dictated by the gross features of the quark mass spectrum: large mass splitting and small mixing angles between the third family and the first two.

The model suggests patterns of masses which differ from conventional assumptions. For example, up or down squarks (or sleptons) of the first two families will be approximately degenerate and, in general, different from third generation masses. Moreover, under the simplest assumptions about the scalar potential, the squarks of the first two generations should have masses of order a TeV, while the top squarks (to avoid naturalness problems) and, therefore, the left-handed bottom squark should be comparatively light. The model also suggests that there will be large supersymmetric contributions to $B^o - \bar{B}^o$ and $D^o - \bar{D}^o$ mixing and CP violation in the K and B systems.

I would like to thank Michael Dine and Rob Leigh for a stimulating and fun collaboration, and the organizers of this workshop for their hospitality.

References

1. H.P. Nilles, *Phys. Rep.* **110** (1984) 1.
2. F. Gabiani and A. Masiero, *Nucl. Phys.* **B322** (1989) 235.
3. J.S. Hagelin, S. Kelley, and T. Tanaka, MIU-THP-92/59.
4. Y. Nir and N. Seiberg, *Phys. Lett.* **309B** (1993) 337.
5. For a recent attempt along these lines, see M. Dine and A. Nelson, UCSD preprint UCSD/PTH 93-05 (1993).
6. S.K. Soni and H.A. Weldon, *Phys. Lett.* **126B** (1983) 215.
7. H. Georgi, *Phys. Lett.* **169B** (1986) 231; L.J. Hall, V.A. Kostelecky and S. Raby, *Nucl. Phys.* **B267** (1986) 415.
8. M. Dine, A. Kagan and S. Samuel, *Phys. Lett.* **243B** (1990) 250.

9. L. Ibanez and D. Lust, *Nucl. Phys.* **B382** (1992) 305; B. de Carlos, J.A. Casas and C. Munoz, CERN-TH. 6436/92 and *Phys. Lett.* **299B** (1993) 234.
10. V.S. Kaplunovsky and J. Louis, CERN-TH. 6809/93 UTTG-05-93.
11. R. Barbieri, J. Louis and M. Moretti, CERN-TH 6856/93.
12. L. Hall, J. Lykken and S. Weinberg, *Phys. Rev.* **D27** (1983) 2359.
13. M. Dine, R. Leigh, A.L. Kagan, SLAC preprint SLAC-PUB-6147, to be published in Phys. Rev. D.
14. G.'t Hooft, in ***Recent Developments in Gauge Theories***, G. 't Hooft *et. al,*, Eds., Plenum (New York) 1980.
15. M. Dine, N. Seiberg, X.G. Wen and E. Witten, *Nucl. Phys.* **B278** (1986) 769; L. Dixon, in *Superstrings, Unified Theories, and Cosmology 1987*, G. Furlan, J.C. Pati, D.W. Sciama, E Sezgin and Q. Shafi, eds., World Scientific (Singapore, 1988).
16. M. Dine and N. Seiberg, *Nucl. Phys.* **B306** (1988) 137.
17. M. Dine, N. Seiberg and E. Witten, *Nucl. Phys.* **B289** (1987) 589.
18. R. Barbieri and G.F. Giudice, *Nucl. Phys.* **B306** (1988) 77; G.G. Ross and R.G. Roberts, *Nucl. Phys.* **B377** (1992) 571; B. de Carlos and J.A. Casas, CERN-TH. 6835/93.
19. H. Georgi, *plb* **297B** (1992) 353; T. Ohl, G. Ricciardi and E.H. Simmons, Harvard preprint HUTP-92/A053.
20. L.I. Bigi and F. Gabbiani, *Nucl. Phys.* **B352** (1991) 309.
21. A. Pomarol, *Phys. Rev.* **D47** (1992) 273; R. Garisto and G. Kane, TRI-UMF preprint TRI-PP-93-1, (1993); L. Hall and S. Weinberg, Texas preprint UTTG-22-92.

THE ELECTROWEAK PHASE TRANSITION IN MINIMAL SUPERGRAVITY MODELS

HEATH POIS
Center for Theoretical Physics, Department of Physics, Texas A&M University
College Station, TX 77843-4242, USA
Astroparticle Physics Group, Houston Advanced Research Center (HARC)
The Woodlands, TX 77381, USA

ABSTRACT

We explore the electroweak phase transition in minimal supergravity models via analysis of the one-loop Higgs potential that includes finite temperature effects. We have searched for the allowed parameter space where the phase transition is sufficiently first order to avoid washout of baryon number after the phase transition via unsuppressed anomalous Electroweak sphaleron processes. This requirement imposes strong constraints on the Higgs sector, and we find $m_h \lesssim 60\,\mathrm{GeV}$ (at one loop). With respect to weak scale baryogenesis, we find however that minimal supergravity is *not* phenomenologically acceptable, since present Higgs searches already rule out $m_h \lesssim 60\,\mathrm{GeV}$ in these models. Thus, if supergravity (with two weak scale Higgs doublets) is to allow for baryogenesis, it *must* also provide for B-L production at or above the weak scale. Our results therefore indicate that that the field/coupling content of minimal supergravity is insufficient to account for the baryon asymmetry of the universe (BAU).

1. Introduction

It is now well known that finite temperature effects can considerably alter the vacuum symmetry of a gauge theory [1], and can lead to restoration or anti-restoration [2,3] of a particular gauge (or global) symmetry due to the interaction of the theory with a plasma at high temperatures. In addition, the non-trivial structure of the electroweak (EW) vacuum naturally admits the possibility of unsuppressed baryon number violation ($\Delta B \neq 0$) at the weak scale via finite temperature non-perturbative sphaleron transitions [4,5,6]. It was further suggested that these baryon number violating effects could actually lead to baryogenesis at the weak scale [7], ($\mathcal{O}(10^2\,\text{GeV})$), since the necessary conditions of (i) C and CP-violation, (ii) B-violation and (iii) thermal non-equilibrium could in principle be satisfied. In the standard scenario for the electroweak phase transition involving the decay of the false vacuum, thermal non-equilibrium requires the transition to be first order. This crucial requirement has led to a recent reappraisal [8] of the EW phase transition beyond the classic treatment of Dolan and Jackiw [9] and Weinberg [2].

Recently, several new weak-scale mechanisms have been proposed which can apparently account for the observed ratio $n_B/n_\gamma \sim 10^{-10}$. Some of these mechanisms are rather economical, and involve *e.g.* two higgs doublets [10,11,12,13], supersymmetry [14], left-right models [15], new heavy Majorana neutrino decays [16], and a CP-violating neutrino mass matrix [17]. This is in marked contrast to GUT-scale mechanisms for baryogenesis, where the characteristic scale is $\mathcal{O}(10^{16}\,\text{GeV})$. Assuming that a GUT scale exists, it not obvious that baryogenesis should be at the same scale. It is certainly possible in a GUT theory for the BAU to arise at the weak scale, and we explore this possibility here.

Irrespective of the details of baryogenesis, if the non-perturbative B,L-violating transitions are in thermal equilibrium above the weak scale (which is true for $200\,\text{GeV} \lesssim T \lesssim 10^{12}\,\text{GeV}$), then $B = \kappa(B-L)$. (The constant κ is of $\mathcal{O}(1)$ and depends on the specific field content of the model.). Thus, one requires either i) $(B-L) \neq 0$ as an *initial* value, or ii) $(B-L)$ production at some intermediate scale between M_U and the weak scale. The requirement relevant for weak scale baryogenesis is that the non-perturbative transitions are out of equilibrium *below* the weak scale. In the SM (or any other $B-L$-conserving theory), an upper bound to the Higgs mass(es) ensues, since the quartic coupling is bounded from above. This ensures an adequate finite-temperature vacuum expectation value at the critical temperature, $v(T_c)$. In the SM, this translates into the limit $m_H \lesssim 45(37)\,\text{GeV}$ (hereafter, Higgs mass limits in parenthesis represent the $\sqrt{\frac{2}{3}}$ reduction due to the higher order effects discussed in [8]) which is in conflict with the LEP result of $m_H \gtrsim 60\,\text{GeV}$ [18].

In the minimal supersymmetric model (MSSM), now that one-loop corrections to the Higgs masses are regarded as essential in certain regions of parameter space, the Higgs masses depend explicitly on the 21 parameters of the MSSM. Thus, any constraints to m_h (the lightest CP-even Higgs) at the EW phase transition also depends explicitly on these many parameters. In order to simplify matters, the allowed parameter space of a SUSY model which employs a single SUSY breaking parameter has been explored in Ref. [19]. A more general case has been considered in Ref. [20]. The upper limit to m_h depends strongly on $\tan\beta$ and m_t. Overall, the least restrictive limit is $m_h \lesssim 65(53)\,\mathrm{GeV}$, corresponding to $m_t = 200\,\mathrm{GeV}$ [19].

Here we show that if one considers the EW phase transition in completely realistic minimal supergravity models, the constraints of unification combined with the experimental and consistency constraints restricts $m_h \lesssim 60(50)\,\mathrm{GeV}$ in order to avoid a washout of the BAU after the EW phase transition [21]. This is just slightly more restrictive than the MSSM, and is not surprising since the total field content is the same as the MSSM. Given the experimental *model independent* LEP limit of $m_h > 43\,\mathrm{GeV}$ [18], it might appear that minimal supergravity models are still barely viable. We argue however that the resulting Higgs spectrum is very SM-like, and leads to the much more restrictive limit $m_h \gtrsim 60\,\mathrm{GeV}$. Therefore, unless there is an additional source of $B-L$ production at or above the weak scale, the BAU would not survive, and would be completely washed out in this class of supergravity unified models.

2. The EW Phase Transition and Weak Scale $\Delta B \neq 0$

Issues related to both weak scale $\Delta B \neq 0$ and the EW phase transition have been discussed recently and extensively in the literature [8,22]. We confine ourselves here to a brief discussion of the issues relevant to our calculations and conclusions. At finite temperature, the SM scalar Higgs potential takes the following form at one-loop [22]:

$$V_T(\phi,T) = D(T^2 - T_2^2)\phi^2 - ET\phi^3 + \frac{1}{4}\lambda_T\phi^4 \tag{1}$$

where ϕ is the real, neutral component of the scalar Higgs field and T_2, E, D, λ_T are calculable parameters that depend on the matter content (see Refs. [22], the first Ref. of [8], or Ref. [20] for the details). Finally, $T_c \simeq T_2$ is the critical temperature; to a good approximation, it can be obtained from[1]

$$\frac{\partial^2 V_T(\phi,T_c)}{\partial^2\phi}|_{\phi=0} \simeq 0 \tag{2}$$

[1] This condition effectively determines the *spinodal* point, which is different than the standard critical temperature condition $V_T(T_c,0) = V_T(T_c,v_c)$. The difference is expected to be small for the range of Higgs masses we consider here [23].

The finite temperature vacuum expectation value $v(T)$ is obtained from the usual $\frac{\partial V_T}{\partial \phi} = 0$ condition, and at the critical temperature T_c,

$$v(T_c) = \frac{3ET_c}{\lambda_T}. \tag{3}$$

One thus sees the crucial role of the cubic term in obtaining a non-zero $v(T_c)$, necessary for the first-order phase transition where the true/false vacuum is separated by a barrier.

Due to the non-trivial EW vacuum and the chiral nature of the theory, unsuppressed, topology changing $\Delta(B+L) \neq 0$ transitions become possible, particularly at high temperatures, where the probability to go *over* the classical barrier is enhanced enormously. A detailed calculation [24] shows that baryon number is safe from 'washout' provided that

$$\frac{M_{sph}(T_c)}{T_c} \geq 45 \rightarrow \frac{v(T_c)}{T_c} \equiv R_c \gtrsim 1.3. \tag{4}$$

The last inequality is obtained from the specific SM sphaleron solution [4,5]. From Eqn. (3) one can easily see that as λ_T grows, R_c is obviously diminished, thus the zero temperature Higgs mass m_H has a natural upper limit. There is an overall asymptotic decrease in R_c with increasing m_H, and by requiring $R_c \gtrsim 1.3$, the SM limit $m_H \lesssim 45\,\mathrm{GeV}$ results. We now consider the EW constraints on the Higgs sector in minimal supergravity models.

3. Minimal Supergravity at Finite Temperature

Minimal supergravity models can be regarded as $SU(3) \times SU(2) \times U(1)$ models with the minimal three generations and two Higgs doublets of matter representations at the EW scale (along with superpartners), and are assumed to unify into a larger gauge group ($SU(5), SO(10)$ or $E6$) at a unification mass of $M_U \approx 10^{16}$ GeV (see talk by S. Kelley in these proceedings). The five dimensional parameter space of this model can be described in terms of three universal soft-supersymmetry breaking parameters at M_U [25]: $m_{1/2}, m_0, A$; the top-quark mass m_t, and finally the ratio of Higgs vacuum expectation values $\tan\beta = v_2/v_1$. We re-define two of the independent soft-SUSY breaking parameters as $\xi_0 = m_0/m_{1/2}$, $\xi_A = A/m_{1/2}$. The sign of the superpotential Higgs mixing term μ is also undetermined. In all previous analysis, we have assumed that the soft-SUSY breaking parameters are real, and that all irreducible phases are zero. To consider the possibility of weak-scale baryogenesis (which requires non-negligible CP-violation), we should include the phases ϕ_B, ϕ_A in the parameter count [26].

Several consistency and phenomenological constraints restrict the range of the model parameters, such as the requirement of one-loop radiative EW symmetry breaking [27,28], a potential bounded from below, $m_{\tilde{q}}^2, m_{\tilde{l}}^2 > 0$ ($\tilde{q}(\tilde{l})$ correspond to the squark (slepton) fields) and the CDF, LEP experimental constraints to $m_{\tilde{g},\tilde{l},\tilde{q}}, m_{\tilde{\chi}^+,\tilde{\chi}^0}, m_{h,A}$ (see Ref. [27]). After all of these constraints have been imposed, one is left with a bounded region in the $m_t, \tan\beta$ parameter space for a given $\xi_0, \xi_A, m_{1/2}$.

Regarding the breaking of the EW symmetry, spontaneously broken supergravity models achieve this goal by inducing radiative corrections to the parameters in the higgs potential. This has the effect of dynamically generating vacuum expectation values for the real, neutral components of the two higgs doublets. At zero temperature, the tree-level potential involving the neutral Higgs fields is [27]:

$$V_0 = (\mu^2 + m_{H_1}^2)h_1^2 + (\mu^2 + m_{H_2}^2)h_2^2 + 2B\mu h_1 h_2 + \frac{(g_2^2 + g'^2)}{8}(h_1^2 - h_2^2)^2 \quad (5)$$

where $g' = \sqrt{\frac{3}{5}}g_1$ and g_2 are the $U_Y(1)$ and $SU_L(2)$ gauge couplings; $h_1 = \phi_3, h_2 = \phi_7$ are the real, neutral components of the H_1, H_2 complex Higgs doublet fields respectively (and contain the eight real degrees of freedom $\phi_{i=1,\dots 8}$), μ is the Higgs mixing term in the superpotential, B is a soft-SUSY breaking parameter. Furthermore, $M_W^2 = \frac{g_2^2}{2}v^2, M_Z^2 = \frac{(g_2^2+g'^2)}{2}v^2$, and $m_t = \frac{1}{\sqrt{2}}\lambda_t v \sin\beta, m_b = \frac{1}{\sqrt{2}}\lambda_b v \cos\beta$, where $\lambda_{t,b}$ are the usual top and bottom quark Yukawa couplings, and $v^2 = v_1^2 + v_2^2$ (where $v = 246\,\text{GeV}$).

In the one-loop approximation $V = V_1 = V_0 + \Delta V$, where [23,29]

$$\Delta V = \frac{1}{64\pi^2}\text{STr}\,\mathcal{M}^4\left(\ln\frac{\mathcal{M}^2}{Q^2} - \frac{3}{2}\right) \quad (6)$$

in the $\overline{MS}$ scheme, and the supertrace is defined as $\text{STr}\,f(\mathcal{M}) = \sum_j (-1)^{2j}(2j+1)\text{Tr}\,f(\mathcal{M}_j)$. $\mathcal{M}_j$ are the higgs-field dependent spin $j = 0, 1/2, 1$ mass matrices, and Q is the renormalization scale. We obtain the one-loop corrected higgs boson masses numerically from the standard mass matrix:

$$M_{ij}^2 = \frac{1}{2}\left(\frac{\partial^2 V}{\partial\phi_i\partial\phi_j}\right)_{\langle\phi_3\rangle=v_1,\langle\phi_7\rangle=v_2} \tag{7}$$

Initially, all of the parameters in V need to be specified at zero temperature. For a given point in the five-dimensional parameter space $(m_t, \tan\beta,\ \xi_0, \xi_A, m_{1/2})$ we numerically solve for μ and B from the minimization conditions for the scalar potential; for a given $\tan\beta$ and M_Z at zero temperature (*i.e.*, $v_1(T=0)$ and $v_2(T=0)$) we find the values of μ and B which solve the following conditions:

$$\left(\frac{\partial V}{\partial h_{1,2}}\right)_{\langle h_1\rangle=v_1,\langle h_2\rangle=v_2} = 0 \tag{8}$$

where V is the zero-temperature scalar Higgs potential. At finite temperature, each fermion/boson of species i adds the following standard term to the effective potential:

$$\Delta V_T = \eta_i(\frac{T^4}{2\pi^2})F_\pm(y_i),\ y_i = m_i(h_1,h_2)/T, \tag{9}$$

and η_i is the multiplicity for each boson/fermion. In our calculation, we have included the following fields in the supertrace appearing in Eqn. (6): the third generation quarks and squarks $t, b(12), \tilde{t}_{1,2}(6),\ \tilde{b}_{1,2}(6)$, the gauge bosons $W(6), Z(3)$, and the Higgs bosons $h, H, A, H^\pm(1)$ (the numbers in parenthesis specify the multiplicity η_i). The functions F_-(bosons), F_+(fermions) are given by the well-known high ($y_i < 1$) and low ($y_i > 1$) temperature expansions. In order to find T_c, we employ $V_T = V_1 + \Delta V_T$ in eqn.(7) and require

$$Det\left(\frac{\partial^2 V_T}{\partial h_i\partial h_j}\right)_{\langle h_1\rangle=0,\langle h_2\rangle=0} \simeq 0 \tag{10}$$

for $i,j = 1,2$, in analogy to Eqn. (2). Having determined T_c, $V_T(T_c)$ is then minimized with respect to h_1, h_2, using the following conditions in order to determine $v_1(T_c), v_2(T_c)$:

$$\left(\frac{\partial V_T}{\partial h_1}\right)_{\substack{\langle h_1\rangle=v_1(T_c)\\ \langle h_2\rangle=v_2(T_c)}},\left(\frac{\partial V_T}{\partial h_2}\right)_{\substack{\langle h_1\rangle=v_1(T_c)\\ \langle h_2\rangle=v_2(T_c)}} \simeq 0 \tag{11}$$

4. Results

We have scanned over the $\xi_0, \xi_A, m_{1/2}, \tan\beta, m_t$ parameter space in an effort to find points for which $R_c > 1.3$. For the realistic supergravity case, the parameters in the Higgs potential are obviously more constrained compared to the generic MSSM. In the case of m_3 (the Higgs mixing term), once the initial values for $\tan\beta$, $\xi_0, \xi_A, m_{1/2}, m_t$ are given, μ, B (and thus m_3) are determined. As a result, m_3 is *not* a free parameter. At this point, it remains to be determined how significant this constraint is in restricting the value for m_h^{max}. We find that the allowed region of $\tan\beta$ for which $v(T_c)/T_c \geq 1.3$ corresponds to $\tan\beta \lesssim 2.0$. Fig. 1 shows a scatter plot of R_c versus m_h for a representative set of distinct minimal supergravity models. We have varied $0 \leq \xi_0 \leq 5$, $1.7 \leq \tan\beta \leq 2.0$, and fixed $\xi_A = 0$. The values for $m_{1/2}, m_t$ are allowed to vary. For $\tan\beta = 2.0$, the tree-level perturbative unitarity constraint $\lambda_t \lesssim 5$ at all scales restricts $m_t \lesssim 170\,\mathrm{GeV}$. We find that increasing $\tan\beta, \xi_0, m_{1/2}$ drives R_c to smaller values quite rapidly. Since $m_{\tilde{q}} \sim m_{1/2}\sqrt{6+\xi_0^2}$, increasing $\xi_0, m_{1/2}$ effectively increases the squark mass and lowers R_c; this behavior was also noted in [19]. Also shown in Fig.1 is a set of points (circles with crosses) which correspond to $m_t = 168\,\mathrm{GeV}$, with $\tan\beta = 2$, $\xi_0 = 3, \xi_A = 0$; only $m_{1/2}$ is allowed to vary.

In order to maximize the value for m_h for the region of parameter space that we explore, $\tan\beta$, $m_{\tilde{q}}$, m_t should all be as large as possible. We have therefore checked that the points near the endpoint of the curve for $\tan\beta = 2.0$, $m_t = 168\,\mathrm{GeV}$ corresponds to the extremal region in the $(m_t, m_{1/2}, \tan\beta, \xi_0)$ parameter space, and to the maximal values of m_h. One can see from the figure that m_h^{max} achieves a maximum near $m_h \simeq 58 - 60\,\mathrm{GeV}$.

Changing the value of ξ_A has little effect on the result, and leads to shifts of $\Delta m_h^{max} \simeq 2\,\mathrm{GeV}$. For the $\mu < 0$ possibility, we find quantitatively similar results. Higher order effects are expected to reduce the upper limit to $m_h^{max} \lesssim 50\,\mathrm{GeV}$. For all points considered in the complete parameter space, m_h grows with increasing $m_{1/2}$, but m_h^{max} never exceeds $60\,\mathrm{GeV}$.

Overall, we find that $R_c > 1.3$ is only possible for $\tan\beta \lesssim 2.0$, and for $\xi_0 \lesssim 5$. In this region, m_h can be as large as $\sim 60\,\mathrm{GeV}$, and one may be tempted to conclude that there is a small, but allowed window. However, in all points we considered, a decoupling phenomenon is driven by the radiative breaking of the EW symmetry combined with a sufficiently large gluino mass, $m_{\tilde{g}}$. This naturally

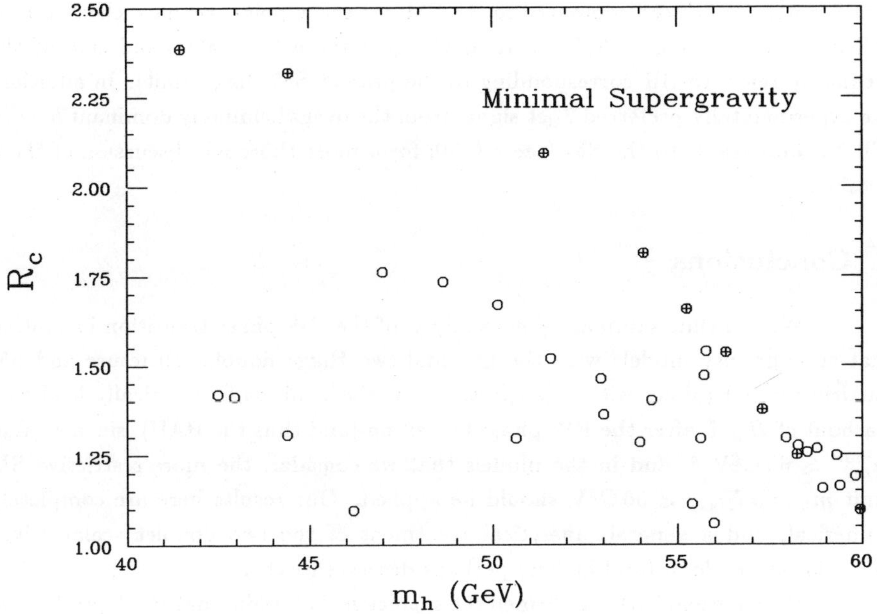

Fig. 1 shows a scatter plot of $R_c \equiv v_c(T_c)/T_c$ versus the lightest CP-even h Higgs for representative minimal supergravity models for $0 \leq \xi_0 \leq 5, \xi_A = 0$ and $1.7 \leq \tan\beta \leq 2.0$. The values for $m_{1/2}, m_t$ are allowed to vary; $m_t \lesssim 170\,\mathrm{GeV}$ due to tree-level perturbative unitarity constraints. The circles with crosses correspond to the case where $m_t = 168\,\mathrm{GeV}$. Increasing $\tan\beta, \xi_0$ causes a systematic downward shift in the scatter plot. For every point shown, the corresponding value for the lightest Higgs is $m_h \lesssim 60\,\mathrm{GeV}$, and is therefore experimentally excluded.

leads to large m_A values [30] and results in a SM-like Higgs sector at low energies, since the hZZ ($\sim \sin(\alpha-\beta)$), $hb\bar{b}$ ($\sim -\sin\alpha/\cos\beta$) couplings $\to 1$ as $m_A \to \infty$. The effect is enhanced even more for small $\tan\beta$. Due to these SM-like couplings, we expect the much more restrictive SM Higgs experimental limit to apply here. The reason is the following: for all points that we consider, we find the value of $\sin(\alpha-\beta) \gtrsim 0.98$ with the coupling $hb\bar{b} > 1$. Therefore, h production is not

significantly suppressed compared to the SM; for every point, we have checked that $BR(Z \rightarrow h^{max} Z^*) \gtrsim BR(Z \rightarrow H_{SM} Z^*)_{expt}$ (where the right hand side of the inequality refers to BR corresponding to the present SM Higgs limit). In addition, the experimentally preferred 2-jet signal from the overwhelmingly dominant $h \rightarrow b\bar{b}$ will be comparable to the SM (see ref [30] for a more thorough discussion of these details).

5. Conclusions

We can thus summarize our analysis of the EW phase transition by stating that supergravity models with the minimal two Higgs doublet structure and SM particle content (along with superpartners) at the weak scale generically lead to a washout of $B + L$ after the EW phase transition (and thus the BAU), since we find $m_h^{max} \lesssim 60\,\text{GeV}$ [2], and in the models that we consider, the more restrictive SM limit $m_h = m_{H_{SM}} \gtrsim 60\,\text{GeV}$ should be applied. Our results here are completely numerical, and a general, analytical treatment of the two doublet scalar Higgs potential at one-loop (and higher) perhaps deserves further study.

We have explicitly confirmed the suspicion that additional $B - L$ production is necessary in order to generate a BAU in the context of supergravity unified models at or above the weak scale. This will require additional fields or couplings such as explicit R-parity breaking, or heavy ν^c (see Ref. [31] for details) in order to generate an acceptable BAU. Although minimal supergravity does contain additional sources of CP-violation in the soft-SUSY breaking sector, we argue that supersymmetric baryogenesis at the weak scale is no longer possible since the scenario is most likely already experimentally excluded by SM Higgs searches. In the case of the minimal $SU(5)$ supergravity model, the same conclusions apply; although new sources of CP-violation arise in the GUT-scale sector, $\Delta(B - L) = 0$ and a GUT scale BAU would also be washed out. Higher dimensional, non-renormalizable interactions which lead R-parity breaking may be required [32,33]; this may arise from as yet unknown 'Planck slop' effects. In the no-scale flipped $SU(5)$ model, a natural mechanism for baryogenesis exists, and arises in the neutrino sector, linking together baryogenesis, solar neutrino physics and dark matter issues.

---

[2] Higher order effects may lower this upper limit to $m_h^{max} \lesssim 50\,\text{GeV}$.

References

[1] D. A. Kirzhnits and A. D. Linde, Phys. Lett. B **42** (1972) 471.

[2] S. Weinberg, Phys. Rev. D **9** (1974) 3357.

[3] A. Masiero, T. Yanagida and D. V. Nanopoulos, Phys. Lett. B **138** (1984) 91; T. W. Kephart, T. J. Weiler and T. C. Yuan, Nucl. Phys. B **330** (1990) 705.

[4] F. R. Klinkhamer and N. S. Manton, Phys. Rev. D **30** (1984) 2212.

[5] N. S. Manton, Phys. Rev. D **28** (1983) 2019.

[6] V. A. Kuzmin, V. A. Rubakov and M. E. Shaposhnikov, Phys. Lett. B **155** (1985) 36.

[7] M. Shaposhnikov, JETP Lett.**44** (1986) 465; Nucl. Phys. B **287** (1987) 757; Nucl. Phys. B **299** (1988) 797; L. McLerran, Phys. Rev. Lett. **62** (1989) 1075.

[8] M. Dine, R. G. Leigh, P. Huet, A. Linde and D. Linde, Phys. Rev. D **46** (1992) 550; C. G. Boyd, D. E. Brahm and S. D. H. Hsu, CALT-68-1795, HUTP-92-A027, EFI-92-22; D. E. Brahm and S. D. H. Hsu, CALT-68-1705, HUTP-91-A063; M. E. Carrington, Phys. Rev. D **45** (1992) 2933; J. R. Espinosa, M. Quiros and F. Zwirner, Phys. Lett. B **291** (1992) 115.

[9] L. Dolan and R. Jackiw, Phys. Rev. D **9** (1974) 3320.

[10] L. McLerran, M. Shaposhnikov, N. Turok and M. Voloshin, Phys. Lett. B **256** (1991) 451.

[11] M. Dine, P. Huet and R. Singleton, Nucl. Phys. B **375** (1992) 625.

[12] A. G. Cohen, D. B. Kaplan and A. E. Nelson, Phys. Lett. B **263** (1991) 86.

[13] N. Turok and J. Zadrozny, Nucl. Phys. B **369** (1992) 729.

[14] A. G. Cohen and A. E. Nelson, Phys. Lett. B **297** (1992) 111; M. Dine, P. Huet, R. Singleton and L. Susskind, Phys. Lett. B **257** (1992) 351; A. Masiero and A. Riotto, Phys. Lett. B **289** (1992) 73; G. F. Giudice, A. Masiero, M. Pietroni and A. Riotto, CERN-TH 6656/92, DFPD 92/TH/43, SISSA-127/92/AP.

[15] R. N. Mohapatra and X. Zhang, Phys. Rev. D **46** (1992) 5331.

[16] M. Fukugita and T. Yanagida, Phys. Lett. B **174** (1986) 45.

[17] A. G. Cohen, D. B. Kaplan and A. E. Nelson, Nucl. Phys. B **349** (1992) 727.

[18] J.-F. Grivaz, Invited talk at the workshop on 'Ten Years of SUSY Confronting Experiment', CERN, Geneva, 7-9 September 1992, LAL 92-59.

[19] S. Myint, Phys. Lett. B **287** (1992) 325.

[20] G. F. Giudice, Phys. Rev. D **45** (1992) 3177.

[21] D. V. Nanopoulos and H. Pois, CTP-TAMU-87/92

[22] G. Anderson and L. Hall, Phys. Rev. D **45** (1992) 2685.

[23] M. Sher, Phys. Rep. **179** (1989) 273.

[24] A. I. Bochkarev and M. E. Shaposhnikov, Mod. Phys. Lett. A **2** (1987) 417.

[25] L. Hall, J. Lykken and S. Weinberg, Phys. Rev. D **27** (1983) 2359; P. Nath, R. Arnowitt and A. H. Chamseddine, Nucl. Phys. B **227** (1983) 121.

[26] J. Ellis, S. Ferrara and D. V. Nanopoulos, Phys. Lett. B **114** (1982) 231; M. Dugan, B. Grinstein and L. Hall, Nucl. Phys. B **255** (1985) 413

[27] S. Kelley, J. L. Lopez, D. V. Nanopoulos, H. Pois and K. Yuan, Nucl. Phys. B **398** (1993) 3.

[28] R. Arnowitt and P. Nath, Phys. Rev. D **46** (1992) 3981.

[29] S. Coleman and E. Weinberg, Phys. Rev. D **7** (1973) 1888.

[30] J. L. Lopez, D. V. Nanopoulos, H. Pois, Xu Wang and A. Zichichi, Phys. Lett. B **306** (1993) 73.

[31] J. Ellis, D. V. Nanopoulos and K. Olive, Phys. Lett. B **300** (1993) 121

[32] B. A. Campbell, S. Davidson, J. Ellis and K. A. Olive, Phys. Lett. B **256** (1991) 457.

[33] R. Arnowitt and P. Nath, Phys. Lett. B **299** (1993) 58.

ONE LOOP CONSTRAINTS ON SUPERGRAVITY MODELS

GYE T. PARK
Center for Theoretical Physics, Department of Physics, Texas A&M University
College Station, TX 77843-4242, USA
and
Astroparticle Physics Group, Houston Advanced Research Center (HARC)
The Woodlands, TX 77381, USA

1. Introduction

Despite the fact that there is still no single experimental evidence for the existence of any supersymmetric partner to the Standard Model (SM) particles, there are some remarkable *indirect* results which indicate that superpartners may be operative at the $\mathcal{O}(1\,\text{TeV})$ scale, and involve several predictions and constraints which could be regarded as much more than coincidence. The most dramatic one is the high-precision unification of the running gauge couplings at superhigh energies in the minimal $SU(5)$ supergravity model, only when virtual sparticles come into play at the TeV scale and modify the running of the gauge couplings [1]. Virtual supersymmetric effects have long been known to be a possible menace to supersymmetric models because of the potentially large one-loop induced flavor-changing-neutral-current (FCNC) processes such as $K-\bar{K}$ mixing and CP violation in the K system [2], requiring high degeneracy of squark masses. Supergravity models with universal soft-supersymmetry breaking naturally tame these effects. More recently, one-loop supersymmetric contributions to FCNC process $b\to s\gamma$ have been shown to be a possible deep probe of supersymmetric models [3].

In this work we explore the possibilities that experimentally testable virtual effects through one-loop electroweak (EW) radiative corrections as well as an interesting FCNC process $b\to s\gamma$ can indeed constrain a class of supergravity models. In particular, we explore the minimal $SU(5)$ [4] and the no-scale flipped $SU(5)$ supergravity models [7] which can be considered to be prototype traditional versus string-inspired supergravity models. This work is based on the very recent phenomenological studies of the supergravity models[5, 6]. Here, we present in the context of the minimal and flipped $SU(5)$ supergravity models a complete study of one-loop EW radiative corrections[6], incorporating the most recent global fits to the precision measurements from LEP as well as a study of $b\to s\gamma$[5], incorporating the recently announced improved CLEO bound on $BR(b\to s\gamma)$[37].

In the following section, we briefly discuss the overall motivation and structure of the minimal and flipped $SU(5)$ models that we study here. In Sec. 3, we present the study of EW radiative corrections. In Sec. 4 we discuss the new analysis of $b\to s\gamma$. And we finally conclude in Section 5.

2.The minimal and flipped SU(5) supergravity models

The minimal [4] and no-scale flipped [7] $SU(5)$ supergravity models that we consider here can be regarded as prototypes for realistic traditional versus string-inspired supergravity unified models. At low energy, both contain (i) the SM $SU(3) \times SU(2) \times U(1)$ gauge symmetry which is radiatively broken at the weak scale, and (ii) the three SM generations and two Higgs doublets of matter representations at the EW scale (along with their superpartners). There are however, several crucial differences between the two models: (i) each unifies into a larger (but different) gauge group, namely $SU(5)$ versus $SU(5) \times U(1)$, (ii) the unification scale in the minimal $SU(5)$ model is $M_U \approx 10^{16}\,\mathrm{GeV}$, whereas in the flipped case, the (string) unification scale is $M_U \approx 10^{18}\,\mathrm{GeV}$, close to the Planck scale. Unification at this higher scale is due to the effects of additional, vector-like representations $(Q, \bar{Q}, D, \bar{D})$ that naturally fit into the $\mathbf{10}, \overline{\mathbf{10}}$ flipped $SU(5)$ representations. At the unification scale the heavy field content is also quite different, and leads to different conclusions regarding proton decay: in the minimal $SU(5)$ model proton decay is highly constraining [10, 11], whereas in the flipped model it is not. Finally, (iii) the pattern of soft-supersymmetry breaking at the unification scale is quite different: in the flipped case, the sole source of SUSY breaking is due to a universal gaugino mass ($m_{1/2}$) as is typical in unified no-scale supergravity models [21], whereas in the minimal $SU(5)$ model, universal scalar (m_0) and trilinear (A) contributions must also be included.

The constraints of gauge and Yukawa unification along with the SUSY breaking assumptions and the satisfaction of all the consistency and phenomenological constraints on these models lead to a restricted five-dimensional parameter space in this class of models [8]. Besides the three SUSY breaking parameters ($m_{1/2}, \xi_0 \equiv m_0/m_{1/2}, \xi_A \equiv A/m_{1/2}$; $\xi_0 = \xi_A = 0$ in the flipped model) one also has the top-quark mass (m_t), and the ratio of Higgs vacuum expectation values ($\tan\beta = v_2/v_1$) defined at the EW scale. The sign of the superpotential Higgs mixing term μ is also undetermined, and both cases ($\pm|\mu|$) must be considered. Primarily as a result of the quite different pattern of SUSY breaking, the low-energy predictions for the squark, slepton, chargino and neutralino mass spectra are quite distinct, and lead to strikingly different phenomenology in the two models (see Refs. [9, 10, 11, 12, 13, 14, 22] for detailed discussions).

As described in detail in Ref. [8], our strategy for studying these models involves a discrete sampling of the ($m_{1/2}, \xi_0, \xi_A, \tan\beta, m_t$) parameter space over their allowed domain. Several consistency and phenomenological constraints restrict the range of the model parameters and yield an allowed region in parameter space. See Refs. [11] and [7] for the determination of the parameter spaces in the minimal and flipped $SU(5)$ supergravity models respectively. In the remainder of this work we explore the constraints arising from one-loop corrections.

3. One-loop EW radiative corrections and the $\epsilon_{1,2,3}$ parameters

Recently, several schemes have been introduced which effectively parametrize the EW vacuum polarization corrections [24, 25, 26]. One can easily show that an expansion of the vacuum polarization tensors to order q^2, results in three independent physical parameters. Alternatively, one can show from an effective field theory point of view, that there are three additional terms in the lagrangian. In the (S, T, U) scheme [25], a SM reference value for $m_t, m_{H_{SM}}$ is used, and the deviation from this reference is calculated and is considered to be "new" physics. This scheme is only valid to lowest order in q^2, and is therefore not applicable to a theory with new, light ($\sim M_Z$) particles. In the two supergravity models we explore here, each point in parameter space is actually a distinct model, and a SM reference point is not meaningful. For these reasons, we choose to use the scheme of Refs. [26, 27, 19]

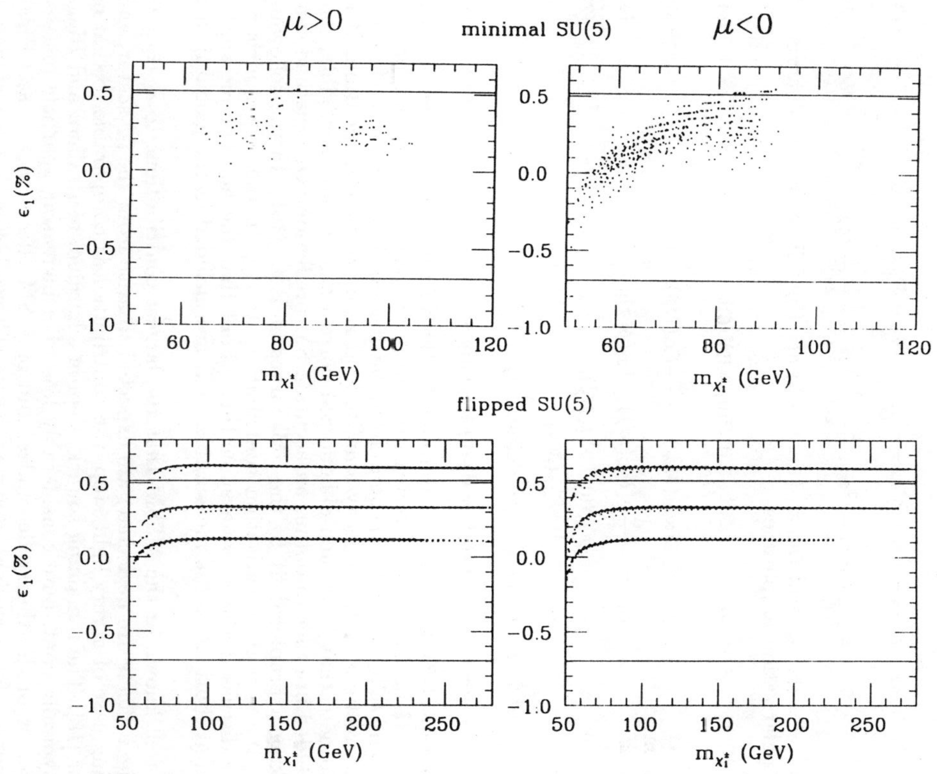

Fig. 1: The total contribution to ϵ_1 as a function of the lightest chargino mass $m_{\chi_1^\pm}$ for the minimal $SU(5)$ supergravity model(top row) and for the no-scale flipped $SU(5)$ supergravity model(bottom row). Points between the two horizontal solid lines are allowed at 90% CL.

where the contributions are *absolute* and valid to higher order in q^2. This $\epsilon_{1,2,3}$ scheme is therefore more applicable to a global fit of the supergravity models we consider here. Regardless of the scheme used, all of the global fits to the three physical parameters are *entirely consistent* with the SM at 90% CL.

In the $\epsilon_{1,2,3}$ scheme these parameters can be written as follows [19, 28],

$$\epsilon_1 = e_1 - e_5 - \frac{\delta G_{V,B}}{G} - 4\delta g_A, \tag{1}$$

$$\epsilon_2 = e_2 - s^2 e_4 - c^2 e_5 - \frac{\delta G_{V,B}}{G} - \delta g_V - 3\delta g_A, \tag{2}$$

$$\epsilon_3 = e_3 + c^2 e_4 - c^2 e_5 + \frac{c^2 - s^2}{2s^2}\delta g_V - \frac{1+2s^2}{2s^2}\delta g_A, \tag{3}$$

where $s^2 = 1 - c^2 = \sin^2\theta_W$ and the e_i, $(i = 1 \to 5)$ are the following combinations of the vacuum polarization amplitudes

$$e_1 = \frac{\alpha}{4\pi s^2 M_W^2}[\Pi_T^{33}(0) - \Pi_T^{11}(0)], \tag{4}$$

$$e_2 = F_{WW}(M_W^2) - \frac{\alpha}{4\pi s^2}F_{33}(M_Z^2), \tag{5}$$

$$e_3 = \frac{\alpha}{4\pi s^2}[F_{3Q}(M_Z^2) - F_{33}(M_Z^2)], \tag{6}$$

$$e_4 = F_{\gamma\gamma}(0) - F_{\gamma\gamma}(M_Z^2), \tag{7}$$

$$e_5 = M_Z^2 F'_{ZZ}(M_Z^2), \tag{8}$$

and the $q^2 \neq 0$ contributions $F_{ij}(q^2)$ are defined by

$$\Pi_T^{ij}(q^2) = \Pi_T^{ij}(0) + q^2 F_{ij}(q^2). \tag{9}$$

The $\delta g_{V,A}$ in Eq. (1-3) are the contributions to the vector and axial-vector form factors at $q^2 = M_Z^2$ in the $Z \to l^+l^-$ vertex from proper vertex diagrams and fermion self-energies, and $\delta G_{V,B}$ in Eq. (1-2) comes from the one-loop box, vertex and fermion self-energy corrections to the μ-decay amplitude at zero external momentum. It is important to note that these non-oblique SM corrections are non-negligible, and must be included in order to obtain an accurate SM prediction. Our method consists of making a graphical fit to the SM curves in Ref. [19]. In the following we calculate the vacuum polarization contributions to $\epsilon_{1,2,3}$ in the two models we consider. As discussed above, we assume throughout that the non-oblique supersymmetric contributions to the measurables that are included in the global fit are negligible.

It is well known in the MSSM that the largest contributions to ϵ_1 (*i.e.*, $\delta\rho$ if q^2-dependent effects are neglected) are expected to arise from the $\tilde{t}$-$\tilde{b}$ sector, and in the limiting case of a very light stop, the contribution is comparable to that of the t-b sector [18]. The remaining squark, slepton, chargino, neutralino, and Higgs sectors all typically contribute considerably less. For increasing sparticle masses, the heavy sector of the theory decouples, and only SM effects with a *light* Higgs survive. However, for very light chargino, as we will see below, a Z-wavefunction renormalization threshold effect can introduce a substantial q^2-dependence in the calculation, thus modifying significantly the standard $\delta\rho$ results. For completeness, we include the complete vacuum polarization contributions from the Higgs sector, the supersymmetric chargino-neutralino and sfermion sectors, and also the corresponding contributions in the SM. Our analytical expressions for the $\Pi_T^{ij}(q^2)$ agree with those given in Refs. [17, 18] as well as other existing references [31].

In Fig. 1 we show the calculated values of ϵ_1 versus the lightest chargino mass ($m_{\chi_1^\pm}$), for the sampled points in two supergravity models. In the flipped model three representative values of m_t were used, $m_t = 100, 130, 160\,\mathrm{GeV}$, whereas in the minimal model , several other values for m_t in the range $90\,\mathrm{GeV} \leq m_t \leq 160\,\mathrm{GeV}$ were sampled. In both models, but most clearly in the flipped model one can see how quickly the sparticle spectrum decouples as $m_{\chi_1^\pm}$ increases, and the value of ϵ_1 asymptotes to the SM value appropriate to each value of m_t and for a *light* ($\sim 100\,\mathrm{GeV}$) Higgs mass. This result is in agreement with Ref. [22] where we have shown that the Higgs sector is SM-like for virtually all points in these models. The threshold effect of $\chi_1^\pm$ is manifest as $m_{\chi_1^\pm} \to \frac{1}{2}M_Z$ and is especially visible for $\mu < 0$ in both models. This systematic, negative contribution comes from the F' term in e_5 which is considered to be a Z-wave-function renormalization effect [19]. Note from Eq. (1-3) that all ϵ's are affected similarly.

Recent values for $\epsilon_{1,2,3}$ obtained from a global fit to the LEP (*i.e.*, $\Gamma_l, A_{FB}^{l,b}, A_{pol}^{\tau}$) and M_W/M_Z measurements are [28],

$$\epsilon_1 = (-0.9 \pm 3.7)10^{-3}, \quad \epsilon_2 = (9.9 \pm 8.0)10^{-3}, \quad \epsilon_3 = (-0.9 \pm 4.1)10^{-3}. \tag{10}$$

For ϵ_1 it is clear that virtually all the sampled points in the minimal $SU(5)$ supergravity model are within the LEP bounds at 90% CL (denoted by the two horizontal solid lines in the figures). Since several values for $90 \leq m_t \leq 160\,\mathrm{GeV}$ were sampled, the trends for fixed m_t are not very clear from the figure. Nonetheless, the points just outside the LEP bounds correspond to $m_t = 160\,\mathrm{GeV}$, which are therefore excluded at the 90% CL. We note that we have imposed the improved experimental constraint $m_h > 60\,\mathrm{GeV}$ [22], which has the effect of removing many of the points corresponding to very light $\chi_1^\pm$, particularly for $\mu > 0$. We have also imposed $m_{\chi_1^\pm} > 50\,\mathrm{GeV}$ as an accuracy cut due to the threshold effect mentioned above.

In the flipped model, the upper bound on m_t depends sensitively on the chargino mass. For example, for $m_t = 160\,\mathrm{GeV}$, only light chargino masses would be acceptable at 90% CL. In fact, we have scanned the region $130\,\mathrm{GeV} \leq m_t \leq 190\,\mathrm{GeV}$ in increments of $5\,\mathrm{GeV}$ and obtained the maximum values for $m_{\chi_1^\pm}$ allowed by the experimental value for ϵ_1 at 90% CL. We immediately find that there is a strong correlation between m_t and $m_{\chi_1^\pm}$: as m_t rises, the upper limit to $m_{\chi_1^\pm}$ falls, and vice versa. In particular, for $m_t \leq 150\,\mathrm{GeV}$ all values of $m_{\chi_1^\pm}$ are allowed, while one could have m_t as large as $160\,(175)\,\mathrm{GeV}$ for $\mu > 0\,(\mu < 0)$ if the chargino mass were light enough($m_{\chi_1^\pm} < 60\,\mathrm{GeV}$). Moreover, since $m_t > 160\,\mathrm{GeV}$ requires $m_{\chi_1^\pm} < 70\,\mathrm{GeV}$, upper bounds on many other sparticle masses follow, since all scale with $m_{1/2}$. In particular, the bounds on $m_{\tilde{t}_1}$ and $m_{\tilde{g}}$ can be as light as $180\,\mathrm{GeV}$ and $240\,\mathrm{GeV}$,respectively for $\mu > 0$. Note that with $100\,\mathrm{pb}^{-1}$ of integrated luminosity, the Tevatron may be able to probe up to $m_{\tilde{g}} \approx m_{\tilde{q}} \approx 250\,\mathrm{GeV}$ through the missing p_T+ jets signal [33]. Thus, the full gluino and squark mass range (for the $\mu > 0$) may be accessible at the Tevatron if the top quark is not seen.

Although we do not present the figures for ϵ_2 and ϵ_3 here,they do not constrain the models at the present level of experimental accuracy.As is well known, $\epsilon_{2,3}$ depend on the top-quark mass only as $\ln(m_t/M_Z)$, and are therefore more sensitive to new physics. This weak dependence on m_t makes these variables potentially good probes of new physics, especially if the top-quark is not within the reach of the Tevatron. For the future, it is expected that 2×10^6 Z-events/experiment at LEPI will lower the uncertainties for $\epsilon_{1,3}$ to $\pm 2.0 \times 10^{-3}$ [28]. Depending on the new central values, one may be able to make definitive statements about constraints on the parameter spaces of these models.

4. Probing Supergravity Models with the $b \to s\gamma$ Microscope

With a recent renewed surge of interest on the flavor-changing-neutral-current (FCNC) $b \to s\gamma$ decay, prompted by the CLEO bound $\mathrm{BR}(b \to s\gamma) < 8.4 \times 10^{-4}$ at 90% CL [34], most recently it has been pointed out [35, 36] that the charged Higgs ($H^{\pm}$) contributions in two-Higgs-doublet models (like the minimal supersymmetric standard model (MSSM)) would violate the CLEO bound if $m_{H^\pm}$ is too light. The recently announced improved CLEO bound of $\mathrm{BR}(b \to s\gamma) < 5.4 \times 10^{-4}$ at 95% CL [37], leads to yet larger excluded domains in the charged Higgs parameter space [33]. In these papers it remained an open question whether inclusion of the full set of supersymmetric contributions to the $b \to s\gamma$ amplitude would change significantly the quoted bounds, and thus the outlook for supersymmetric Higgs searches at LEPII. The full supersymmetric contributions to $b \to s\gamma$ have been known for some time [38] and have been recently given in a justifiably simplified form in Ref. [39]. The latter paper argues that the bound on $m_{H^\pm}$ in Refs. [35, 36] could well be evaded in the full supersymmetric calculation since in the unbroken supersymmetric limit the $b \to s\gamma$ amplitude vanishes.

We show below that the *new* CLEO bound does not yet constrain the minimal $SU(5)$ model, while some as-yet-mild constraints are imposed on the flipped $SU(5)$ model, where a new phenomenon can drastically suppress the $b \to s\gamma$ amplitude. We present the results for $\mathrm{BR}(b \to s\gamma)$ in these two models and show that improved sensitivity could probe them in ways not possible at present collider experiments.

We use the following expression for the branching ratio $b \to s\gamma$ [39]

$$\frac{\mathrm{BR}(b \to s\gamma)}{\mathrm{BR}(b \to ce\bar{\nu})} = \frac{6\alpha}{\pi} \frac{\left[\eta^{16/23}A_\gamma + \frac{8}{3}(\eta^{14/23} - \eta^{16/23})A_g + C\right]^2}{I(m_c/m_b)\left[1 - \frac{2}{3\pi}\alpha_s(m_b)f(m_c/m_b)\right]}, \tag{11}$$

where $\eta = \alpha_s(M_Z)/\alpha_s(m_b)$, I is the phase-space factor $I(x) = 1 - 8x^2 + 8x^6 - x^8 - 24x^4 \ln x$, and $f(m_c/m_b) = 2.41$ the QCD correction factor for the semileptonic decay. The A_γ, A_g are the coefficients of the effective $bs\gamma$ and bsg penguin operators evaluated at the scale M_Z. Their simplified expressions are given in the Appendix of Ref. [39], where the gluino and neutralino contributions have been justifiably neglected [38] and the squarks are considered degenerate in mass, except for the $\tilde{t}_{1,2}$ which are significantly split by m_t. This is a fairly good approximation to the actual result obtained in the two supergravity models we consider below, since $m_{\tilde{q}} > 200\,\mathrm{GeV}$ in these models. To include at least partially the QED corrections to the $b \to s\gamma$ operator, in Eq. (11) we take $\alpha = \alpha(m_b) = 1/131.2$. We use the 3-loop expressions for α_s and choose Λ_{QCD} to obtain $\alpha_s(M_Z)$ consistent with the recent measurements at LEP. For the numerical evaluation of Eq. (11) we take $\alpha_s(M_Z) = 0.118$, $\mathrm{BR}(b \to ce\bar{\nu}) = 10.7\%$, $m_b = 4.8\,\mathrm{GeV}$ and $m_c/m_b = 0.3$.

In Fig. 2 (top row) we present $\mathrm{BR}(b \to s\gamma)$ for the minimal $SU(5)$ supergravity model, plotted against one of the light sparticle masses in the model, namely the lightest chargino mass. We find

$$2.3\,(2.6) \times 10^{-4} < \mathrm{BR}(b \to s\gamma)_{minimal} < 3.6\,(3.3) \times 10^{-4}, \tag{12}$$

for $\mu > 0\,(\mu < 0)$, which are all within the new CLEO bound. Since these numbers are within the SM range, we obtain their ratios to the corresponding SM contribution,

$$0.90\,(0.97) < \mathrm{BR}(b \to s\gamma)_{minimal}/\mathrm{BR}(b \to s\gamma)_{SM} < 1.20\,(1.13), \tag{13}$$

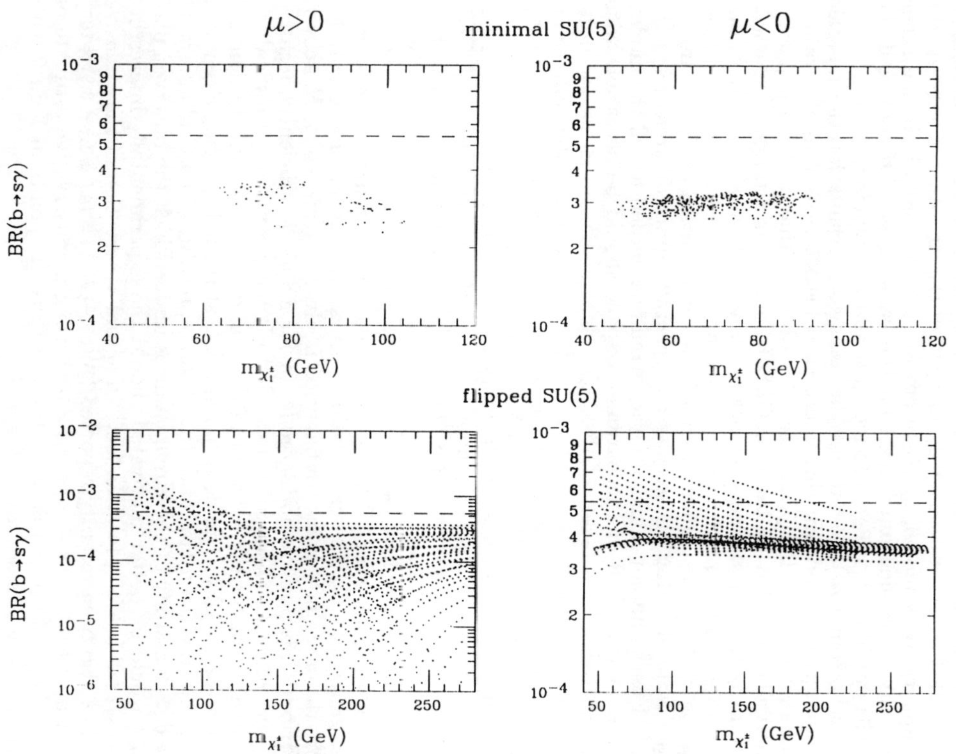

Fig. 2: The calculated values of BR($b \to s\gamma$) versus the chargino mass in the minimal $SU(5)$ supergravity model(top row) and in the no-scale flipped $SU(5)$ supergravity model(bottom row). The horizontal dashed line represents the new CLEO upper bound.

for $\mu > 0\,(\mu < 0)$. This implies that $\mathrm{BR}(b \to s\gamma)$ would need to be measured with better than 20% accuracy to start disentangling the minimal $SU(5)$ supergravity model from the SM. Moreover, for $\mu < 0$ there is a band of points which will be difficult to tell apart (requires $<$ 1% accuracy). It is interesting to remark that an analogous set of plots versus m_t instead, does not reveal any particular structure in m_t, thus an independent determination of m_t is not likely to constrain the possible values of $\mathrm{BR}(b \to s\gamma)$. This is in sharp contrast with the SM, where such a measurement would determine $\mathrm{BR}(b \to s\gamma)$, up to QCD uncertainties. In other words, in the minimal $SU(5)$ supergravity model, a measurement of $\mathrm{BR}(b \to s\gamma)$ within the range of Eq. (13), is unlikely to shed much light on the value of m_t. Of course, if the measurement falls outside this range, the model would be excluded.

In Fig. 2(bottom row) we present the analogous results for the no-scale flipped $SU(5)$ supergravity model(in this figure $m_t = 100, 130, 160\,\mathrm{GeV}$.). The results are strikingly different than in the prior case. One observes that part of the parameter space is actually excluded by the new CLEO bound, for a range of sparticle masses. However, light charged Higgs masses are not excluded altogether. It can be as light as $150\,\mathrm{GeV}$. Perhaps the most surprising feature of the results is the strong suppression of $\mathrm{BR}(b \to s\gamma)$ which occurs for a good portion of the parameter space for $\mu > 0$. Note also that the results are generally quite different from the corresponding SM ones and not much accuracy would be needed to tell the models apart. Moreover, at least half of the parameter space should be accessible with an increase in sensitivity by a factor of two. The dependence on m_t is also quite significant in this model, and for $m_t = 100\,\mathrm{GeV}$ and $\mu < 0$, the calculated $\mathrm{BR}(b \to s\gamma)$ values are quite large and are starting to be constrained: note the set or nearly parallel lines intersecting the CLEO bound for $\mu < 0$.

The large suppression for $\mu > 0$ in the flipped model requires further explanation. What happens is that in Eq. (11), the A_γ term nearly cancels against the QCD correction factor C; the A_g contribution is small. The A_γ amplitude receives three contributions: from the W-t loop (*i.e.*, the SM contribution), from the $H^\pm$-t loop, and from the $\chi^\pm_{1,2}$-$\tilde{q}$ loop. The first two contributions are always negative, whereas the last one can have either sign. When the chargino-squark contribution is positive and in the range $0.4 - 0.6$, the cancellation occurs. No cancellation occurs for $\mu < 0$ since there the chargino-squark contribution is always below ≈ 0.15. Moreover, when this contribution is positive and large (> 1.2 for $\mu > 0$) or sufficiently negative (< -0.05 for $\mu < 0$), the CLEO bound is exceeded. It has proven quite difficult to determine a simple reason for the behavior of the chargino-squark contribution and its sign-of-μ dependence. All we can do is to remark that the lightest stop eigenstate $\tilde{t}_1$ is lighter for $\mu > 0$ and it can approach the $\chi^\pm_2$ mass, resulting in an enhancement in A_γ. A more detailed study of the circumstances under which the observed phenomenon occurs would be better motivated should the experimentally interesting range of $\mathrm{BR}(b \to s\gamma)$ fall below $\sim 10^{-4}$. In fact, an actual observation of the exclusive $B \to K^*\gamma$ mode has also been recently reported ($\mathrm{BR}(B \to K^*\gamma) = (4.5 \pm 1.5 \pm 0.9) \times 10^{-5}$ [37]), which indicates that $\mathrm{BR}(b \to s\gamma) > 6.0 \times 10^{-5}$ using a conservative estimate of the ratio of exclusive to inclusive decay rates and thus the points in Fig. 2 with very suppressed $\mathrm{BR}(b \to s\gamma)$ values are disfavored.

5. Conclusions

In this work, we have explored the minimal $SU(5)$ and the no-scale flipped $SU(5)$ supergravity model contributions to the three physical one-loop EW correction parameters, and have compared these results to the latest global fit to the LEP precision measurements. We include the *complete* sparticle spectrum, and also include q^2-dependent effects which are important for a light chargino spectrum. This effect is due to a Z-boson wavefunction renormalization threshold effect, and systematically lowers the values of $\epsilon_{1,2,3}$. As we have shown, the effect is most significant for $m_{\chi_1^\pm} <$ 70 GeV, where light $\chi_1^\pm$ can reduce the value for ϵ_1 to fall within the experimental limits for m_t as large as 175 GeV in the flipped model. However, this region for $\chi_1^\pm$ is quite restrictive, and should be thoroughly explored at LEP II [13]. Moreover, in this case the Tevatron should be able to explore at least half of the parameter space of the flipped model through gluino and squark production. We also conclude through a new analysis on $b \to s\gamma$ that the $b \to s\gamma$ microscope can probe the parameter spaces of the two models in ways not possible with present high-energy collider experiments. With further improvements in sensitivity to the inclusive $b \to s\gamma$ mode, it should not be long before these models begin to show their true colors.

Acknowledgments

This work has been supported by World-Laboratory Fellowship.

6. References

1. J. Ellis, S. Kelley and D. V. Nanopoulos, Phys. Lett. B **249** (1990) 441, Phys. Lett. B **260** (1991) 131, Nucl. Phys. B **373** (1992) 55; P. Langacker and M.-X. Luo, Phys. Rev. D **44** (1991) 817; F. Anselmo, L. Cifarelli, A. Peterman, and A. Zichichi, Nuovo Cim. **104A** (1991) 1817 and **105A** (1992) 581.
2. J. Ellis and D. V. Nanopoulos, Phys. Lett. B **110** (1982) 44.
3. S. Bertolini, F. Borzumati, A. Masiero, and G. Ridolfi, Nucl. Phys. B **353** (1991) 591; V. Barger, M. Berger, and R. J. N. Phillips, Phys. Rev. Lett. **70** (1993) 1368; J. Hewett, Phys. Rev. Lett. **70** (1993) 1045; R. Barbieri and G. Giudice, CERN-TH.6830/93.
4. For reviews see: R. Arnowitt and P. Nath, *Applied N=1 Supergravity* (World Scientific, Singapore 1983); H. P. Nilles, Phys. Rep. **110** (1984) 1.
5. J. L. Lopez, D. V. Nanopoulos, and Gye T. Park, Texas A & M University preprint CTP-TAMU-16/93 (to appear in Phys. Rev. D).
6. J. L. Lopez, D. V. Nanopoulos, G. T. Park, H. Pois, and K. Yuan Texas A & M University preprint CTP-TAMU-19/93.
7. J. L. Lopez, D. V. Nanopoulos, and A. Zichichi, CERN-P-TH.6667/92, CTP-TAMU-68/92.
8. S. Kelley, J. L. Lopez, D. V. Nanopoulos, H. Pois, and K. Yuan, Texas A & M University preprint CTP-TAMU-16/92 and CERN-TH.6498/92 (to appear in Nucl. Phys. B).
9. J. L. Lopez, D. V. Nanopoulos, and A. Zichichi, Phys. Lett. B **291** (1992) 255.
10. J. L. Lopez, D. V. Nanopoulos, and H. Pois, Phys. Rev. D **47** (1993) 2468.
11. J. L. Lopez, D. V. Nanopoulos, H. Pois, and A. Zichichi, Phys. Lett. B **299** (1993) 262.
12. J. L. Lopez, D. V. Nanopoulos, X. Wang, and A. Zichichi, CERN/PPE/92-194, CTP-TAMU-76/92.
13. J. L. Lopez, D. V. Nanopoulos, H. Pois, X. Wang, and A. Zichichi, CERN-

PPE/93-16, CTP-TAMU-89/92.
14. J. L. Lopez, D. V. Nanopoulos, X. Wang, and A. Zichichi, CERN-PPE/93-64, CTP-TAMU-15/93.
15. E. Eliasson, Phys. Lett. B **147** (1984) 67; S. Lim, *et. al.*, Phys. Rev. D **29** (1984) 1488; J. Grifols and J. Sola, Nucl. Phys. B **253** (1985) 47; B. Lynn, *et. al.*, in Physics at LEP, eds. J. Ellis and R. Peccei, CERN Yellow Report CERN86-02, Vol. 1; R. Barbieri et al., Nucl. Phys. B **341** (1990) 309.
16. A. Bilal, J. Ellis, and G. Fogli, Phys. Lett. B **246** (1990) 459.
17. M. Drees, K. Hagiwara, and A. Yamada, Phys. Rev. D **45** (1992) 1725.
18. M. Drees and K. Hagiwara, Phys. Rev. D **42** (1990) 1709.
19. R. Barbieri, M. Frigeni, and F. Caravaglios, Phys. Lett. B **279** (1992) 169.
20. G. Altarelli, R. Barbieri, and F. Caravaglios, CERN-TH.6770/93.
21. For a review see, A. B. Lahanas and D. V. Nanopoulos, Phys. Rep. **145** (1987) 1.
22. J. L. Lopez, D. V. Nanopoulos, H. Pois, X. Wang, and A. Zichichi, CERN-PPE/93-17, CTP-TAMU-05/93 (to appear in Phys. Lett. B).
23. Particle Data Group, Phys. Rev. D **45** (1992) S1.
24. D. Kennedy and B. Lynn, Nucl. Phys. B **322** (1989) 1; D. Kennedy, B. Lynn, C. Im, and R. Stuart, Nucl. Phys. B **321** (1989) 83.
25. M. Peskin and T. Takeuchi, Phys. Rev. Lett. **65** (1990) 964; W. Marciano and J. Rosner, Phys. Rev. Lett. **65** (1990) 2963; D. Kennedy and P. Langacker, Phys. Rev. Lett. **65** (1990) 2967.
26. G. Altarelli and R. Barbieri, Phys. Lett. B **253** (1990) 161.
27. G. Altarelli, R. Barbieri, and S. Jadach, Nucl. Phys. B **369** (1992) 3.
28. R. Barbieri, CERN Report No. CERN-TH.6659/92 (unpublished).
29. S. Sarantakos, A. Sirlin, and W. J. Marciano, Nucl. Phys. B **217** (1983) 84.
30. A. Sirlin, Phys. Rev. D **22** (1980) 971.
31. C. S. Lim, T. Inami, and N. Sakai, Phys. Rev. D **29** (1984) 1488; B. A. Kniehl, Nucl. Phys. B **352** (1991) 1.
32. D. Levinthal, F. Bird, R. Stuart, and B. Lynn, Z. Phys. C, **53** (1992) 617.
33. See, for example, V. Barger, talk presented at the HARC workshop "Recent Advances in the Superworld", The Woodlands, Texas, April 1993.
34. M. Battle, *et. al.* (CLEO Collab.) in Proceedings of the joint Lepton-Photon and Europhysics Conference on High-Energy Physics, Geneva 1991.
35. V. Barger, M. Berger, and R. J. N. Phillips, Phys. Rev. Lett. **70** (1993) 1368.
36. J. Hewett, Phys. Rev. Lett. **70** (1993) 1045.
37. E. Thorndike, CLEO Collab., talk given at the 1993 Meeting of the American Physical Society, Washington D.C., April 1993.
38. S. Bertolini, F. Borzumati, A. Masiero, and G. Ridolfi, Nucl. Phys. B **353** (1991) 591.
39. R. Barbieri and G. Giudice, CERN-TH.6830/93.

Superstrings and Phenomenology

SUPERSTRING and BEYOND: FRACTIONAL SUPERSTRING

S.-H. Henry Tye
Newman Laboratory of Nuclear Studies
Cornell University
Ithaca, NY 14853-5001

Abstract

We comment on the present status of the superstring theory and its phenomenological implications. We then argue why it is important to search for consistent string theories beyond the superstring theory. In particular, we consider some promising possibilities, namely, fractional superstrings. More specifically, we present the $K = 4$, or spin-4/3 fractional superstring case. This theory is characterized by a chiral algebra involving a spin-4/3 current on the world-sheet in addition to the energy-momentum tensor. Scattering amplitudes of this theory are described in some detail. They satisfy both spurious state decoupling and cyclic symmetry (duality). Examples of such amplitudes are calculated using an explicit $c = 5$ realization of the spin-4/3 current algebra. This representation has a 3-dimensional flat Minkowski space-time picture, with fermions and bosons only in its spectrum.

1 Introduction

String theory[1] is the only known theory with the potential for describing all matter and forces in nature in a unified way. In particular, the superstring theory and the closely related heterotic string theory entail many structures, including gravity, gauge fields and chiral fermions, that are central to the present understanding of our universe. However, their critical space-time dimension is ten, and there are numerous mechanisms to reduce the number of observable space-time dimensions. Although the number of models in ten dimensions is very limited, namely the SO(32) and the $E_8 X E_8$ models[2], the number of models with four flat space-time dimension is huge[3]. Furthermore, there is no known compelling reason why the superstring theory should have only four large space-time dimensions. While it is important to search for dynamical and/or symmetry reasons explaining how our universe could be realized in the heterotic/superstring framework, it is also important to examine the four-dimensional models that were constructed to determine their phenomenological viability. This analysis can serve at least two purposes:

(1) To find models, or a unique model, that is compatible with nature. This is a trial and error procedure. Due to the lack of understanding of the dynamics of superstring theory, the analysis is subtle and phenomenological tests can be misleading at times. However, this is the best we can do at the moment, and efforts should be dedicated to this approach.

(2) Recently, supersymmetry/supergravity phenomenology has received a lot of attention, in part due to the strong indication (from the unification of the strong and electroweak couplings at grand unification scale) of supersymmetric particles at TeV scale, accessible by future experiments. In supersymmetry/supergravity phenomenology, the minimal SUSY $SU(5)$ GUT has five parameters, while the number of parameters explodes in any other models. Superstring models can, and did, provide inspirations to cut down the parameter space of SUSY model-building. There are structures that are quite natural from the string point of view, but look rather arbitrary from the SUSY point of view. Flip SU(5) is such a case, while zeroes in Yukawa couplings is another.

All these points are fully addressed in this workshop by other speakers. So I shall limit myself to only one comment on phenomenology, namely, the cosmological constant problem in string theory. A four-dimensional superstring model with spacetime supersymmetry (i.e. supergravity) automatically has a zero cosmological constant to all orders in perturbation theory. On the other hand, a non-supersymmetric string model has a non-zero cosmological constant through string loop corrections. Such a cosmological constant is typically a hundred orders of magnitude too big. I like to argue that this does not happen in superstring theory, where spacetime supersymmetry is broken spontaneously. The argument is based on two observations: (1) in string theory, a non-zero cosmological constant will destroy the factorization property in multi-loop string diagrams, thus invalidates the consistency of string perturbation expansions; (2) starting with a consistent quantum theory, spontaneous/dynamical symmetry breaking will occur only in a self-consistent way. These observations imply the following: if a superstring model with spacetime supersymmetry is a consistent quantum theoy, then, after supersymmetry breaking, it must remain a consistent theory, i.e., the resulting cosmological constant in flat Minkowski spacetime must remain precisely zero, since a consistent string perturbative expansion must exist in the new non-supersymmetric vacuum. Of course, this argument sheds no light on whether or not (or how) supersymmetry breaking takes place in string theory.

With the success/failure of the superstring theory mentioned above, it is important to search for other possible string theories, beyond the superstring theory, that are phenomenologically viable. String theories are characterized by the local symmetries of two-dimensional conformal field theories on the string world-sheet. The bosonic string is invariant under diffeomorphisms and local Weyl rescalings on the world-sheet; whereas the superstring is characterized by a locally supersymmetric version of these symmetries. This involves a spin-3/2 supercurrent, in addition to the energy-momentum tensor, on the string world-sheet. It is natural to ask whether other symmetries on the world-sheet can give rise to consistent string theories, which have lower space-time dimensions. Recently, a proposal for a large class of new string theories, called fractional superstrings (FSS), was advanced [4, 5]. Since fractional-spin fields exist in two-dimensional theories, one can imagine new local symmetries on the world-sheet involving fractional-spin currents (replacing the spin-3/2 supercurrent of the superstring). We shall call this symmetry the fractional superconformal (FSC) symmetry. Evidence has been presented[5, 6] that fractional superstrings with spin 4/3, 6/5, and 10/9 currents on the world-sheet have potentially interesting phenomenologies in lower space-time dimensions. These models, if exist, seem to have many phenomenologically interesting features, such as the presence of space-time supergravity and Yang-Mills fields.

This talk will focus exclusively on the spin-4/3 fractional superstring and its tree scattering amplitudes[7]. In the course of the discussion we shall simply state many of the properties; for proofs and more details, see Ref[7, 8]. Classically, the spin-4/3 algebra is the constraint algebra arising from gauge-fixing the local world-sheet symmetry. Quantum mechanically, the constraints generate physical state conditions which pick out the propagating degrees of freedom from the larger string state space. Although the classical world-sheet gauge symmetry giving rise to a spin-4/3 constraint algebra is not understood at present, we can make progress by taking the constraint algebra itself as a starting point, and checking the consistency of the resulting string theory by constructing unitary scattering amplitudes for the physical states. This is known as the old covariant approach.

In Section 2, the spin-4/3 FSC algebra is reviewed. In Section 3, the generalized commutation relations for what turns out to be the space-time boson sector is given. Properties of a vertex operator and the physical state conditions are presented. In Section 4, scattering amplitudes of N massless vector particles are constructed. Their duality and cyclic symmetry properties are explicitly demonstrated. An explicit $c = 5$ represntation of the world-sheet symmetry is described in Section 5. This representation has a 3-dimensional flat Minkowski space-time interpretation, with fermions and bosons in its spectrum. Since the critical central charge $c = 10$, this representation is sub-critical; that is, it can be consistent only at the tree level. The argument for tree-level unitarity (i.e., the no-ghost theorem) is based on the recently derived Kac determinant[8]. The fermionic sector is discussed in Section 6. Section 7 summarizes the properties of the FSS at the critical central charge. We conclude with some remarks in Section 8.

2 The spin-4/3 fractional superconformal chiral algebra

Before describing the scattering amplitudes, we must first define in more detail the spin-4/3 fractional superstring. The motivating idea behind the construction of this string is to replace the spin-1/2 world-sheet fermions ψ^μ appearing in the ten-dimensional superstring with spin-1/3 world-sheet fields ϵ^μ. The world-sheet supersymmetry of the superstring is then replaced with a world-sheet "fractional supersymmetry" which relates coordinate boson fields X^μ not to fermions but rather to the field ϵ^μ. The fractional supersymmetry is generated by a generalization of the supercurrent, a new chiral current G of the form $G(z) \sim \epsilon^\mu \partial X_\mu + \cdots$, whose conformal dimension is 4/3.

The fractional current, $G(z)$, and the energy-momentum tensor, $T(z)$, together generate the fractional superconformal (FSC) chiral algebra, encoded in the singular terms of the operator product expansions[9, 10]:

$$\begin{aligned} T(z)T(w) &= \frac{1}{(z-w)^4}\left\{\frac{c}{2} + 2(z-w)^2T(w) + (z-w)^3\partial T(w)\right\} , \\ T(z)G(w) &= \frac{1}{(z-w)^2}\left\{\frac{4}{3}G(w) + (z-w)\partial G(w)\right\} , \\ G(z)G(w) &= \frac{1}{(z-w)^{8/3}}\left\{\frac{3c}{4} + 2(z-w)^2T(w)\right\} \\ &\quad + \frac{\lambda}{(z-w)^{4/3}}\left\{G(w) + \frac{1}{2}(z-w)\partial G(w)\right\} . \end{aligned} \tag{1}$$

The first OPE just states that $T(z)$ obeys the Virasoro algebra, while the second implies that $G(z)$ is a dimension-4/3 chiral primary field. We will take this algebra as the constraint algebra generating the physical state conditions for the spin-4/3 fractional string. Although one can imagine other chiral algebras involving dimension-4/3 currents, we have chosen the above FSC algebra to define the spin-4/3 fractional superstring because this algebra is known to have a sensible representation theory.[7, 9] In particular, it is known that associativity fixes λ as a function of the central charge c:

$$\lambda^2 = \frac{8-c}{6}. \tag{2}$$

An important feature of the FSC algebra is the appearance of cuts in the GG OPE. Since there are two different cuts on the right hand side, upon continuation of a correlation function involving $G(z)G(w)$ along a contour interchanging z and w it is not consistent for the correlator to pick up a simple phase. This situation is described by saying that the current G satisfies "nonabelian braid relations." For such currents, their OPEs alone do not define the chiral algebra; they must be supplemented by the current braid relations.[10, 7] These braid relations can be described in a simple way. Since only two fractional cuts appear in the GG OPE, one can split the G current into two pieces, $G(z) = G^+(z) + G^-(z)$, which satisfy abelian braid relations[9]:

$$\begin{aligned} G^\pm(z)G^\pm(w) &= \frac{\lambda}{(z-w)^{4/3}}\left\{G^\mp(w) + \frac{1}{2}(z-w)\partial G^\mp(w)\right\} , \\ G^\pm(z)G^\mp(w) &= \frac{1}{(z-w)^{8/3}}\left\{\frac{3c}{8} + (z-w)^2T(w)\right\} . \end{aligned} \tag{3}$$

Under interchange of z and w (along a prescribed path, say a counterclockwise switch) the only consistent phase that G^+ or G^- can pick up with itself is $e^{4i\pi/3}$. The phase that develops upon interchange of G^+ with G^- can be taken to be $e^{2i\pi/3}$. This serves to define the braid relations satisfied by the full FSC current $G(z)$. It is worth emphasizing that the split algebra currents $G^\pm$ do *not* generate the physical state conditions for the spin-4/3 superstring; only their combination $G = G^+ + G^-$ does. The split algebra of Eq. (3) is only introduced as a technical tool to describe the braid relations of the fractional supercurrent $G(z)$.

The specification of the FSC algebra, Eq. (1), along with the braid relations of G defines the spin-4/3 fractional superstring. The rest of this talk will outline the first step towards showing that this string in fact exists, by constructing tree scattering amplitudes which are consistent with the physical state conditions following from the FSC algebra. We will see by explicit calculation in an example that these scattering amplitudes have a sensible space-time interpretation.

3 Physical state conditions and FSC highest-weight modules

The first step in constructing tree scattering amplitudes is to derive the algebra of the constraints from the FSC algebra. The physical state conditions are imposed by the requirement that the positive (annihilation) modes of the FSC currents vanish when acting on physical states. Deriving the constraint algebra is thus equivalent to deriving from the OPEs and braid relations the commutation relations satisfied by the modes of the currents. This is a technically complicated step the results of which I will simply state; for a fuller derivation see Ref.[7].

The split algebra, Eq. (3), was studied by Zamolodchikov and Fateev[9], who noted a $\mathbf{Z}_3$ symmetry which organizes its representation theory. In particular, the currents G^+ and G^- can be assigned $\mathbf{Z}_3$ charges $q = 1$ and -1, respectively, while the energy-momentum tensor T (as well as the identity) have charge $q = 0$.

Since the split algebra is abelianly braided, the arguments of Ref.[9, 11] can be directly applied to derive the mode expansions and generalized commutation relations following from Eq. (3). The mode expansions of T, G^+ and G^- are defined by

$$\begin{aligned} T(z)\chi_q(0) &= \sum_n z^{-n-2} L_n \chi_q(0) , \\ G^\pm(z)\chi_q(0) &= \sum_n z^{n\mp q/3} G^\pm_{-1-n-(1\mp q)/3}\chi_q(0) , \end{aligned} \tag{4}$$

where χ_q is an arbitrary state with $\mathbf{Z}_3$ charge q. This means, in particular, that the L_n's have integral moding, while the G_r's are moded by one-third integers. The mode expansion for the full fractional superconformal current can be built from the split algebra pieces: $G_r = G_r^+ + G_r^-$. The modes of the $G^\pm$ currents and the energy-momentum tensor T satisfy the commutation relations

$$\begin{aligned} [L_m, L_n] &= (m-n)L_{m+n} + \frac{c}{12}(m^3 - m)\delta_{m+n} , \\ \left[L_m, G_r^\pm\right] &= \left(\frac{m}{3} - r\right) G^\pm_{m+r} , \end{aligned} \tag{5}$$

and the generalized commutation relations

$$\begin{aligned} &\sum_{\ell=0}^{\infty} c_\ell^{(-2/3)} \left[G^\pm_{\pm\frac{q}{3}+n-\ell} G^\pm_{\frac{2\pm q}{3}+m+\ell} - G^\pm_{\pm\frac{q}{3}+m-\ell} G^\pm_{\frac{2\pm q}{3}+n+\ell} \right] = \\ &\qquad\qquad \frac{\lambda}{2}(n-m) G^\mp_{\frac{2\pm 2q}{3}+n+m} , \\ &\sum_{\ell=0}^{\infty} c_\ell^{(-1/3)} \left[G^+_{\frac{1+q}{3}+n-\ell} G^-_{-\frac{1+q}{3}+m+\ell} + G^-_{-\frac{2+q}{3}+m-\ell} G^+_{\frac{2+q}{3}+n+\ell} \right] = \\ &\qquad\qquad L_{n+m} + \frac{3c}{16}\left(n+1+\frac{q}{3}\right)\left(n+\frac{q}{3}\right)\delta_{n+m} , \end{aligned} \tag{6}$$

when acting on a state with $\mathbf{Z}_3$ charge q. The $c_\ell^{(\alpha)} = (-1)^\ell \binom{\alpha}{\ell}$ are fractional binomial coefficients. Because of the infinite sum on the left hand side, the mode algebra in Eq. (6) is not a graded Lie algebra, but a new algebraic structure on the string world-sheet.

The physical state conditions require that a physical state $|\phi\rangle$ be annihilated by the positive modes of T and G, and be eigenstates of their zero modes:

$$\begin{aligned} (L_n - v\delta_{n,0})|\phi\rangle &= 0 , \quad 0 \le n \in \mathbf{Z} , \\ (G_r - \beta\delta_{r,0})|\phi\rangle &= 0 , \quad 0 \le r \in \mathbf{Z}/3 . \end{aligned} \tag{7}$$

Here v and β are "intercepts", normal ordering constants in the definitions of T and G. All the positively-moded constraints can be generated from those of the set $\{L_1, L_2, G_{1/3}, G_{2/3}, G_1, G_{4/3}\}$. A state $|s\rangle$ obeying the zero-mode conditions in Eq. (7) is called a spurious state if it is orthogonal to all physical states. Such a state can be written as

$$|s\rangle = \sum_{n>0} L_{-n}|\chi_n\rangle + \sum_{r>0} G_{-r}|\psi_r\rangle \;, \tag{8}$$

in terms of some other states $|\chi_n\rangle$ and $|\psi_r\rangle$. All states not satisfying the physical state conditions must have a spurious component. A physical state can itself be spurious, in which case it is a null state, and should decouple from all scattering amplitudes. Thus, the decoupling of all spurious states in the scattering amplitudes of physical states is a test of whether Eq. (6) is a sensible constraint algebra.

The physical state conditions imply that the physical state $|\phi\rangle$ is a highest weight state of the FSC algebra with highest weight v. The properties of the highest weight modules can be derived from the commutation relations Eqs. (5–6), and it turns out that they fall into two classes.[9] The highest weight state of an S (singlet) module has $\mathbf{Z}_3$ charge $q = 0$, the allowed modings of the fractional current acting on it are $G_{n-(1/3)}$, and its descendants have dimensions $v + n$ and $v + \frac{1}{3} + n$ where n is an integer. The highest weight state of a D (doublet) module is doubly degenerate in the split algebra with $\mathbf{Z}_3$ charges $q = \pm 1$, allowed current modings G_n and $G_{n-(2/3)}$, and descendants of dimensions $v+n$ and $v+\frac{2}{3}+n$. In the D-module, associativity fixes the G_0 intercept β in terms of v and c.

We will see later, by way of an example, that the space-time bosons fall into two classes. In analogy to the superstring case, we can define a G-parity, $G = (-1)^{q+1}$. It turns out that the G-parity-odd sector contains the tachyonic state as its lighest particle; we shall call this sector as the Tachyonic sector, or simply the T-sector. The lightest state in the G-parity-even sector is the massless vector particle (or graviton for a closed string). So we shall call the G-parity-even sector as the Vector- or V-sector. As mentioned earlier, there is also a third sector consisting of highest weight modules that are double-valued with respect to the fractional supercurrents, called the R-sector. The appropriate moding of the currents in this sector is half-integral, and the algebra obeyed by the modes is different from Eq. (6). The physical states in this sector correspond to space-time fermions. We shall discuss this sector in Section 6. In the remaining of this section, we shall construct the vertex operator for the lowest state in the D-module. This vertex operator correponds to the massless vector particle. As we shall see, the corresponding vertex operator for the tachyon does not exist in the old covariant formalism.

The main properties of D-modules are summarized by the OPEs of the current G with the highest weight vertex operator $W_D(z)$ of dimension-v and its first descendant operator $V_D(z)$ of dimension $v + \frac{2}{3}$ [7]:

$$\begin{aligned} G(z)W_D &= \left(\sqrt{v - \frac{c}{24}}\right) \frac{W_D}{z^{4/3}} + \frac{V_D}{z^{2/3}} + \ldots \\ G(z)V_D &= \left(v + \frac{c}{12} + \frac{1}{12}\sqrt{(8-c)(24v-c)}\right) \frac{1}{z^2} \left\{W_D + \frac{z}{3v}\partial W_D\right\} \\ &\quad + \left(\frac{\sqrt{24v-c} - \sqrt{8-c}}{36v}\right) \frac{\widetilde{W}_D}{z} + \ldots \end{aligned} \tag{9}$$

For ease of writing, we have inserted the D-module vertex operators W_D *etc.* at the origin of the complex plane and have dropped their arguments. The first OPE fixes the normalization of the V_D vertex, while the second one introduces the new Virasoro (though not FSC) primary $\widetilde{W}_D = (L_{-1} - 3\sqrt{6}vG_{-1})W_D$ of dimension $v + 1$. Note that the fractional current is single-valued with respect to V_D, but is nonabelianly braided with respect to W_D. These OPEs play an important role in the development of consistent tree scattering amplitudes for D-sector physical states.

4 Tree scattering amplitudes and spurious state decoupling

In what follows, we will construct open fractional superstring tree scattering amplitudes. Closed string scattering amplitudes at tree level are easily formed by combining two open string amplitudes using a level-matching condition for left- and right-movers[12]. The construction we use will be closely analogous to that of open Ramond-Neveu-Schwarz superstring tree amplitudes in the "old covariant" formalism. (For a detailed explanation of this formalism, see, *e.g.*, Chapter 7 of Ref.[1].)

The world-sheet in an open string tree scattering process is conformally equivalent to a unit disc with vertex operators $V(x)$ representing the asymptotic scattering states inserted at points on the boundary. Since we must be able to integrate these vertex operators over their insertion positions, they must be dimension-one operators in the two-dimensional world-sheet theory. We can conformally map the disk to the complex upper half-plane, fixing the positions of three of the vertex insertions at ∞, 1, and 0 on the real axis, with the remaining insertions at points $1 < x_i < \infty$. This picture is suitable for writing the string amplitude as a correlator of operators in the open string Fock space with a radial-ordering prescription:

$$\mathcal{A}_N = \int \frac{dx_3 \cdots dx_{N-1}}{x_3 \cdots x_{N-1}} \langle V_N | V_{N-1}(x_{N-1}) \cdots V_2(1) | V_1 \rangle \ , \tag{10}$$

where the "in" and "out" states are the insertions at $x_1 = 0$ and $x_N = \infty$, and the integration is over all x_i preserving the order $1 < x_3 < \cdots < x_{N-1} < \infty$.

From our previous mapping to the disk, it is clear that $\mathcal{A}_N$ should be invariant under cyclic permutations of the vertex ordering. This means that after passing the V_N vertex to the right through all the other vertices in Eq. (10), the value of $\mathcal{A}_N$ must be unchanged. This condition on the braiding of the vertex operators can be satisfied if two such operators commute when zero space-time momentum is flowing into either vertex. The representation theory of the spin-4/3 split algebra implies that commuting operators can only have $\mathbf{Z}_3$ charge $q = 0$. Thus, in particular, the V_D (but not the W_D) operator in the D-module is a potential candidate for the vertex insertions in Eq. (10). For this also to be a dimension-one operator implies that the D-module intercept must be $v = 1/3$.

A vertex insertion at x can be rewritten as $V(x) = x^{L_0} V(1) x^{-L_0}$, and the positions of the insertions explicitly integrated over to give the amplitude in the form

$$\mathcal{A}_N = \langle V_N | V_{N-1}(1) \tilde{\Delta} \ldots \tilde{\Delta} V_2(1) | V_1 \rangle \tag{11}$$

where the propagator is $\tilde{\Delta} = (L_0 - 1)^{-1}$. For scattering of D-module states, the vertices in Eq. (11) correspond to the V_D descendant states in a FSC highest-weight module. We can convert Eq. (11) to a different "picture" involving the highest-weight states W_D using the general properties of the D modules. In particular, by taking appropriate fourier components of the

D-module OPE in Eq. (9), it follows that

$$[G_r, V_D(1)] = \left(L_0 + r - \frac{1}{3}\right) W_D(1) - W_D(1)\left(L_0 - \frac{1}{3}\right) \quad \text{for all } r \in \mathbf{Z}/3\,. \tag{12}$$

In deriving this formula we have used $v = 1/3$ and the relation $[L_0, W_D(1)] = (1/3)W_D(1) + \partial W_D(1)$ which is true for any dimension-1/3 Virasoro primary W_D. Note that when $v = 1/3$ the dimension $1+v$ descendant $\widetilde{W}_D$ (which has no analog in the superstring) decouples from the OPE in Eq. (9). It is this unexpected decoupling at precisely the physical value of the intercept which allows us to derive Eq. (12), and in turn construct sensible tree scattering amplitudes.

We can now replace the "in" state $|V_D\rangle$ with a physical state using $|V_D\rangle = G_{-2/3}|W_D\rangle$ which follows from the first OPE in Eq. (9). The $G_{-2/3}$ mode can be commuted to the left using Eq. (12) as well as the relation

$$G_r(L_0 - a - r)^{-1} = (L_0 - a)^{-1} G_r \ , \tag{13}$$

following from Eq. (5). Acting on the "out" state, $\langle V_D|G_{-2/3} = \langle W_D|$, which is a consequence of the second OPE in Eq. (9) with $v = 1/3$. The extra insertions coming from the right-hand side of Eq. (12) vanish by a "cancelled propagator" argument, since setting $r = -2/3$ in Eq. (12) gives factors of $L_0 - 1$ and $L_0 - 1/3$ which cancel the propagators to the left and right, respectively.[1] Thus, the final form we find for the amplitude is

$$\mathcal{A}_N = \langle W_N|V_{N-1}(1)\Delta \ldots \Delta V_2(1)|W_1\rangle \ , \tag{14}$$

where the propagator in this picture is $\Delta = (L_0 - \frac{1}{3})^{-1}$.

Now we can investigate the crucial issue of spurious state decoupling in our amplitudes. If we start with physical states defined as highest-weight vectors of FSC modules, will they scatter only to other physical states? For this to be true, only physical states must contribute to residues of poles in amplitudes when an internal propagator goes on-shell. Suppose we fix the external momenta such that some state $|s\rangle$ in the string Fock space at momentum $\kappa = k_{M+1} + \cdots + k_N$ is on-shell: $(L_0 - 1/3)|s\rangle = 0$. If we factorize

[1] Tree amplitudes with cancelled propagators are holomorphic in the Mandelstam invariant of the cancelled propagator channel, and thus vanish if the amplitudes have Regge asymptotic behavior. We will see in the next section that they do have this soft high energy behavior.

the amplitude in Eq. (14) by inserting a sum over a complete set of states of momentum κ at the propagator between V_{M+1} and V_M, then the $|s\rangle\langle s|$ term in the sum will contribute a pole in momentum space. The requirement of spurious state decoupling is that if $|s\rangle$ is spurious, its contribution to the residue of the pole should vanish:

$$\langle s|V_M(1)\Delta\cdots\Delta V_2(1)|W_1\rangle = 0 \ . \tag{15}$$

To prove this, consider one term, $\langle\psi|G_r$ with $r > 0$, in the presentation of $\langle s|$ as a sum of descendant states, Eq. (8). (The L_n descendant pieces can be shown to decouple by a similar argument.) The G_r mode can be commuted to the right in Eq. (15) using Eqs. (12–13). The insertions coming from the right-hand side of Eq. (12) again vanish by a cancelled propagator argument. Finally, the G_r mode acting on the "in" state $|W_1\rangle$ vanishes by the physical state conditions Eq. (7), thus proving spurious state decoupling.

Our prescription for constructing dual N-point tree amplitudes of V-sector states satisfying spurious state decoupling can be extended to include two G-odd-sector states by simply replacing the "in" and "out" W_D states in Eq. (14) with T-sector physical states W_S :

$$\mathcal{A}_N = \langle W_S, N|V_{N-1}(1)\Delta\ldots\Delta V_2(1)|W_S, 1\rangle \ , \tag{16}$$

where the propagator in this picture is again taken to be $\Delta = (L_0 - \frac{1}{3})^{-1}$.

The argument for spurious state decoupling then goes through unchanged. However, it turns out that there is no appropriate dimension-1 commuting vertex operator in the T-sector to play the role of the V_D vertices. Thus, we cannot prove cyclic symmetry of the amplitudes with two T-sector states, nor can we extend the prescription to include scattering of three or more T-sector states.

Note that upon factorizing the V-sector scattering amplitude in Eq. (14) on any propagator, we can never obtain a T-sector intermediate state. The reason is simply that the W_D "in" state is G-parity odd, while the V_D vertices are G-parity even. Thus, by the conservation of the G-parity, only the G-parity odd intermediate states can contribute, whereas the T-sector physical states W_S have even G-parity. This selection rule means that it is consistent at tree level to drop the T-sector altogether, a desirable feature since it will turn out that the T-sector contains the tachyon. This is completely analogous to the GSO-projection in superstring[13].

5 A c=5 free-field representation of the FSC algebra

The scattering amplitude construction we have presented so far has depended only on general properties of the highest-weight modules of the FSC algebra. In this section we will flesh out this construction with an explicit conformal field theory which forms a representation of the FSC algebra at central charge $c = 5$. In sec. 7, we will argue that the critical central charge of the FSC algebra is $c = 10$, so the representation presented below is sub-critical. As is the case with sub-critical representations of the bosonic and superstring, tree amplitudes are perfectly well-behaved. The restriction to the critical central charge is expected only to appear once loop amplitudes are included.

The $c = 5$ representation is constructed from five free (massless) scalar fields on the world-sheet. Three of them, $X^\mu(z)$, $\mu = 0, 1, 2$, are interpreted as the string coordinate fields, and obey the standard OPE

$$X^\mu(z)X^\nu(w) = -\eta^{\mu\nu}\ln(z-w) \ , \tag{17}$$

where $\eta^{\mu\nu}$ is the three-dimensional Minkowski metric with signature $(-++)$. This X^μ CFT has a global SO(2,1) Lorentz symmetry. The remaining two fields, $\varphi^i(z)$, $i = 1, 2$, are compactified on a triangular lattice—the su(3) root lattice. In a basis in which the φ^i boundary conditions are diagonalized, $\varphi^i = \varphi^i + 2\pi$, their OPEs read

$$\varphi^i(z)\varphi^j(w) = -g^{ij}\ln(z-w) \ , \qquad g^{ij} = \frac{1}{3}\begin{pmatrix} 2 & -1 \\ -1 & 2 \end{pmatrix} \ . \tag{18}$$

The vertex operator $V_{\mathbf{m}} = \exp\{im_j\varphi^j\}$ has dimension $\Delta_{\mathbf{m}} = \frac{1}{2}m_i g^{ij} m_j$ for integer m_i. A triplet of dimension-one fields, $U^\mu = \{[V_{(2,1)} + V_{(-2,-1)}], [V_{(-1,1)} + V_{(1,-1)}], [V_{(1,2)} + V_{(-1,-2)}]\}$, generates an SO(2,1)$_2$ Kac-Moody algebra, organizing all the fields in the φ^i CFT in SO(2,1) representations.†[0] Some important fields of low conformal dimension Δ are:

$$\Delta = 1/3: \quad \epsilon_\mu = \left\{V_{(1,0)}, V_{(0,1)}, V_{(-1,-1)}\right\} \ , \ \epsilon^\dagger_\mu = \left\{V_{(-1,0)}, V_{(0,-1)}, V_{(1,1)}\right\}$$

[0] **Appropriate, though standard, cocycles must be added to the φ^i CFT to realize this symmetry.**

$$\Delta = 1 \quad : \qquad W_{\mu\nu} = \left\{ \tfrac{i}{2}\left[V_{(2,1)} - V_{(-2,-1)}\right], \tfrac{i}{2}\left[V_{(1,-1)} - V_{(-1,1)}\right], \right.$$
$$\left. \tfrac{i}{2}\left[V_{(-1,-2)} - V_{(1,2)}\right], i\partial\varphi^j \right\}$$
$$\Delta = 4/3 : \qquad s = \tfrac{1}{3}\left[V_{(-2,0)} + V_{(0,-2)} + V_{(2,2)}\right] ,$$
$$s^\dagger = \tfrac{1}{3}\left[V_{(2,0)} + V_{(0,2)} + V_{(-2,-2)}\right] . \tag{19}$$

The dimension-1/3 vector fields, ϵ_μ and $\epsilon^\dagger_\mu$, are the analogs of the dimension-1/2 ψ^μ fields in the flat-space superstring representation. The scalars, s and $s^\dagger$, and the spin-2 (symmetric, traceless) $W_{\mu\nu}$ have no superstring analogs.

A current obeying the FSC algebra Eq. (1) is given by

$$G = \frac{1}{\sqrt{2}}\left[(\epsilon_\mu - \epsilon^\dagger_\mu)\partial X^\mu - \frac{3}{2}(s + s^\dagger)\right] . \tag{20}$$

It is manifestly an SO(2,1) scalar Virasoro primary dimension-4/3 field. A splitting of the current satisfying the split algebra in Eq. (3) exists with $\mathbf{Z}_3$ charge assignments $q = 1$ for ϵ_μ and s, $q = -1$ for $\epsilon^\dagger_\mu$ and $s^\dagger$, and $q = 0$ for X^μ, U^μ, and $W^{\mu\nu}$. The simplest S-sector vertex is $W_S = \exp\{ik \cdot X\}$. The physical state conditions, Eq. (7), applied to W_S imply that $k^2 = 2v$. We will show in the next section that a consistent choice for the S-sector intercept is $v = 1/3$, which implies that W_S is a tachyon.

The simplest D-sector vertex is $W_D = (\zeta \cdot \epsilon + \zeta' \cdot \epsilon^\dagger)\exp\{ik \cdot X\}$. The physical state conditions acting on W_D constrain ζ_μ and ζ'_μ to be expressible in terms of a single polarization vector ξ_μ, such that $k^2 = k \cdot \xi = 0$, and

$$W_D = \left[\xi^\mu(\epsilon_\mu - \epsilon^\dagger_\mu) - \varepsilon^{\mu\nu\rho}\xi_\mu k_\nu(\epsilon_\rho + \epsilon^\dagger_\rho)\right] e^{ik\cdot X} . \tag{21}$$

Here $\varepsilon^{\mu\nu\rho}$ is the completely antisymmetric tensor in three dimensions. W_D represents a massless vector particle. In a closed string, we could match a left-moving and a right-moving version of W_D to form the usual graviton, dilaton, and antisymmetric tensor fields. The V_D descendant of W_D, defined by the first OPE in Eq. (9), is

$$V_D = -\sqrt{2}\left[\xi^\mu \partial X_\mu - \varepsilon^{\mu\nu\rho}\xi_\mu k_\nu U_\rho - ik^\mu \varepsilon^{\nu\rho\sigma}\xi_\rho k_\sigma W_{\mu\nu}\right] e^{ik\cdot X} . \tag{22}$$

Note that upon making a gauge transformation $\delta\xi^\mu \propto k^\mu$, one finds $\delta V_D \propto \partial(\exp\{ik \cdot X\})$, a spurious state which decouples by the arguments of the last section. The operators ∂X_μ, U_μ, and $W_{\mu\nu}$ appearing in V_D are all single-valued fields on the world-sheet—they have no cuts in their OPEs with any other field—a property shared by all operators with $\mathbf{Z}_3$ charge $q = 0$. Therefore, only simple commutators, as opposed to the generalized commutators appearing in the constraint algebra Eq. (6), are needed to evaluate an N-point amplitude. Consider, for example, the coupling of three massless vector particles, given by $\mathcal{A}_3 = \langle W_D(k_3, \xi_3)|V_D(k_2, \xi_2; 1)|W_D(k_1, \xi_1)\rangle$. One finds

$$\begin{aligned} \mathcal{A}_3 = \quad & i2\sqrt{2}\Big[(k_1 \cdot \xi_3)(\xi_2 \cdot \xi_1) + (k_2 \cdot \xi_1)(\xi_3 \cdot \xi_2) + \\ & +(k_3 \cdot \xi_2)(\xi_1 \cdot \xi_3) - 3(\xi_1 \cdot k_2)(\xi_2 \cdot k_3)(\xi_3 \cdot k_1)\Big] \, . \end{aligned} \tag{23}$$

The first three terms are precisely the expected Yang-Mills coupling; the last term represents a non-linear correction to the Yang-Mills action which is higher-order in the string tension, and therefore is suppressed at energies far below the Planck scale. The non-linear term also appears in the three-vector coupling in the bosonic string, where it has coefficient $+1$ instead of -3; in the superstring no such term appears in the three-point coupling (though string correction terms do appear in higher-point functions).

One can calculate higher-point amplitudes in a similar way. While the details are not illuminating, the main features of these amplitudes are easily understood. Because the contribution from the $e^{ik\cdot X}$ pieces essentially factorizes, it is clear that one will obtain gamma-function factors similar to those that appear in the Veneziano amplitude for the bosonic string. The remaining part of the vertices can only contribute factors which are polynomial in the momenta. Thus, the fractional superstring amplitudes have the soft high-energy Regge behavior characteristic of bosonic and superstring amplitudes.

6 The R-sector and Space-time Fermions

Let us consider the fermions in the FSS. Recall the su(3) lattice in the $c = 5$ representation discussed in the last section. The R-sector contains the twist fields in the (two-dimensional) $\mathbf{Z}_2$ orbifold of the su(3) lattice; as we shall see, this is only half of the fields in the R-sector. These fields in the R-sector appear upon orbifolding the φ^i in the CFT by a 180° rotation of its lattice. The twist fields of this $\mathbf{Z}_2$ orbifold transform in the spinor representation of

SO(2,1), and the zero modes of the ϵ^{μ} fields acting on these states satisfy a three-dimensional Clifford algebra. The R-sector consists of highest weight modules that are double-valued with respect to the fractional currents,$G(z)$. The modings of $G^{\pm}$ currents in this case are given by the following OPEs

$$G^{\pm}(z)\chi_R(0) = 2^{\frac{1}{3}} \sum_{n=-\infty}^{\infty} z^{n/2-4/3} G^{\pm}_{n/2}\chi_R(0), \tag{24}$$

where the overall normalization has been chosen for later convenience. These mode expansions can be inverted to give

$$G^{\pm}_{n/2}\chi_R(0) = 2^{-\frac{4}{3}} \oint_{\gamma} \frac{dz}{2\pi i}\, z^{1/3+n/2} G^{\pm}(z)\chi_R(0). \tag{25}$$

Here, γ is a contour encircling the origin twice, where $\chi_R(0)$ is inserted.

Note that, for $c < 8$, the linear combinatons

$$G_{n/2} = G^{-}_{n/2} + (-1)^n G^{+}_{n/2}, \tag{26}$$

have the following hermiticity properties

$$(G_{n/2})^{\dagger} = G_{-n/2}. \tag{27}$$

The operators $G_{n/2}$ satisfy the following generalized commutation relation:

$$\sum_{\ell=0}^{\infty} D^{(\ell)}_{(\frac{1}{3},-\frac{1}{3})} \left[G_{\frac{n-\ell}{2}} G_{\frac{m+\ell}{2}} + G_{\frac{m-\ell}{2}} G_{\frac{n+\ell}{2}} \right] =$$
$$\frac{\lambda}{2}(-1)^{n+m} G_{\frac{n+m}{2}} + [(-1)^n + (-1)^m] \left\{ L_{\frac{n+m}{2}} + \frac{3c}{8}\left(\frac{n^2}{8} - \frac{5}{48}\right)\delta_{n+m} \right\}, \tag{28}$$

where $D^{(\ell)}_{(\alpha,\beta)}$ are binomial coefficients

$$(1-x)^{\alpha}(1+x)^{\beta} = \sum_{\ell=0}^{\infty} D^{(\ell)}_{(\alpha,\beta)} x^{\ell}. \tag{29}$$

The physical state conditions require that a physical state $|R\rangle$ be annihilated by the positive modes of T and G, and be eigenstates of their zero modes:

$$\begin{aligned} (L_n - v_R\delta_{n,0})|R\rangle &= 0\,, \quad 0 \le n \in \mathbf{Z}\,, \\ (G_{n/2} - \beta\delta_{r,0})|R\rangle &= 0\,, \quad 0 \le n \in \mathbf{Z}\,. \end{aligned} \tag{30}$$

Note that $G_{n/2}$ is $\mathbf{Z}_2$ invariant when n is even, but is $\mathbf{Z}_2$ odd when n is odd. So, for odd n, $G_{n/2}$ moves a $\mathbf{Z}_2$ invariant state to a $\mathbf{Z}_2$ odd state and vice versa. The $\mathbf{Z}_2$ odd states can be built from the τ fields that are present in the $\mathbf{Z}_2$ twisting of the bosons φ^i :

$$\partial\varphi(z)\sigma(w) = (z-w)^{1/2}\tau(w) + \cdots \tag{31}$$

where σ has conformal dimension 1/8 and τ has conformal dimension 5/8. Our prescription for constructing dual N-point tree amplitudes of V-sector states satisfying spurious state decoupling can be extended to include two R-sector states by simply replacing the "in" and "out" W_D states in Eq. (14) with R-sector physical states :

$$\mathcal{A}_N = \langle R_N|V_{N-1}(1)\Delta \ldots \Delta V_2(1)|R_1\rangle \ , \tag{32}$$

where the propagator in this picture is again taken to be $\Delta = (L_0 - v_R)^{-1}$, as discussed in Section 5.

The argument for spurious state decoupling then goes through unchanged. Since we do not have an appropriate dimension-1 commuting vertex operator in the R-sector to play the role of the V_D vertices, we cannot prove cyclic symmetry of the amplitudes with two R-sector states, nor can we extend the prescription to include scattering of three or more R-sector states. This problem is similar to the problem in superstring in the old covriant formalism. As is obvious, V-sector particle-scattering does not produce R-sector states. This is interpreted as a space-time spin-statistics selection rule.

7 Tree-level unitarity, Extra spurious states, and Critical central charge

Spurious state decoupling by itself does not imply that tree amplitudes are unitary. One must also prove the no-ghost theorem, i.e.,that there are no negative-norm physical states for the specific representation of the constraint algebra under consideration. In Ref.[8], the Kac determinants for the spin-4/3 FSC algebra are explicitly constructed. Using these Kac determinants (for the $S-$, the $D-$ and the $R-$modules), we argue that the $c=5$ representation discussed above is free of negative-norm states. The line of argument is similar to that of Brower and Thorn[14] for the bosonic string. The readers can see Ref.[8] for details.

In the bosonic and superstrings, for representations with one time-like (space-time) dimension, a non-negative physical state space occurs up to a maximum value of the central charge. As one passes through this critical value of the central charge the norm of some physical states change sign, implying that at the critical central charge there are extra null states. The presence of these extra spurious states is a clear indication of a "stringy gauge symmetry". Using the Kac determinant, it is easy to find the critical central charge for FSS, $c = 10$, and to search for the existence of extra sets of zero-norm physical states and their respective intercepts in a representation-independent way. Here, we shall give only the results. Details can be found in Ref.[8].

Since a null physical state is spurious, consider the simplest (V-sector) spurious state $|\psi_1\rangle = G_{-1/3}|\widetilde{\psi}\rangle$, where $\widetilde{\psi}$ is an arbitrary state. The physical state conditions Eq. (7) are satisfied by ψ_1 only when the intercept $v = 1/3$, irrespective of the value of the central charge. This is the value of the V-sector intercept determined above by the requirement of spurious state decoupling; when $\widetilde{\psi} = e^{ik\cdot X}$, ψ_1 is just the longitudinal component of the massless vector state in Eq. (21).

To determine the critical central charge we must consider the more complicated general V-sector spurious state, at level 4/3. Applying the physical state conditions, one finds a null state with the same intercept v only when $c = 10$[4, 15]. One can check in the $c = 5$ sub-critical representation of the last section that this same state is in fact a positive-norm state, as expected.

One can also search for sets of T-sector spurious string states. If the intercept of the T-sector is the same as that of the V-sector, then again we can find two sets of spurious states, although the second set is at level three, uncomfortably high. Of course, similar to the superstring case, we expect the T-sector states to be GSO-projected out. It is interesting to point out another possibility here. If the intercept of the T-sector can be different from that of the V-sector, there are two sets of spurious states in the T-sector at intercept $v_T = 0$, in which case the ground state becomes massless, and need not be projected out. In the R-sector, there are again two sets of spurious states, for intercept $v_R = 3/8$. So a consistent FSS at the critical central charge $c = 10$ can have both fermions and bosons.

8 Discussion and outlook

The tree-level considerations discussed so far leave us with a certain amount of arbitrariness in constructing spin-4/3 fractional superstrings. In particular, we are free to include or not the T-sector states and/or the R-sector states; we can couple left- and right-moving theories at will on the world-sheet in type II and heterotic-like constructions; and the choice of CFT representation of the spin-4/3 FSC algebra is constrained by tree unitarity only to have central charge less than or equal to its critical value $c = 10$. The inclusion of string loop amplitudes should remove much of this arbitrariness. As is the case with the bosonic and superstrings, one expects that loop amplitudes will only be consistent at the critical central charge, and modular invariance will determine which left- and right-moving sectors, at which values of their intercepts, can be consistently coupled together. As we mentioned earlier, the constraints for FSS is even tighter than superstring, in the sense that the FSC algebra is non-linear. One difficulty in constructing a critical ($c = 10$) representation of the FSC algebra is its non-linearity. The fact that the FSC structure constant λ is a non-linear function of the central charge c, Eq. (2), means that the FSC algebra as a whole is non-linear: the tensor product of two representations of this algebra is not itself a representation. In particular, the tensor product of two copies of the $c = 5$ representation described above will not make a $c = 10$ representation of the FSC algebra.

Once given a $c = 10$ representation of the FSC algebra, the construction of loop amplitudes may still not be an easy task. One can imagine "sewing" tree amplitudes in the "old covariant" formalism described above to form one-loop amplitudes by a suitable generalization of the sewing procedure for the bosonic string. Such an amplitude would not only have to be unitary, but also modular invariant. An indication that this may not be too much to expect may be the existence of the modular-invariant fractional string partition functions proposed in Ref.[5, 6]. At higher loops it seems likely that a clearer understanding of the "fractional moduli" describing the sewing of tree amplitudes will be necessary.

Finally, what are the space-time features of critical fractional superstrings? This, also, is an open question, since its answer depends on the resolution of the world-sheet issues outlined above. So far, only hints of possibly new space-time structures have been gleaned from the fractional superstring partition function [5, 6, 16].

Acknowledgements

This talk reports on joint works with Philip Argyres and Zurab Kakushadze described in Ref[7, 8]. It is a pleasure to thank K. Dienes and J. Grochocinski for useful discussions and collaborations. This work was supported in part by the National Science Foundation.

References

[1] See *e.g.* M.B. Green, J.H. Schwarz and E. Witten, *Superstring Theory*, Cambridge University Press (1987).

[2] D.J. Gross, J.A. Harvey, E. Martinec, and R. Rohm, *Nucl. Phys.* **B256** (1985) 253.

[3] See e.g., P. Candelas, G. Horowitz, A. Strominger, and E. Witten,*Nucl. Phys.* **B258** (1985) 46; L. Dixon, J. Harvey, C. Vafa, and E. Witten, *Nucl. Phys.* **B261** (1985) 285; H. Kawai, D.C. Lewellen, and S.-H.H. Tye, *Nucl. Phys.* **B288** (1987) 1; D. Senechal, *Phys. Rev.* **D39** (1989) 3717.

[4] P.C. Argyres, A. LeClair and S.-H.H. Tye, *Phys. Lett.* **253B** (1991) 306.

[5] P.C. Argyres and S.-H.H. Tye, *Phys. Rev. Lett.* **67** (1991) 3339.

[6] K.R. Dienes and S.-H.H. Tye, *Nucl. Phys.* **B376** (1992) 297; P.C. Argyres, K.R. Dienes and S.-H.H. Tye, *New Jacobi-like Identities for Parafermion Characters*, to appear in *Comm. Math. Phys.*

[7] P.C. Argyres and S.-H.H. Tye, *Tree Scattering Amplitudes for the K=4 Fractional Superstring*, Cornell preprint CLNS 92/1176, to appear.

[8] P.C. Argyres, Z. Kakushadze, and S.-H.H. Tye, *Kac Determinant and Fractional Superstring*, Cornell preprint, 1993.

[9] A.B. Zamolodchikov and V.A. Fateev, *Theor. Math. Phys.* **71** (1987) 451.

[10] P.C. Argyres, J. Grochocinski and S.-H.H. Tye, *Nucl. Phys.* **B367** (1991) 217, and **B391** (1993) 409.

[11] A.B. Zamolodchikov and V.A. Fateev, *Sov. Phys. J.E.T.P.* **62** (1985) 215; *Sov. Phys. J.E.T.P.* **63** (1986) 913.

[12] H. Kawai, D.C. Lewellen, and S.-H.H. Tye, *Nucl. Phys.* **B269** (1986) 1.

[13] F. Gliozzi, J. Scherk and D. Olive, *Nucl. Phys.* **B122** (1977) 253.

[14] R.C. Brower and C.B. Thorn, *Nucl. Phys.* **B31** (1971) 163.

[15] P.C. Argyres, E. Lyman and S.-H.H. Tye, *Phys. Rev.* **D46** (1992) 4533.

[16] P. Frampton and J. Liu, *Phys. Rev. Lett.* **70** (1993) 130; G.B. Cleaver and P.J. Rosenthal, *Aspects of Fractional Superstrings,* Caltech preprint CALT-68-1756.

[17] P. Goddard and A. Schwimmer, *Phys. Lett.* **206B** (1988) 62.

NON-PERTURBATIVE STRING PHENOMENOLOGY

JORGE L. LOPEZ

Center for Theoretical Physics, Department of Physics, Texas A&M University College Station, TX 77843-4242, USA

ABSTRACT

A phenomenological analysis is presented of possible non-quantum-mechanical and CPT-violating phenomena in the $K-\bar{K}$ system in the context of new generic non-perturbative string theory effects. Casting the problem in the density matrix formalism, one introduces three new parameters (α, β, γ) which then appear in explicit predictions for the $K \to 2\pi$ rate and the semileptonic decay charge asymmetry δ. Intermediate- and long-time measurements are then used to determine upper bounds on these parameters.

1 Introduction

All of perturbative string phenomenology is based on a core two-dimensional conformal field theory (CFT), which is believed to describe, *e.g.*, every vacuum state of the heterotic string in four dimensions. Therefore, a complete classification of CFTs would be equivalent to an enumeration of all possible string vacua. However, quantum consistency imposes severe constraints on the possibly allowed CFTs. String perturbation theory consists of a sum over topologies of the two-dimensional manifold where the CFT lives in, *i.e.*, a sum over surfaces with increasing number of handles. A scattering process is described by the insertion on this two-dimensional surface of a vertex operator for each external particle, and by the average over the positions of these operators (up to topological deformations of the surface under modular transformations). Even though a complete classification of all CFTs is presently unavailable, large classes of CFTs have been indentified, as well as some important connections among them. A few of these classes (*i.e.*, a few classes of string vacua) have been explored in detail to see if the known low-energy phenomenology can be reproduced. One such string vacuum, supplemented by various assumptions about the low-energy effective theory of string (most notably about supersymmetry breaking) is called a "string model". String phenomenology is based mostly on results obtained at tree-level in the genus expansion (*i.e.*, when the surface mentioned above is a sphere), although important one-loop and all-orders results are known.

Crucial to our understanding of the connection between string theory and the low-energy world is the specification of the low-energy effective theory of string. String model-building and the various known model-independent results depend on a prescription to truncate the infinite tower of string massive states when dealing with energy scales below the string scale of $\sim 10^{18}$ GeV. Naively such theory would just be a generic supergravity theory with calculable coefficients for the higher-derivative terms and specific forms for the three basic functions: the superpotential, the Kähler potential, and the gauge kinetic function. Even though this expectation is realized in practice, a string property called duality or target-space modular invariance is seen to pervade all three functions. This property has its origin in the dependence of the low-energy effective theory on the so-called moduli fields which parametrize the deformations of the compactified space. In fact, it has been realized that supersymmetry breaking in the context of string-inspired supergravities can be usefully classified by the relative magnitudes of the F-terms of the various moduli fields.

From the phenomenological point of view, an important question is whether there any other string theory symmetries (like duality) which may be generically present in the low-energy effective theory. If so, it may be possible to derive new (hopefully model-independent) tests or predictions of string theory. Beyond duality, no such symmetries are apparent in string perturbation theory. However, an analogous duality symmetry involving the dilaton field instead, has been conjectured to exist at the non-perturbative level (since it relates the string coupling to its inverse) [1].

2 Quantum Gravity and the $\not S$ Matrix

It has been known for some time that quantum coherence in the presence of evaporating black holes may be in jeopardy [2]: the large amount of information carried by the particles accreting to form the black hole is locked-in since the evaporation process is conjectured to be thermal in nature, and as such it cannot carry away any information. It is hoped that this puzzle should have an answer within our only working theory of quantum gravity, namely string theory. There has been a significant amount of theoretical activity in this field lately, with several alternatives put forward to explain this puzzle, although not all invoking the physics of string. A particularly simple and appealing suggestion has been proposed by Ellis, Mavromatos, and Nanopoulos [3]. These authors conjecture that it may be possible for black holes to evaporate and still not lead to the loss of information.* The idea is simple: a Planck size black hole is represented by a massive state of a critical string theory in two-dimensional space-time. Solutions to the equations of motion of such theories can be shown to correspond to the gravitational metric induced by a black hole [4]. In these two-dimensional theories there is a set of "topological quantum numbers" associated with each state in the spectrum. These are special states of the theory which exist only for fixed values of the energy and momentum. The evaporation process is depicted as the "decay" of a massive black hole into less massive ones [5]. This cascade decay would eventually produce the lightest possible

*It should be stressed that all these ideas properly apply to models of black holes in two space-time dimensions. The relevance of these studies to four-dimensional physical systems remains unclear.

states which would be equivalent to an evaporation phenomenon. The important point is that the emitted states carry away all of the information stored in the black hole, since it is encoded in the topological quantum numbers. These quantum numbers are usually not measured, although one could envision gedanken experiments which could in principle measure them [6]. In fact, when one sums over the normally unobserved quantum numbers, the original Hawking entropy [2] is recovered [7] – a rather remarkable result. Moreover, the picture of evaporation described above appears different from that proposed by Hawking, although detailed studies of the evaporation process would be required to give a more precise description of the main differences.

The type of "open" system represented by the black hole is usually described in terms of a density matrix formalism, which allows the time-evolution of pure states into mixed states. In field theory language one considers a superscattering operator $\not{S}$ [8], which in the usual case factorizes as follows

$$\not{S} = S^{\dagger} S, \qquad \rho_{out} = \not{S} \rho_{in}, \tag{1}$$

where ρ is the density matrix. At very short distances, where quantum gravity effects matter, there may arise non-factorizable contributions to $\not{S}$ which would lead to a loss of information [8]. Intuitively, if all states in the system are accounted for, then $\not{S} = S^{\dagger} S$. However, one gets

$$\not{S} = S^{\dagger} S + \delta \not{S} \tag{2}$$

if only a "convenient" (incomplete) set of states is used, *i.e.*, such that the $\delta\not{S}$ effects are only relevant near the Planck scale. In practice, topological states (those spread over all space and time) cannot be incorporated in an S-matrix formalism and the $\delta\not{S}$ contribution appears unavoidable.

How large can the $\delta\not{S}$ effects be? Dimensionally one would expect $\mathcal{O}(m_p^2/M_{Pl})^n$, where $m_p \sim 1\,\mathrm{GeV}$ is the low-energy scale and $n \geq 1$. For $n = 1$, one gets $\sim 10^{-18}\,\mathrm{GeV}$. Is there a system sensitive enough to this tiny effects? Yes, the $K - \bar{K}$ system, since there $\Delta M_K \sim \Delta\Gamma_K \sim 10^{-15}\,\mathrm{GeV}$ and well measured observables could show deviations from conventional ($\delta\not{S} = 0$) expectations [9].

It has been argued that in extreme quantum gravity scenarios, CPT invariance is not likely to hold. In fact, detailed string theory arguments in the context of the two-dimensional string theories mentioned above indicate that CPT violation may be a generic property of the low-energy theory of string [9]. Unfortunately, a complete operator language describing this phenomenon is at present still lacking, although great strides in this direction have been recently made [10]. In what follows we will assume that such effects are generically present and will study ways of testing for them in the $K - \bar{K}$ system.

3 The $K-\bar{K}$ Mixing Analysis in the Density Matrix Formalism

The time-evolution equation for the density matrix in the presence of $\delta\not{S} \neq 0$ is given by [11]

$$\dot{\rho} = i\,[\rho, H] + \delta\not{H}, \tag{3}$$

where $\delta\not{H}$ is the open system term, which generally contains CPT-conserving and CPT-violating terms. The $K-\bar{K}$ two-state problem can be recast in this language and reduces to one 2×2 complex matrix linear differential equation, which depends on the usual $K-\bar{K}$ parameters. In the language of Ref. [11], this equation also depends on three new CPT-violating parameters (α, β, γ). One obtains

$$\dot{\rho}_{11} = -\Gamma_L\rho_{11} + \gamma\rho_{22} - 2\chi \mathrm{Re}\,\rho_{12} - 2\beta \mathrm{Im}\,\rho_{12}, \tag{4}$$

$$\dot{\rho}_{12} = -(\Gamma + i\Delta m)\rho_{12} - 2i\alpha \mathrm{Im}\,\rho_{12} + (\chi - i\beta)(\rho_{11} - \rho_{22}), \tag{5}$$

$$\dot{\rho}_{22} = -\Gamma_S\rho_{22} + \gamma\rho_{11} + 2\chi \mathrm{Re}\,\rho_{12} + 2\beta \mathrm{Im}\,\rho_{12}, \tag{6}$$

with $\Gamma_L = (5.17\times 10^{-8}\,\mathrm{s})^{-1}$ and $\Gamma_S = (0.8922\times 10^{-10}\,\mathrm{s})^{-1}$ the inverse long and short K-meson lifetimes, $\Gamma = (\Gamma_S + \Gamma_L)/2$, $|\Delta\Gamma| \approx \Gamma_S$, and $\Delta m = 0.5351\times 10^{10}\,\mathrm{s}^{-1} = 3.522\times 10^{-15}\,\mathrm{GeV}$ the $K-\bar{K}$ mass difference. Moreover

$$\chi = \mathrm{Im}\,M_{12} = \tfrac{1}{2}\frac{|\Delta\Gamma||\epsilon|}{\cos\phi}, \quad \epsilon = |\epsilon|e^{i\phi}, \quad \tan\phi = \frac{\Delta m}{\frac{1}{2}|\Delta\Gamma|}, \tag{7}$$

with $|\epsilon| = 2.265\times 10^{-3}$ and $\phi \approx 43.7°$.

At present there are three experiments running which are studying the traditional $K-\bar{K}$ mixing and CP-violating parameters, and could in principle observe CPT violations as well. These are: (i) the E731 experiment at Fermilab, where a K_L beam goes through a regenerator and gives a coherent mixture of K_S and K_L, and the $\pi^+\pi^-$ and $\pi^0\pi^0$ decay products are studied [12]; (ii) the NA31 experiment at CERN which is similar to the E731 experiment but no regenerator is used [13]; and (iii) the new CPLEAR experiment at CERN in which K^0 and $\bar{K}^0$ are produced simultaneously and symmetrically in $p\bar{p} \to K^-\pi^+K^0, K^+\pi^-\bar{K}^0$, and the time-dependent decay asymmetries are studied [14].

In the density matrix formalism, every observable is computed as $\langle \mathcal{O}\rangle = \mathrm{Tr}\,\mathcal{O}\rho/\mathrm{Tr}\,\rho$ for a suitably chosen operator $\mathcal{O}$. In the case of the $K \to 2\pi$ decay one takes [11]

$$\mathcal{O}_{2\pi} = \begin{array}{c} \\ |K_2\rangle \\ |K_1\rangle \end{array}\!\!\begin{array}{c} |K_2\rangle \quad |K_1\rangle \\ \begin{pmatrix} 0 & 0 \\ 0 & 1 \end{pmatrix} \end{array}, \tag{8}$$

where

$$|K_1\rangle = \frac{|K^0\rangle + |\bar{K}^0\rangle}{\sqrt{2}}, \qquad |K_2\rangle = \frac{|K^0\rangle - |\bar{K}^0\rangle}{\sqrt{2}}, \tag{9}$$

are the CP-even and CP-odd eigenstates which decay into 2π and 3π final states respectively. In the present discussion "direct" CP violation is neglected since $|\epsilon'/\epsilon| \lesssim$

10^{-3}. The rate of decay into 2π is then given by

$$R_{2\pi} = \frac{\rho_{22}}{\rho_{11} + \rho_{22}}. \tag{10}$$

Another observable of interest is the semileptonic decay charge asymmetry δ

$$\delta = \frac{\Gamma(K \to \pi^- l^+ \nu_l) - \Gamma(K \to \pi^+ l^- \bar{\nu}_l)}{\Gamma(K \to \pi^- l^+ \nu_l) + \Gamma(K \to \pi^+ l^- \bar{\nu}_l)}, \tag{11}$$

where $l = e, \mu$. In this case the relevant operators are [11]

$$\mathcal{O}_{\pi^- l^+ \nu_l} = \begin{pmatrix} 1 & 1 \\ 1 & 1 \end{pmatrix}, \qquad \mathcal{O}_{\pi^+ l^- \bar{\nu}_l} = \begin{pmatrix} 1 & -1 \\ -1 & 1 \end{pmatrix}, \tag{12}$$

and one obtains

$$\delta = 2\mathrm{Re}\,\rho_{12}/(\rho_{11} + \rho_{22}). \tag{13}$$

4 The Standard ($\alpha = \beta = \gamma = 0$) Case

With the initial condition

$$\rho(0) = \frac{1}{2}\begin{pmatrix} 1 & 1 \\ 1 & 1 \end{pmatrix}, \tag{14}$$

corresponding to a pure $|K^0\rangle$ beam, one can solve the differential equations (Eqs. (4–6)) in the standard case (*i.e.*, no CPT violation $\alpha = \beta = \gamma = 0$) to obtain the expected result

$$\rho_{11}(t) \approx e^{-\Gamma_L t} \approx 1, \tag{15}$$

$$\rho_{22}(t) \approx e^{-\Gamma_S t} + 2|\epsilon| e^{-\Gamma t} \cos(\Delta m t + \phi) + |\epsilon|^2 e^{-\Gamma_L t}. \tag{16}$$

From Eq. (10) it is clear that the expression for ρ_{22} above gives the $R_{2\pi}$ rate. Note that since we are neglecting ϵ' effects we cannot distinguish between $\pi^+\pi^-$ and $\pi^0\pi^0$ final states. The formula for $R_{2\pi}$ approaches an asymptotic region after several K_S lifetimes and also exhibits an interference region. These features are shown in Fig. 1, where the dotted lines represent the $|\epsilon|^2 e^{-\Gamma_L t}$ and $e^{-\Gamma_S t}$ solutions which would exist in the absence of mixing.

The asymptotic region provides a determination of $|\epsilon|$, whereas the interference region allows the measurement of the angle ϕ. The latest experimental data on the interference region comes from the E731 experiment and is shown in Fig. 2. The plotted curve is the "cosine" extracted from Eq. (16) by subtracting out $e^{-\Gamma_S t} + |\epsilon|^2 e^{-\Gamma_L t}$ and dividing by $2|\epsilon| e^{-\Gamma t}$. Note that there are no obvious deviations from the expected cosine curve.

With the same initial condition as in Eq. (14), one obtains for the semileptonic decay asymmetry

$$\delta = 2|\epsilon| \cos\phi (e^{-\Gamma_L t} + e^{-\Gamma_S t}) + e^{-\Gamma t} \cos(\Delta m t). \tag{17}$$

In Fig. 3 we show the time-dependence of δ which exhibits an asymptotic region ($\delta \to \delta_L = 2|\epsilon| \cos\phi$) as well as an interference region which is being used by the CPLEAR collaboration to measure Δm [14].

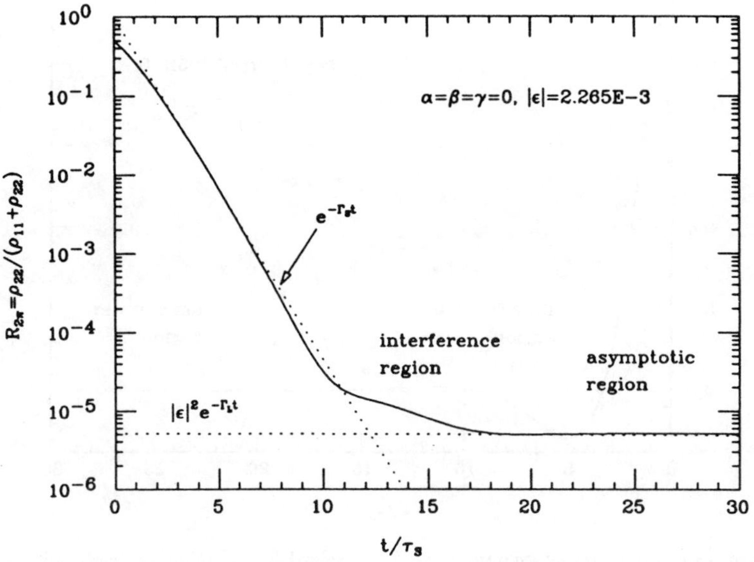

Figure 1: The rate for $K \to 2\pi$ in the standard $K - \bar{K}$ mixing scenario with no CPT violation as a function of time in units of the K_S lifetime (τ_s). The dotted lines represent the functions $e^{-\Gamma_S t}$ and $|\epsilon|^2\, e^{-\Gamma_L t}$. The curve asymptotes to $|\epsilon|^2$.

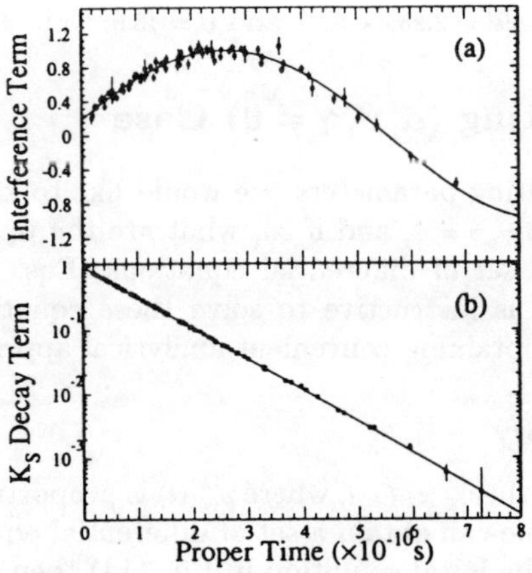

Figure 2: The latests experimental determination of the "cosine" interference term in $K \to 2\pi$ decays. Figure from Ref. [12].

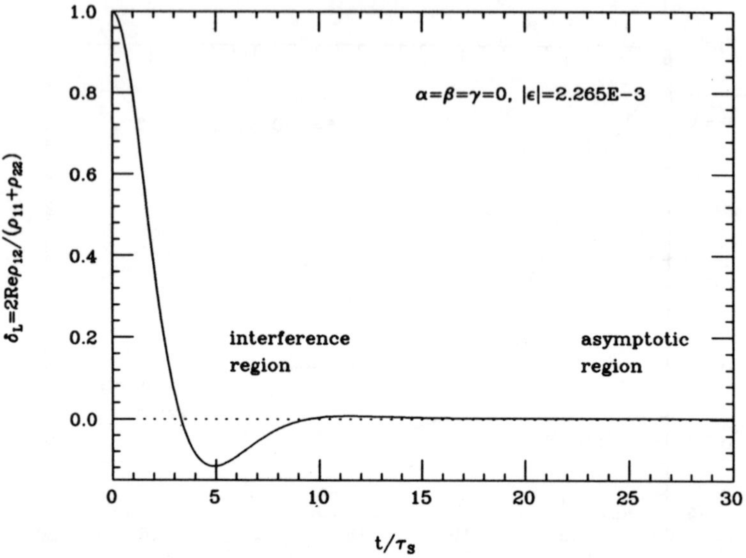

Figure 3: The semileptonic decay asymmetry in the standard $K-\bar{K}$ mixing scenario with no CPT violation as a function of time in units of the K_S lifetime (τ_s). The curve asymptotes to $2|\epsilon|\cos\phi$.

Experimentally one obtains [15]

$$\sqrt{R_{2\pi}} = 2.265(1 \pm 0.0102) \times 10^{-3}, \qquad \delta_L = 3.27(1 \pm 0.0367) \times 10^{-3}, \tag{18}$$

which are compatible with $|\epsilon| = 2.265 \times 10^{-3}$ and $\phi = 43.68° \pm 0.14°$.

5 The CPT-violating ($\alpha, \beta, \gamma \neq 0$) Case

Turning on the CPT violating parameters one would like to know if the best fit to the data is given by $\alpha = \beta = \gamma = 0$, and if so, what are the upper bounds on these parameters. The coupled set of differential equations (Eqs. (4-6)) can be solved numerically. However, it is instructive to solve these equations in perturbation theory (in $\alpha, \beta, \gamma, |\epsilon|$) thus obtaining convenient analytical approximations.

5.1 Perturbation theory

Writing $\rho_{ij}(t) = \rho_{ij}^{(0)}(t) + \rho_{ij}^{(1)}(t) + \rho_{ij}^{(2)}(t) + \cdots$, where $\rho_{ij}^{(k)}(t)$ is proportional to $\alpha^{p_\alpha}\beta^{p_\beta}\gamma^{p_\gamma}|\epsilon|^{p_\epsilon}$, with $p_\alpha + p_\beta + p_\gamma + p_\epsilon = k$, one can obtain a set of differential equations at each order in perturbation theory. The inital condition in Eq. (14) then implies the following zero-th order results

$$\rho_{11}^{(0)}(t) = e^{-\Gamma_L t}, \quad \rho_{22}^{(0)}(t) = e^{-\Gamma_S t}, \quad \rho_{12}^{(0)}(t) = e^{-(\Gamma + i\Delta m)t}. \tag{19}$$

For higher orders the typical equation is of the form

$$\dot{\rho}_{ij}^{(k)}(t) = -A\rho_{ij}^{(k)}(t) + \sum{}' \rho_{kl}^{(k-1)}(t) \tag{20}$$

where $\sum'$ excludes the ρ_{ij} term. Multiplying by the integrating factor e^{At} one obtains

$$\frac{d}{dt}\left[e^{At}\rho_{ij}^{(k)}(t)\right] = e^{At}\sum{}'\rho_{ij}^{(k-1)}(t) \tag{21}$$

which can be straightforwardly integrated in terms of the known functions at $(k-1)$-th order, and the initial condition $\rho_{ij}^{(k)}(0) = 0$, for $k \geq 1$.

Following the above straightforward (although tedious) procedure, one obtains expressions for the relevant quantities (recall that $R_{2\pi} \approx \rho_{22}$ and $\delta \approx 2\mathrm{Re}\,\rho_{12}$):

$$\begin{aligned} \rho_{22}(t) &= \left[1 - \hat{\gamma} - 2|\epsilon|\cos\phi - 4\hat{\beta}\sin\phi\cos\phi\right] e^{-\Gamma_S t} \\ &\quad +2|\epsilon| e^{-\Gamma t}\cos(\Delta m t - \phi) - 4\hat{\beta}\cos\phi\, e^{-\Gamma t}\sin(\Delta m t - \phi) \\ &\quad +\hat{\gamma} e^{-\Gamma_L t} + \sum_{i=1}^{7} c_i R_i(t), \end{aligned} \tag{22}$$

where we have defined

$$\hat{\alpha} = \alpha/|\Delta\Gamma|, \quad \hat{\beta} = \beta/|\Delta\Gamma|, \quad \hat{\gamma} = \gamma/|\Delta\Gamma|, \tag{23}$$

and the second-order time-dependent functions $R_i(t)$ are given by

$$R_1(t) = e^{-\Gamma_L t} - e^{-\Gamma_S t} \tag{24}$$

$$R_2(t) = t\, e^{-\Gamma_S t} \tag{25}$$

$$R_3(t) = e^{-\Gamma t}\cos(\Delta m t) - e^{-\Gamma_S t} \tag{26}$$

$$R_4(t) = e^{-\Gamma t}\sin(\Delta m t) - e^{-\Gamma_S t} \tag{27}$$

$$R_5(t) = e^{-\Gamma t}\cos(\Delta m t - \phi) - e^{-\Gamma_S t}\cos\phi \tag{28}$$

$$R_6(t) = e^{-\Gamma t}\sin(\Delta m t - \phi) + e^{-\Gamma_S t}\sin\phi \tag{29}$$

$$R_7(t) = e^{-\Gamma t}\left[\frac{|\Delta\Gamma| t}{2\cos\phi}\cos(\Delta m t - \phi) - \cos(\Delta m t - 2\phi)\right] + e^{-\Gamma_S t}\cos 2\phi \tag{30}$$

with coefficients

$$c_1 = |\epsilon|^2 - 4\hat{\beta}|\epsilon|\sin\phi - 4\hat{\beta}^2\cos^2\phi + \hat{\gamma}(\hat{\gamma} + 2|\epsilon|\cos\phi - 4\hat{\beta}\sin\phi\cos\phi) \tag{31}$$

$$c_2 = |\Delta\Gamma|(|\epsilon|^2 + 4\hat{\beta}|\epsilon|\sin\phi - 4\hat{\beta}^2\cos^2\phi) \tag{32}$$

$$c_3 = -4\hat{\gamma}|\epsilon|\cos\phi \tag{33}$$

$$c_4 = 8\hat{\beta}\hat{\gamma}\cos^2\phi \tag{34}$$

$$c_5 = -4|\epsilon|^2\cos\phi + 16\hat{\beta}^2\cos^3\phi \tag{35}$$

$$c_6 = 4|\epsilon|\hat{\alpha}/\tan\phi + 16|\epsilon|\hat{\beta}\cos^2\phi \tag{36}$$

$$c_7 = -4\hat{\alpha}|\epsilon|\cos\phi + 8\hat{\alpha}\hat{\beta}\cos^2\phi \tag{37}$$

Analogously one can determine the first-order contributions to $\rho_{12}(t)$,

$$\begin{aligned} \delta(t) = 2\mathrm{Re}\,\rho_{12}(t) &= \delta_L e^{-\Gamma_L t} + \delta_S e^{-\Gamma_S t} + 2e^{-\Gamma t}(1 - \hat{\alpha}|\Delta\Gamma| t - 2|\epsilon|\cos\phi)\cos(\Delta m t) \\ &\quad +2e^{-\Gamma t}(2\hat{\alpha}/\tan\phi + 4\hat{\beta}\cos^2\phi)\sin(\Delta m t), \end{aligned} \tag{38}$$

where

$$\delta_L = 2|\epsilon|\cos\phi - 4\hat{\beta}\sin\phi\cos\phi, \quad (39)$$
$$\delta_S = 2|\epsilon|\cos\phi + 4\hat{\beta}\sin\phi\cos\phi. \quad (40)$$

5.2 Bounds on β and γ

In the expressions for all the ρ-matrix elements we can identify three regimes (see Figs. 1,3): (i) the short-time regime: $0 \le t/\tau_s \lesssim 5$, (ii) the intermediate-time regime: $5 \lesssim t/\tau_s \lesssim 20$, and (iii) the long-time regime: $20 \lesssim t/\tau_s \ll \tau_L/\tau_s \approx 580$. The standard $K_L \to 2\pi$ measurements are carried out in the long-time regime, and in this case $\rho_{22}(t)$ simplifies considerably (*i.e.*, taking $\Gamma, \Gamma_S \to \infty$, $\Gamma_L \to 0$)

$$\begin{aligned} R_{2\pi} &\to \hat{\gamma} + c_1 \\ &= \hat{\gamma} + |\epsilon|^2 - 4\hat{\beta}|\epsilon|\sin\phi - 4\hat{\beta}^2\cos^2\phi + \hat{\gamma}(\hat{\gamma} + 2|\epsilon|\cos\phi - 4\hat{\beta}\sin\phi\cos\phi) \end{aligned} \quad (41)$$

Comparing this prediction with the experimental measurement, one can define an effective $|\epsilon|$ parameter

$$\begin{aligned} |\epsilon_{eff}| &= \left[\hat{\gamma} + |\epsilon|^2 - 4\hat{\beta}|\epsilon|\sin\phi - 4\hat{\beta}^2\cos^2\phi + \hat{\gamma}(\hat{\gamma} + 2|\epsilon|\cos\phi - 4\hat{\beta}\sin\phi\cos\phi)\right]^{1/2} \\ &= 2.265(1 \pm 0.0102) \times 10^{-3} \end{aligned} \quad (42)$$

This relation has the very interesting possible implication that CP violation may not exist (*i.e.*, $|\epsilon| = 0$) and yet the experimental value of $R_{2\pi}$ could still be obtained through CPT-violating interactions. Unfortunately, as we now show, this possibility is quickly ruled out by intermediate-time measurements. Note that since the standard result ($|\epsilon|^2$) does not appear until second order, the value of the first-order contribution ($\hat{\gamma}$) is bound to be severely constrained (*i.e.*, $\hat{\gamma} < |\epsilon|^2 \sim 10^{-6}$).

In the intermediate-time regime, interference effects are important and have been observed experimentally. In this region we can effectively take $\Gamma_S \to \infty$ and $\Gamma_L \to 0$, with the $e^{-\Gamma t}$ and oscillatory factors giving the time dependence. In practice, to first order the "cosine" term is obtained by dividing the interference terms by $2|\epsilon_{eff}|e^{-\Gamma t}$, to get

$$\frac{|\epsilon|}{|\epsilon_{eff}|}\cos(\Delta m t - \phi) - \frac{2\hat{\beta}}{|\epsilon_{eff}|}\cos\phi\sin(\Delta m t - \phi) \equiv C\cos(\Delta m t - \phi - \delta\phi), \quad (43)$$

with

$$\tan\delta\phi = -\frac{2\hat{\beta}\cos\phi}{|\epsilon|}, \qquad C = \frac{|\epsilon|}{|\epsilon_{eff}|}\frac{1}{\cos\delta\phi}. \quad (44)$$

Since there is no evidence that the "cosine" term has an anomalous amplitude (see Fig. 2) one must have $C \approx 1$. Moreover, the phase $\phi + \delta\phi$ is not observed to deviate from the "superweak" phase ϕ by more than 2° [12], *i.e.*, $|\delta\phi| \lesssim \pi/90$, which implies

$$|\hat{\beta}| \lesssim \frac{\pi|\epsilon|}{180\cos\phi} \approx 5.466 \times 10^{-5}\frac{|\epsilon|}{|\epsilon_{eff}|}. \quad (45)$$

The above upper bound on $\delta\phi$ also implies

$$0.9994\,\mathcal{C} < |\epsilon|/|\epsilon_{eff}| < \mathcal{C}, \tag{46}$$

which substituted back in Eq. (45) gives

$$|\hat{\beta}| < 5.466 \times 10^{-5} \times \mathcal{C} \lesssim 5.7 \times 10^{-5} \qquad \text{(intermediate} - \text{time data)} \tag{47}$$

if we assume that the "cosine" amplitude is measured experimentally to within 5%, *i.e.*, $\mathcal{C} \approx 1 \pm 0.05$. This assumption on $\mathcal{C}$ also implies

$$\frac{|\epsilon_{eff}|}{|\epsilon|} \approx 1 \pm 0.05. \tag{48}$$

Therefore, the intermediate-time data indicate that one must have a "full strength" $|\epsilon|$ and the other contributions to $|\epsilon_{eff}|$ must be no larger than allowed by the above ratio.

Going back to the asymptotic value of $R_{2\pi}$ in Eq. (42), factoring out the leading $|\epsilon|$ dependence, one obtains an expression for $|\epsilon_{eff}|/|\epsilon|$,

$$\frac{|\epsilon_{eff}|}{|\epsilon|} \approx 1 + \tfrac{1}{2}\frac{\hat{\gamma}}{|\epsilon|^2} - 2\frac{|\hat{\beta}|}{|\epsilon|}\sin\phi - 2\frac{\hat{\beta}^2}{|\epsilon|^2}\cos^2\phi + \frac{\hat{\gamma}}{|\epsilon|}\left(\frac{\hat{\gamma}}{|\epsilon|} + \frac{2}{|\epsilon|}\cos\phi - 4\frac{\hat{\beta}}{|\epsilon|}\sin\phi\cos\phi\right), \tag{49}$$

which when constrated with Eq. (48) implies

$$\hat{\gamma} < 0.1|\epsilon|^2 \approx 5 \times 10^{-7} \tag{50}$$

although this result cannot really be trusted since $|\epsilon'| \sim 10^{-6}$ effects have been neglected. As well one obtains an upper bound on $|\hat{\beta}|$, $|\hat{\beta}| < 0.05|\epsilon|/2\sin\phi \approx 8.2 \times 10^{-5}$ which is weaker than the one deduced above on the basis of the intermediate-time data. Note however that if the "cosine" amplitude is determined more precisely, then the bound on $|\hat{\beta}|$ obtained in the present way could easily become more stringent than the one in Eq. (47).

5.3 Bounds on α

In order to involve the parameter α one can consider second-order intermediate-time contributions to $R_{2\pi}$. This analysis will result in a generalization of Eqs. (43,44) as follows

$$2|\epsilon|e^{-\Gamma t}\left[A\cos(\Delta mt - \phi) + B\sin(\Delta mt - \phi)\right] \equiv 2|\epsilon|e^{-\Gamma t}\,\mathcal{C}\cos(\Delta mt - \phi - \delta\phi) \tag{51}$$

with

$$\tan\delta\phi = \frac{B}{A}, \qquad \mathcal{C} = \frac{A}{\cos\delta\phi}\,. \tag{52}$$

Keeping in mind the bounds on $\hat{\gamma}$ and $\hat{\beta}$ obtained above, and that only terms proportional to $e^{-\Gamma t}$ are relevant, one can show that the only non-negligible second-order terms are proportional to $c_{5,6,7}$. Fortunately, $c_{6,7}$ are sensitive to $\hat{\alpha}$. In this manner one obtains the following next-to-leading-order expressions for A and B

$$A = 1 - 2|\epsilon|\cos\phi - \hat{\alpha}(t|\Delta\Gamma| - 2\cos^2\phi), \quad (53)$$

$$B = -2\frac{\hat{\beta}}{|\epsilon|}\cos\phi + 2\hat{\alpha}(1/\tan\phi + \sin\phi\cos\phi). \quad (54)$$

From the discussion in the previous subsection, it is clear that $A \approx 1 \pm 0.05$. In the expression for A, the second term in the parenthesis is much smaller than the first one, since $5 \lesssim t|\Delta\Gamma| \approx t/\tau_s \lesssim 30$ in this regime. Therefore, to stay within the (5%) bounds of the "cosine" amplitude, we must demand

$$|\hat{\alpha}| \lesssim \frac{0.05}{30} \approx 2 \times 10^{-3}. \quad (55)$$

In the expression for B, if $\hat{\beta} > 0$, it would appear that a cancellation of the two terms could yield a sufficiently small $\delta\phi$ for relatively large values of the parameters. However, such a cancellation is only possible for rather small values of the parameters, since we already know that $\hat{\alpha}$ must be small, in fact smaller than the upper bound $|\hat{\beta}|/|\epsilon| \lesssim 0.024$ deduced in the previous subsection. For $\hat{\beta} < 0$, the $\hat{\alpha}$ contribution to B ($B \lesssim \pi/90$) does not affect the bound on $\hat{\beta}$ deduced above. This is not the case for $\hat{\alpha}$ (which must be positive [9]) if $\hat{\beta}$ is close to its maximum allowed value: if $|\hat{\beta}| > 4.5 \times 10^{-5}$ then the bound on $\hat{\alpha}$ will be more stringent than that in Eq. (55).

The presence of $\hat{\alpha}$ is also manifest in the first-order intermediate-time contribution to $\delta(t)$ in Eq. (38). One can rewrite this expression as follows

$$\delta(t) = \delta_L e^{-\Gamma_L t} + \delta_s e^{-\Gamma_S t} + 2e^{-\Gamma t} C_\delta \cos(\Delta m t - \phi_\delta), \quad (56)$$

with

$$\tan\phi_\delta = \frac{B_\delta}{A_\delta}, \qquad C_\delta = \frac{A_\delta}{\cos\phi_\delta}, \quad (57)$$

and

$$A_\delta = 1 - 2|\epsilon|\cos\phi - \hat{\alpha}t|\Delta\Gamma|, \qquad B_\delta = 2\hat{\alpha}/\tan\phi + 4\hat{\beta}\cos^2\phi. \quad (58)$$

It is not hard to see (*i.e.*, by setting the first derivative of the interference term to zero) that the minimum of $\delta(t)$ (see Fig. 3) occurs (approximately) for $\Delta m t - \phi = \pi/2$ in the standard case, and it is shifted to $\Delta m t - \phi - \phi_\delta = \pi/2$ in the CPT-violating case. Thus, if the position of the minimum could be measured with some accuracy (by the CPLEAR collaboration) then one could obtain some experimental information on $\hat{\alpha}$. The bound on $\hat{\alpha}$ derived above already constrains the possible shift in the position of the minimum to $|\phi_\delta| \lesssim 6 \times 10^{-3} \approx 0.3°$.

6 Conclusions

We conclude that new tests or predictions of string theory could be readily accessible at present (E731,NA31,CPLEAR) and near future (DAϕNE) Kaon physics facilities. These experimental probes are essential to determine whether these string theory ideas are relevant to the physical world. However, more effort would need to be invested from the formal side in order to produce specific predictions for *e.g.*, the α, β, γ parameters, as well as any other parameters which may occur in other systems of interest. For all we know, it may turn out that these effects, even though non-zero, are yet completely negligible because too strong Planck mass suppression. On the other hand, it is encouraging to note that the bound on $\hat{\beta}$ obtained above, *i.e.*, $|\hat{\beta}| \lesssim 10^{-5}$, implies an upper bound on $|\beta|$ itself of $|\beta| \lesssim 10^{-19}\,\mathrm{GeV}$ which would be expected to be the largest possible value from naive dimensional analysis (*i.e.*, $(1\,\mathrm{GeV})^2/M_{Pl}$).

Acknowledgements

I would like to thank J. Ellis, N. Mavromatos, and D. Nanopoulos for the on-going collaboration which has led to the above calculations and results. This work has been supported by an SSC Fellowship.

1. A. Font, L. Ibáñez, D. Lüst, and F. Quevedo, Phys. Lett. B **249** (1990) 35; J. Schwarz and A. Sen, Phys. Lett. B **312** (1993) 105.
2. S. W. Hawking, Comm. Math. Phys. **43** (1975) 199; J. D. Bekenstein, Phys. Rev. D **12** (1975) 3077.
3. J. Ellis, N. Mavromatos, and D. V. Nanopoulos, Phys. Lett. B **267** (1991) 465 and Phys. Lett. B **272** (1991) 261.
4. E. Witten, Phys. Rev. D **44** (1991) 314; G. Mandal, A. Sengupta, and S. Wadia, Mod. Phys. Lett. A **6** (1991) 1685.
5. J. Ellis, N. Mavromatos, and D. V. Nanopoulos, Phys. Lett. B **276** (1992) 56.
6. J. Ellis, N. Mavromatos, and D. V. Nanopoulos, Phys. Lett. B **284** (1992) 27,43.
7. J. Ellis, N. Mavromatos, and D. V. Nanopoulos, Phys. Lett. B **278** (1992) 246.
8. S. W. Hawking, Comm. Math. Phys. **87** (1982) 395.
9. J. Ellis, N. Mavromatos, and D. V. Nanopoulos, Phys. Lett. B **293** (1992) 142 and CERN-TH.6755/92 (Dec. 92).
10. J. Ellis, N. Mavromatos, and D. V. Nanopoulos, CERN-TH.6897/93 (May 93).
11. J. Ellis, J. Hagelin, D. V. Nanopoulos, and M. Srednicki, Nucl. Phys. B **241** (1984) 381.
12. See *e.g.*, L. K. Gibbons, *et. al.* (E731 Coll.), Phys. Rev. Lett. **70** (1993) 1199 and references therein.
13. See *e.g.*, H. Burkhardt, *et. al.* (NA31 Coll.), Phys. Lett. B **206** (1988) 169 and references therein.
14. See *e.g.*, C. Guyot, *et. al.* (CPLEAR Coll.) in Proceedings of the XXVIth International Conference on High-Energy Physics, Dallas, August 6-12, 1992,

edited by J. Sanford (AIP Conference Proceedings No. 272), p. 510.
15. Particle Data Group, Phys. Rev. D **45** (1992) S1.

BREAKDOWN OF DUALITY IN (0,2) SUPERSTRING MODELS

JENS ERLER
Department of Physics, David Rittenhouse Laboratory, University of Pennsylvania, 209 South 33rd Street, Philadelphia, PA 19104, U.S.A.

ABSTRACT

After pointing out the role of the compactification lattice for spectrum calculations in orbifold models, I discuss modular discrete symmetry groups for Z_N orbifolds. I consider the Z_7 orbifold as a nontrivial example of a (2,2) model and give the generators of the modular group for this case, which does not contain $[SL(2, \mathbf{Z})]^3$ as had been speculated. I also discuss how to treat cases where quantized Wilson lines are present. I consider in detail an example, demonstrating that quantized Wilson lines affect the modular group in a nontrivial manner. In particular, I show that it is possible for a Wilson line to break $SL(2, \mathbf{Z})$.

1. Introduction

The space of string vacua is locally parametrized by moduli, which are marginal deformations of the underlying conformal field theory (CFT). In the low energy effective theory the moduli correspond to vacuum expectation values of massless scalar fields that have flat potentials to all orders in perturbation theory. An intriguing feature of string compactifications is that the natural parametrization of the moduli space label the points in a redundant way, since all physical quantities are invariant under the action of some discrete group acting on the moduli.

The motivations for studying this modular discrete symmetry groups for compactifications of the heterotic string are manifold. The most obvious one is that its knowledge allows a restriction to a fundamental domain of moduli space, reducing the number of string vacua. Next, being an exact symmetry it should persist in any approximation, and in particular in the low energy effective supergravity theory. This constrains the superpotential, the gauge kinetic functions, etc. Assuming the exactness of this symmetry even after taking into account nonpertubative string effects, gives a powerful tool for the construction of nonpertubative potentials[1]. They can fix the moduli vacuum expectation values (vev's) and lift the vacuum degeneracy. Recent discussions of soft supersymmetry breaking terms and minimal string models were restricted to the modular group $SL(2, \mathbf{Z})$. It will be one of the main results of this talk that this group plays a much smaller role as a symmetry in moduli space as previously believed. Finally, modular symmetries may also play an important role in the context of gauge coupling unification. Here threshold corrections potentially explain the discrepancy between the reduced Planck scale and the unification scale. Again, the relevant threshold terms have to transform appropriately under the given symmetry group[2].

I will discuss these modular symmetries for orbifold models. They represent a simple construction and the orbifold CFT is exactly solvable. This implies that the discrete symmetry groups are exactly derivable, as well, and that the phenomenology of orbifold models can be pushed forward very far. At the same time "quasi realistic" models with 3 fermion generations transforming under $SU(3) \times SU(2) \times U(1)^n$ belong to this class[3]. A natural hierarchy of fermion masses arises due to the fixed point structure of the twisted sectors. Finally, exact deformations of fermionically constructed models can be obtained via the $Z_2 \times Z_2$ orbifold.

2. Symmetric Z_N Orbifolds

The possible point groups leading to $D = 4$, $N = 1$ supersymmetric Z_N orbifolds are well known[4]:

$$Z_3, \quad Z_4, \quad Z_6, \quad Z_6', \quad Z_7, \quad Z_8, \quad Z_8', \quad Z_{12}, \quad Z_{12}'. \tag{1}$$

However, in order to find all the corresponding *models* a classification of the possible compactification lattices is necessary. As an example of how the underlying lattice affects basic properties of the model consider the Z_4 orbifold. It can be realized e.g. using three copies of the root lattice of $SO(4)$ with a twist simultaneously acting as 90° rotations in two copies and as a reflection in the last one:

$$\Lambda_1 = [SO(4)]^3 : \qquad e_{1,3} \to e_{2,4} \quad e_{2,4} \to -e_{1,3} \quad e_{5,6} \to -e_{5,6}.$$

Alternatively, one can utilize two copies of the root lattice of $SU(4)$ with Coxeter twists acting in them:

$$\Lambda_2 = [SU(4)]^2 : \qquad e_{1,4} \to e_{2,5} \quad e_{2,5} \to e_{3,6} \quad e_{3,6} \to -\sum_{i=1}^{3} e_{i,i+3}. \tag{2}$$

It has been shown that for two models to be equivalent, there must be a matrix $M \in GL(d, \mathbf{Z})$ with respect to which the integer valued twist (lattice basis) matrices θ are similar[5], i.e.

$$\exists \quad M \in GL(d, \mathbf{Z}) \quad \text{with} \quad \theta' = M \theta M^{-1}. \tag{3}$$

For the case at hand such a matrix cannot be found. What does that mean for the spectra?

Clearly, in the untwisted sector the lattice Λ (winding modes) and its dual Λ^* (momenta) only affect the massive string modes. The untwisted massless states are the same. Since there are no fixed tori in the first twisted sector, it is completely independent of Λ. The number of generations coming from this sector equals the number of fixed points, which in turn can be determined with help of the Lefschetz fixed point theorem. In our case we find

$$\text{Det}(1 - \theta) = 16. \tag{4}$$

The second twisted sector, however, *does* depend on Λ due to the complex plane which is left fixed under the twist action. The number of generations coming from such a sector is given by the number of fixed tori.

In the case of the $[SO(4)]^3$ lattice the fixed plane is orthogonal to the twisted directions. This implies that the Lefschetz fixed point theorem is again applicable if restricted to the twisted directions. One finds 16 fixed tori. A closer investigation of twist phases reveals that 10 of them correspond to generations and 6 to antigenerations. In contrast, in the $[SU(4)]^2$ model the Lefschetz theorem cannot be employed. Explicit construction of the fixed tori gives rise to four generations and no antigenerations.

The same result can be obtained by computing the one loop partition function. The relevant projections in the two cases are[6]

$$\Lambda_1 : \frac{1}{4}[16 + 4\Delta + 16\Delta^2 + 4\Delta^3], \tag{5}$$

$$\Lambda_2 : \frac{1}{4}[4 + 4\Delta + 4\Delta^2 + 4\Delta^3], \tag{6}$$

with $\Delta = +1$ for generations and $\Delta = -1$ for antigenerations. The degeneracy factors in Eqs. (5) and (6) are different because the twist acts as a rotation in the even, selfdual lattice composed of winding and momentum states: In the orthogonal case of Λ_1 the invariant sublattice is even and selfdual itself and thus the volume factor which one encounters when performing modular world sheet transformations is trivially one. On the other hand, a nontrivial volume factor is found for Λ_2.

¿From Eqs. (5) and (6) it is also obvious, that the Euler number determining the number of net generations is independent of the underlying compactification lattice. This one indeed expects, since this number can be computed by a formula conjectured by Dixon, Harvey, Vafa and Witten[4] and proved by Markushevich, Olshanetskii and Perelomov[7], which only uses twist eigenvalues.

Because of their appealing phenomenology and a considerable confusion in the literature, it was worthwhile to construct all supersymmetric, symmetric Z_N orbifolds with (2,2) world sheet supersymmetry[6]. The resulting 18 models have different massive and *massless* spectra, different modular symmetry groups and consequently different phenomenologies.

3. Duality and Discrete Symmetry Groups

3.1. Overview

To begin the discussion of discrete symmetries in moduli space, consider as a simple example the Z_3 orbifold in two dimensions. The twist acts in the root lattice of $SU(3)$ and the scaling deformations thereof. There is also a continuous antisymmetric tensor field compatible with the twist:

$$g = R^2 \begin{pmatrix} 2 & -1 \\ -1 & 2 \end{pmatrix}, \qquad b_{ij} = \begin{pmatrix} 0 & -b \\ b & 0 \end{pmatrix}. \tag{7}$$

The orbifold radius R and the antisymmetric tensor parameter b, which are the real untwisted moduli of the model, can be combined to the complex parameter

$$\lambda = b + i\sqrt{\mathrm{Det}\, g}, \tag{8}$$

which takes values in the complex upper half plane.

Duality symmetry is the statement that the exchange of winding and momentum quantum numbers,

$$n \to Wm, \qquad m \to {W^{-1}}^T n, \tag{9}$$

and a simultaneous transformation of the background,

$$g \pm b \to {W^{-1}}^T \frac{1}{g \pm b} W^{-1}, \tag{10}$$

lead to the same physical theory. Here the matrix $W \in GL(2, \mathbf{Z})$ has to satisfy the condition[8]

$$\theta W = W {\theta^{-1}}^T. \tag{11}$$

The transformation of the complex modulus λ takes the particularly simple form of an inversion,

$$\lambda \to -1/\lambda. \tag{12}$$

Inclusion of the *axionic* shift symmetry,

$$b \to b + 1 \quad \Rightarrow \quad \lambda \to \lambda + 1, \tag{13}$$

completes the holomorphic symmetry group[9] $PSL(2, \mathbf{Z})$. As shown by Lauer, Mas and Nilles[10] these symmetries are also respected by the correlation functions. Finally, there is a nonholomorphic generator[11],

$$b \to -b \quad \Rightarrow \quad \lambda \to -\bar{\lambda}, \tag{14}$$

which induces an exchange of particles and antiparticles in the twisted sectors.

The results of this example can be summarized in compact form by noting that the global structure of the moduli space is given by

$$\mathcal{M}_{T_2/Z_3} = \frac{\frac{SU(1,1)}{U(1)}}{SU(1,1;\mathbf{Z}) \times Z_2}. \tag{15}$$

Written this way, for the holomorphic part one can state the result in saying that the discrete symmetry group is given by the maximal discrete subgroup of the group characterizing the local structure of the moduli space.

The question arises whether this is a general feature. Consider the local structures of the moduli spaces of the Z_N orbifolds[12]:

$$\mathcal{M}_{Z_3} = \frac{SU(3,3)}{SU(3) \times SU(3) \times U(1)} \tag{16}$$

$$\mathcal{M}_{Z_{4,6}} = \frac{SU(2,2)}{SU(2)\times SU(2)\times U(1)} \times \left[\frac{SU(1,1)}{U(1)}\right]^n \qquad n=1,2 \tag{17}$$

$$\mathcal{M}_{Z_{N\geq 7}} = \left[\frac{SU(1,1)}{U(1)}\right]^n \qquad n=3,4. \tag{18}$$

It has often been assumed that, in fact, the maximal discrete subgroups of the above local structures describe the modular symmetries. Also, extended versions for cases including Wilson lines leading to (0,2) models with broken $E_6 \times E_8$ gauge symmetry are used. For instance, for the Z_3 orbifold one would expect the group $SU(3,3;\mathbf{Z})$, and, similarly, $[SU(1,1;\mathbf{Z})]^3 \cong [PSL(2,\mathbf{Z})]^3$ for Z_7. Actually, this hypothesis has been confirmed for the Z_3 case without Wilson lines[13]. Let us now turn to the more complicated case of the Z_7 orbifold.

3.2. Z_7 Orbifold

It is defined as the Coxeter twist

$$e_i \to e_{i+1} \qquad i=1,\ldots,5 \qquad e_6 \to -\sum_{i=1}^{6} e_i \tag{19}$$

in the $SU(7)$ root lattice and the three metrical deformations thereof. Analogously, three independent parameters of the antisymmetric tensor background are compatible with this twist. Again drastic simplifications occur when using complex moduli, e.g.

$$t_1 = -i\tan(\frac{\pi}{7})[g_1 + \frac{s_1^2}{s_3^2}g_2 + \frac{s_2^2}{s_3^2}g_3] + b_1 + \frac{s_2}{s_1}b_2 + \frac{s_3}{s_1}b_3, \tag{20}$$

$$t_2 = -i\tan(\frac{2\pi}{7})[g_1 + \frac{s_2^2}{s_1^2}g_2 + \frac{s_3^2}{s_1^2}g_3] + b_1 - \frac{s_3}{s_2}b_2 - \frac{s_1}{s_2}b_3, \tag{21}$$

$$t_3 = -i\tan(\frac{3\pi}{7})[g_1 + \frac{s_3^2}{s_2^2}g_2 + \frac{s_1^2}{s_2^2}g_3] + b_1 - \frac{s_1}{s_3}b_2 + \frac{s_2}{s_3}b_3, \tag{22}$$

with the abbreviations $s_k := \sin\frac{2\pi k}{7}$ and $c_k := \cos\frac{2\pi k}{7}$. These moduli are normalized in such a way that one of the axionic shift symmetries ($b_1 \to b_1 + 1$) takes a simple form,

$$T: (t_1, t_2, t_3) \to (t_1+1, t_2+1, t_3+1). \tag{23}$$

Similarly, a matrix W (cf. Eqs. (9) and (10)) can be found[14] such that a $PSL(2,\mathbf{Z})$ structure arises,

$$S: (t_1, t_2, t_3) \to (-1/t_1, -1/t_2, -1/t_3). \tag{24}$$

However, no transformation changing only one *individual* complex modulus can be found, no matter which definition for the t_i one uses[14]. This already reveals that there is no simple $[SL(2,\mathbf{Z})]^3$ symmetry which one would expect on the basis of the aforementioned conjectures. In contrast, a rich structure arises. It is best described by introducing another type of symmetry generators:

Transforming quantum numbers and background according to

$$n \to Vn, \qquad m \to {V^{-1}}^T m, \tag{25}$$

and

$$g \pm b \to\to V^{-1^T}(g \pm b)V^{-1}, \tag{26}$$

respectively, gives rise to a symmetry, presupposing the matrix $V \in GL(2, \mathbf{Z})$ satisfies

$$\theta V = V\theta^p. \tag{27}$$

p is an integer which is in general allowed to be greater than one[15]. However, p is required[11] to have no common divisor with the twist order N. The significance of transformations with $p > 1$ is that they correspond to nonholomorphic transformations (cf. Eq. (14)).

For the Z_7 case we found the transformation

$$R : (t_1, t_2, t_3) \to (-2c_1\bar{t_2}, -2c_2t_3, -2c_3t_1). \tag{28}$$

R^3 yields just complex conjugation. It is interesting that the inclusion of nonholomorphic transformations actually reduces the number of symmetry *generators* since it relates holomorphic ones. For instance the shift symmetries $T_{2(3)} := b_{2(3)} \to b_{2(3)} + 1$ can be generated like $(T = T_1)$,

$$RT_1R^{-1} =: T_3, \tag{29}$$

$$RT_3R^{-1} =: T_2. \tag{30}$$

Similarly $(S = S_1)$,

$$S_2 := R^2S_1R^{-2}, \tag{31}$$

$$S_3 := RS_1R^{-1}. \tag{32}$$

This way more $PSL(2, \mathbf{Z})$ subgroups arise, and in fact, there is an infinity of them. However, no pair of these subgroups mutually commute, again showing that there is no $[SL(2, \mathbf{Z})]^3$.

The three generators R, S and T are sufficient to generate the whole symmetry group. There are many relations between them, e.g.

$$\begin{gathered} S^2 = (ST)^3 = R^6 = \mathbf{1} \\ (TR)^6 = (TR^3)^2 = (SR^3)^2 = \mathbf{1} \\ (SRTR^{-1})^7 = \mathbf{1} \end{gathered} \tag{33}$$

etc.

It is interesting to note that whereas duality is known to relate small and large radii, the combination SR^2SR^{-2} yields a *rescaling symmetry*:

$$(t_1, t_2, t_3) \to (\frac{1}{4c_1^2}t_1, \frac{1}{4c_2^2}t_2, \frac{1}{4c_3^2}t_3). \tag{34}$$

It is of infinite order, transforming two of the moduli to smaller and one of them to larger values.

All holomorphic transformations are of the form

$$t_k \to \frac{a_k + b_k t_k}{c_k + d_k t_k}. \tag{35}$$

The 12 integers appearing in Eq. (35) satisfy three nonlinear relations, generalizing the determinant condition $ad - bc = 1$ in the case of $SL(2, \mathbf{Z})$. They enter the matrix representation of the symmetry group at hand[14].

We checked that the symmetries are respected by the correlation functions, which are now available with their complete moduli dependance[10,16].

4. Discrete Background Fields and (0,2) Models

4.1. Orbicircle ⊗ Circle

In this section I describe how discrete background fields break the modular symmetry group down to a subgroup. The simplest example is the product theory of an orbicircle (a circle with its Z_2 symmetry divided out) with a circle. The twist matrix is given by

$$\theta = \begin{pmatrix} 1 & 0 \\ 0 & -1 \end{pmatrix}, \tag{36}$$

giving rise to the metric

$$g = \begin{pmatrix} R_C^2 & 0 \\ 0 & R_O^2 \end{pmatrix}, \tag{37}$$

characterized by the orbifold radius R_O and the circle radius R_C. In this example the antisymmetric tensor field does not possess a continuous parameter and, in contrast to the two-dimensional Z_3 orbifold discussed earlier, does *not* correspond to a modulus. It only takes discrete values,

$$b_{ij} = \begin{pmatrix} 0 & k \\ -k & 0 \end{pmatrix}, \qquad k \in \frac{1}{2}\mathbf{Z}. \tag{38}$$

Two different cases have to be distinguished. In the first, one has $k \in \mathbf{Z} \cong 0$. This corresponds to a proper direct product theory of the orbicircle and the circle. Thus it is clear that the symmetry transformations consist of the two independent duality involutions

$$R_C \to 1/R_C \qquad \textbf{and/or} \qquad R_O \to 1/R_O, \tag{39}$$

giving rise to $Z_2 \times Z_2$.

The second case has $k \in \mathbf{Z} + 1/2$. The nontrivial antisymmetric tensor field has two effects. Its presence only allows for a *simultaneous* inversion of the radii,

$$R_C \to 1/2R_C \qquad \textbf{and} \qquad R_O \to 1/2R_O, \tag{40}$$

and changes the self-dual point. In other words, it breaks[5] the $Z_2 \times Z_2$ symmetry down to Z_2.

4.2. Z_3 Orbifold $\otimes$ Gauge Lattice

As an example with more phenomenological applicability consider the Z_3 orbifold with a *quantized Wilson line* turned on. In complete analogy to the toy example in the preceding subsection, we have the combination of an orbifold with a torus connected through a discrete background field. The presence of quantized Wilson lines is phenomenologically highly desired. They lift the degeneracy of the twisted sectors' fixed points, thereby reducing the number of fermion generations from 36 to a smaller number and three generation models can be found[3]. At the same time the $E_6 \times E_8$ gauge symmetry is broken to realistic gauge groups. The lesson to be learned here is that they also break the discrete symmetry group in moduli space. Most surprisingly, they break duality symmetry. The Z_2 duality generator has to be replaced by an infinite order generator acting on the background as

$$A : g + b - \Delta_{1/3} \to W^{-1T} \frac{1}{9(g \pm b + \Delta_{1/3})} W^{-1}, \tag{41}$$

where

$$\Delta_{1/3} := \frac{1}{3} \begin{pmatrix} 0 & 1 \\ -1 & 0 \end{pmatrix}. \tag{42}$$

In order to show that the remaining symmetry is in fact a subgroup of $SL(2,\mathbf{Z})$, I introduce the rescaled complex modulus

$$\lambda' = \frac{1}{3}\lambda = \frac{1}{3}(b + i\sqrt{\mathrm{Det\, g}}). \tag{43}$$

The $SL(2,\mathbf{Z})$ transformations w.r.t. λ' are defined as

$$\sigma : \quad \lambda' \to -\tfrac{1}{\lambda'}, \tag{44}$$

$$\tau : \quad \lambda' \to \lambda' + 1. \tag{45}$$

Now one can write

$$A := \tau\sigma\tau, \tag{46}$$

$$B := \tau^3, \tag{47}$$

where B corresponds to the axionic shift symmetry. The $SL(2,\mathbf{Z})$ relations $S^2 = (ST)^3 = 1$ have to be replaced by $(AB)^3 = 1$ with no order two relation.

Thus *quantized Wilson lines break $SL(2,\mathbf{Z})$ to a subgroup*[11,14]. On the other hand, the "canonical" duality transformation is *not* modular and leads to an asymmetric orbifold[5].

5. Conclusions and Outlook

It has become clear that only in a very limited number of cases the modular group $SL(2,\mathbf{Z})$ is realized as a symmetry in moduli space. This has to be compared with the result of an investigation of the mirror manifold of the quintic threefold[17]. As in our case, there is a shift symmetry but no order two generator (duality), which in their case is replaced by an order five generator. The phenomenological investigations using $SL(2, \mathbf{Z})$ or the "maximally discrete symmetry hypothesis" have to be generalized to other groups.

The classification of (2,2) orbifolds necessarily involved the construction of all possible compactification lattices, since they can affect the massless spectra. The list of symmetric orbifolds is now complete and the one of asymmetric orbifolds (point groups *and* lattices) is under way.

5. Acknowledgements

It is a pleasure to thank D. Jungnickel, Albrecht Klemm, J. Lauer, J. Mas, M. Spaliñski, S. Stieberger and especially H. P. Nilles with whom I collaborated on the subjects presented here. I also thank S. Theisen for stimulating discussions. I acknowledge support by Evangelisches Studienwerk and Deutsche Forschungsgemeinschaft.

6. References

1. For a review see S. Ferrara and S. Theisen, *Moduli Spaces, Effective Actions and Duality Symmetry in String Compactifications*, preprint CERN–TH 5652/90 (1990).
2. L. J. Dixon, V. Kaplunovsky and J. Louis, *Nucl. Phys.* **B355** (1991) 649;
J. P. Derendinger, S. Ferrara, C. Kounnas and F. Zwirner, *Nucl. Phys.* **B372** (1992) 145;
P. Mayr and S. Stieberger, *Threshold Corrections to Gauge Couplings in Orbifold Compactifications*, Munich preprint MPI–PH/93–07 (1993).
3. L. E. Ibáñez, J. E. Kim, H. P. Nilles and F. Quevedo, *Phys. Lett.* **191B** (1987) 282.
4. L. Dixon, J. A. Harvey, C. Vafa and E. Witten, *Nucl. Phys.* **B261** (1985) 678 and **B274** (1986) 285;
L. E. Ibáñez, J. Mas, H. P. Nilles and F. Quevedo, *Nucl. Phys.* **B301** (1988) 157.
5. J. Erler, D. Jungnickel and H. P. Nilles, *Phys. Lett.* **276B** (1992) 303.
6. Jens Erler and Albrecht Klemm, *Comment on the Generation Number in Orbifold Compactifications*, Munich preprint MPI–Ph/92–60 and TUM–TH–146/92 (1992) to appear in *Comm. Math. Phys.*
7. D. G. Markushevich, M. A. Olshanetskii and A. M. Perelomov, *Comm. Math. Phys.* **111** (1987) 247.
8. M. Spaliñski, *Nucl. Phys.* **B377** (1992) 339.
9. W. Lerche, D. Lüst and N. P. Warner, *Phys. Lett.* **231B** (1989) 417.

10. J. Lauer, J. Mas and H. P. Nilles, *Phys. Lett.* **226B** (1989) 251 and *Nucl. Phys.* **351B** (1991) 353.

11. J. Erler, *Investigation of Moduli Spaces in String Theories*, Thesis at the Technical University of Munich (in German language), Munich preprint MPI–Ph/92–21 (1992).

12. M. Cvetic, J. Louis and B. Ovrut, *Phys. Lett.* **206B** (1988) 227.

13. S. Ferrara, P. Frè and P. Soriani, *Class. Quant. Grav.* **9** (1992) 1649.

14. J. Erler and M. Spaliñski, *Modular Groups for Twisted Narain Models*, Munich preprint MPI–Ph/92–61 and TUM–TH–147/92 (1992).

15. M. Spaliñski, *Phys. Lett.* **275B** (1992) 47.

16. S. Hamidi and C. Vafa, *Nucl. Phys.* **B279** (1987) 465;
L. J. Dixon, D. Friedan, E. Martinec and S. Shenker, *Nucl. Phys.* **B282** (1987) 13;
J. A. Casas, F. Gomez and C. Munoz, *Phys. Lett.* **251B** (1990) 99;
T. T. Burwick, R. K. Kaiser and H. F. Müller, *Nucl. Phys.* **B355** (1991) 689;
J. Erler, D. Jungnickel, J. Lauer and J. Mas, *Ann. Phys.* **217** (1992) 318;
J. Erler, D. Jungnickel and J. Lauer, *Phys. Rev.* **D45** (1992) 3651;
S. Stieberger, D. Jungnickel, J. Lauer and M. Spaliñski, *Mod. Phys. Lett.* **A7** (1992) 3059;
J. A. Casas, F. Gomez and C. Munoz, *Int. J. Mod. Phys.* **A8** (1993) 455;
J. Erler, D. Jungnickel, M. Spaliñski and S. Stieberger, *Higher Twisted Sector Couplings of Z(N) Orbifolds*, Munich preprint MPI–Ph/92–56 and TUM–TH–142/92 (1992);
D. Jungnickel, *Correlation Functions of Two-dimensional Twisted Conformal Field Theories*, Thesis at the Technical University of Munich (in German language), Munich preprint MPI–Ph/92–27 (1992);
S. Stieberger, *Higher Twisted Sector Couplings and Duality Transformations in Orbifold Compactifications*, Dilpoma thesis at the Technical University of Munich (in German language), Munich preprint TUM–TH–148/92 (1992).

17. P. Candelas, X. C. De la Ossa, P. S. Green and L. Parkes, *Nucl. Phys.* **B359** (1991) 21 and *Phys. Lett.* **258B** (1991) 118.

CLASSICAL STRING SOLITONS

RAMZI R. KHURI
Physics Department, Texas A&M University
College Station, TX 77843, USA

ABSTRACT

We discuss some recent work in the field of classical solitonic solutions in string theory. In particular, we construct instanton and monopole solutions and discuss the dynamics of string-like solitons. Some of the motivation behind this work is that instantons may provide a nonperturbative understanding of the vacuum structure of string theory, while monopoles may appear in string predictions for grand unification. The string-like solitons represent extended states of fundamental strings. The essential role of supersymmetry in both the saturation of the Bogomol'nyi bound and in the cancellation of higher order corrections is emphasized.

1. Introduction

In the past few years, classical solitonic solutions in string theory with higher-membrane structure have been actively investigated. These solutions are static multi-soliton solutions obeying a zero-force condition and saturating a Bogomol'nyi bound between ADM mass and charge. In certain cases, exact solutions of bosonic and heterotic string theory may be constructed, each solution in principle corresponding to an exact conformal field theory (CFT) of the sigma-model. Although the solutions are initially conceived as perturbative expansions in the classical string parameter α' (the inverse string tension), the exact solutions acquire nonperturbative status. Being classical, the solitons are tree-level solutions in the (quantum) topological expansion of the string worldsheet, but are also nonperturbative in the loop parameter e^{ϕ_0}. Therefore, full quantum string-loop extensions of these solutions await an understanding of nonperturbative string theory. Nevertheless, it is possible to use these solitons in nonperturbative calculations (such as vacuum-tunneling) since it is often the case that higher order corrections do not contribute to these effects.

In this work, we discuss the construction of two classes of solitons, instantons and monopoles[1], and examine the dynamics of string-like solitons[2]. Both the monopole and instanton solutions have "fivebrane" structure in $D = 10$ (i.e. they are $5+1$-dimensional objects in $9+1$-dimensional spacetime).

The motivation for the study of these solutions includes the following: Since these solutions represent Planck scale solitons, their existence constitutes a possible test for string theory as a theory of quantum gravity. Another interest in these solutions stems from the application of known field-theoretic techniques (and their stringy analogs) in the physics of solitons and instantons to string theory. For example, the string instanton solutions, through vacuum tunneling computations, may lead to an understanding of the structure of the vacuum in string theory, much in the same manner as instantons are used in field theory. The string monopole solutions may appear in grand unification predictions of string theory, while the macroscopic string solitons may be used to represent extended string states. An especially noteworthy feature of the monopoles shown here is the cancellation between gauge and gravitational singularities in the action, a feature, which, if it survives quantization, promises to shed light on the nature of string theory as a finite theory of quantum gravity. An important point regarding the role of supersymmetry in these solutions is that the supersymmetric solutions represent extremal limits of generalized black hole type solutions. For each class of solutions, the existence of partially unbroken spacetime supersymmetries leads to the saturation of a Bogomol'nyi bound and guarantees the stability of the solitons. In addition, for the heterotic solutions, worldsheet supersymmetry leads to the cancellation of higher order corrections. A different point of view to these solutions is the study of the resultant string-inspired low-energy field theories, which may well capture the essential behaviour of these solitons (e.g. the singularity cancellation in the monopole solutions) without requiring an expansion in the full string theory. Finally, there is the open problem of proving the string/fivebrane duality conjecture[3], and its implications to low energy string theory, the singularity structure of string theory[4] and its relation to the more familiar electric/magnetic duality conjecture of Montonen and Olive[5]

2. Instantons and Monopoles

We first summarize the 't Hooft ansatz for the Yang-Mills instanton, and then write down the tree-level bosonic axionic instanton solution. An exact bosonic solution, with corresponding CFT can be obtained for the special case of a linear dilaton wormhole. An exact multi-instanton solution of heterotic string theory is then obtained by combining the Yang-Mills gauge solution with the bosonic axionic instanton. An exact heterotic multi-monopole can be obtained from the same ansatz via a slight modification of the solution. The notable feature in this case is the cancellation between gauge and gravitational divergences in the effective action.

Consider the four-dimensional Euclidean action

$$S = -\frac{1}{2g^2}\int d^4x \mathrm{Tr} F_{\mu\nu}F^{\mu\nu}, \qquad \mu,\nu = 1,2,3,4. \tag{1}$$

For gauge group $SU(2)$, the fields may be written as $A_\mu = (g/2i)\sigma^a A_\mu^a$ and $F_{\mu\nu} = (g/2i)\sigma^a F_{\mu\nu}^a$ (where σ^a, $a = 1,2,3$ are the 2×2 Pauli matrices). A self-dual solution (but not the most general one) to the equation of motion of this action is given by the 't Hooft ansatz[6]

$$A_\mu = i\overline{\Sigma}_{\mu\nu}\partial_\nu \ln f, \tag{2}$$

where $\overline{\Sigma}_{\mu\nu} = \overline{\eta}^{i\mu\nu}(\sigma^i/2)$ for $i = 1,2,3$, where

$$\begin{aligned} \overline{\eta}^{i\mu\nu} = -\overline{\eta}^{i\nu\mu} &= \epsilon^{i\mu\nu}, \qquad && \mu,\nu = 1,2,3, \\ &= -\delta^{i\mu}, && \nu = 4 \end{aligned} \tag{3}$$

and where $f^{-1}\Box f = 0$. The ansatz for the anti-self-dual solution is similar, with the δ-term in Eq. (3) changing sign. To obtain a multi-instanton solution, one solves for f in the four-dimensional space to obtain

$$f = 1 + \sum_{i=1}^{N} \frac{\rho_i^2}{|\vec{x} - \vec{a}_i|^2}, \tag{4}$$

where ρ_i^2 is the instanton scale size and $\vec{a}_i$ the location in four-space of the ith instanton. Note that this solution has $5N$ parameters, while the most general self-dual solution has $8N - 3$ parameters.

It turns out that there is an analog to the Yang-Mills instanton in the gravitational sector of the string, namely the axionic instanton[7]. In its simplest form, this instanton appears as a solution for the massless fields of the bosonic string. The bosonic sigma model action can be written as

$$I = \frac{1}{4\pi\alpha'} \int d^2x \left(\sqrt{\gamma}\gamma^{ab}\partial_a x^\mu \partial_b x^\nu g_{\mu\nu} + i\epsilon^{ab}\partial_a x^\mu \partial_b x^\nu B_{\mu\nu} + \alpha' \sqrt{\gamma} R^{(2)}\phi \right), \tag{5}$$

where $g_{\mu\nu}$ is the sigma model metric, ϕ the dilaton and $B_{\mu\nu}$ the antisymmetric tensor, and where γ_{ab} is the worldsheet metric and $R^{(2)}$ the two-dimensional curvature. The classical equations of motion of the effective action of the massless fields is equivalent to Weyl invariance of I. A tree-level solution is given by any dilaton function satisfying $e^{-2\phi}\Box\, e^{2\phi} = 0$ with

$$\begin{aligned} g_{\mu\nu} &= e^{2\phi}\delta_{\mu\nu} \qquad && \mu,\nu = 1,2,3,4, \\ g_{ab} &= \delta_{ab} && a,b = 5,...,26, \\ H_{\mu\nu\lambda} &= \pm\epsilon_{\mu\nu\lambda\sigma}\partial^\sigma\phi && \mu,\nu,\lambda,\sigma = 1,2,3,4. \end{aligned} \tag{6}$$

In order to see the instanton structure of this solution, we define a generalized curvature $\hat{R}^i{}_{jkl}$ in terms of the standard curvature $R^i{}_{jkl}$ and $H_{\mu\alpha\beta}$:

$$\hat{R}^i{}_{jkl} = R^i{}_{jkl} + \frac{1}{2}\left(\nabla_l H^i{}_{jk} - \nabla_k H^i{}_{jl}\right) + \frac{1}{4}\left(H^m{}_{jk}H^i{}_{lm} - H^m{}_{jl}H^i{}_{km}\right). \tag{7}$$

One can also define $\hat{R}^i{}_{jkl}$ as the Riemann tensor generated by the generalized Christoffel symbols $\hat{\Gamma}^\mu_{\alpha\beta}$, where $\hat{\Gamma}^\mu_{\alpha\beta} = \Gamma^\mu_{\alpha\beta} - (1/2)H^\mu{}_{\alpha\beta}$. The crucial observation for obtaining higher-loop and even exact solutions is the following. For any solution given by Eq. (6), we can express the generalized curvature in covariant form in terms of the dilaton field as[7]

$$\hat{R}^i{}_{jkl} = \delta_{il}\nabla_k\nabla_j\phi - \delta_{ik}\nabla_l\nabla_j\phi + \delta_{jk}\nabla_l\nabla_i\phi - \delta_{jl}\nabla_k\nabla_i\phi \pm \epsilon_{ijkm}\nabla_l\nabla_m\phi \mp \epsilon_{ijlm}\nabla_k\nabla_m\phi, \tag{8}$$

It easily follows that

$$\hat{R}^i{}_{jkl} = \mp\frac{1}{2}\epsilon_{kl}{}^{mn}\hat{R}^i_{jmn}. \tag{9}$$

So the instanton appears in the gravitational sector of the string in the (anti) self-duality of the generalized curvature. A tree-level multi-instanton solution is therefore given by Eq. (6) with the dilaton given by

$$e^{2\phi} = C + \sum_{i=1}^{N} \frac{Q_i}{|\vec{x} - \vec{a}_i|^2}, \tag{10}$$

where Q_i is the charge and $\vec{a}_i$ the location in the four-space (1234) (the transverse space) of the ith instanton. In the spherically symmetric case $e^{2\phi} = Q/r^2$, we can explicitly solve the higher order equations of motion by rescaling the dilaton and fixing the metric and antisymmetric tensor in lowest order form. For example, the two-loop dilaton is given by $e^{2\phi} = Q/r^{2(1-\frac{\alpha'}{Q})}$. To get an exact solution in this linear dilaton[8] case, we notice that the sigma-model action can be decomposed according to $I = I_1 + I_3$, where for $u = 1/r$

$$I_1 = \frac{1}{4\pi\alpha'} \int d^2x \left(Q(\partial u)^2 + \alpha' R^{(2)}\phi\right) \tag{11}$$

is the action for a Feigin-Fuchs Coulomb gas, a one-dimensional CFT with central charge given by $c_1 = 1 + 6\alpha'(\partial\phi)^2$ and I_3 is the Wess–Zumino–Witten action on an $SU(2)$ group manifold with central charge

$$c_3 = \frac{3k}{k+2} \simeq 3 - \frac{6}{k} + \frac{12}{k^2} + ... \tag{12}$$

where $k = Q/\alpha'$, the level of the WZW model, is an integer, from the quantization condition on the Wess-Zumino term. Thus Q is not arbitrary, but is quantized in units of α'.

We use this splitting to obtain exact expressions for the fields by fixing the metric and antisymmetric tensor field in their lowest order form and rescaling the dilaton to all orders in α'. The resulting expression for the dilaton is

$$e^{2\phi} = \frac{Q}{r^{\sqrt{\frac{4}{1+\frac{2\alpha'}{Q}}}}} \tag{13}$$

for arbitrary α'.

We now turn to the heterotic solution. The tree-level supersymmetric vacuum equations for the heterotic string are given by

$$\begin{aligned} \delta\psi_M &= \left(\nabla_M - \tfrac{1}{4}H_{MAB}\Gamma^{AB}\right)\epsilon = 0, \\ \delta\lambda &= \left(\Gamma^A\partial_A\phi - \tfrac{1}{6}H_{AMC}\Gamma^{ABC}\right)\epsilon = 0, \\ \delta\chi &= F_{AB}\Gamma^{AB}\epsilon = 0, \end{aligned} \tag{14}$$

where ψ_M, λ and χ are the gravitino, dilatino and gaugino fields. The Bianchi identity is given by

$$dH = \alpha'\left(\mathrm{tr}R\wedge R - \frac{1}{30}\mathrm{Tr}F\wedge F\right). \tag{15}$$

The $(9+1)$-dimensional Majorana-Weyl fermions decompose down to chiral spinors according to $SO(9,1) \supset SO(5,1)\otimes SO(4)$ for the $M^{9,1} \to M^{5,1}\times M^4$ decomposition. Let $\mu,\nu,\lambda,\sigma = 1,2,3,4$ and $a,b = 0,5,6,7,8,9$. Then the ansatz

$$\begin{aligned} g_{\mu\nu} &= e^{2\phi}\delta_{\mu\nu}, \\ g_{ab} &= \eta_{ab}, \\ H_{\mu\nu\lambda} &= \pm\epsilon_{\mu\nu\lambda\sigma}\partial^\sigma\phi \end{aligned} \tag{16}$$

with constant chiral spinors $\epsilon_\pm$ solves the supersymmetry equations with zero background fermi fields provided the YM gauge field satisfies the instanton (anti)self-duality condition

$$F_{\mu\nu} = \pm\frac{1}{2}\epsilon_{\mu\nu}{}^{\lambda\sigma}F_{\lambda\sigma}. \tag{17}$$

A perturbative solution representing a supersymmetric fivebrane was first derived by Strominger[9]. In the absence of a gauge sector, the multi-fivebrane solution is identical to the tree-level type II supersymmetric fivebrane solution of Duff and Lu[10], derived in terms of the dual seven-form formulation of supergravity, with $K = e^{-\phi*}H = dA$, where A is the antisymmetric six-form associated with a fivebrane. An exact solution is obtained as follows. Define a generalized connection by

$$\Omega^{AB}_{\pm M} = \omega^{AB}_M \pm H^{AB}_M \tag{18}$$

embedded in an SU(2) subgroup of the gauge group, and equate it to the gauge connection A_μ [11] so that $dH = 0$ and the corresponding curvature $R(\Omega_\pm)$ cancels against the Yang-Mills field strength F. As in the bosonic case, for $e^{-2\phi}\Box\, e^{2\phi} = 0$ with the above ansatz, the curvature of the generalized connection can be written in covariant form in terms of the dilaton as in Eq. (8) from which it follows that both F and R are (anti)self-dual. This solution becomes exact since $A_\mu = \Omega_{\pm\mu}$ implies that all the higher order corrections vanish[7,12,13] The self-dual solution for the gauge connection is then given by the 't Hooft ansatz. An interesting feature of the heterotic solution is that it combines a YM instanton structure in the gauge sector with an axionic instanton structure in the gravity sector. In addition, the heterotic solution has finite action.

Note that the single instanton solution in the heterotic case carries through to higher order without correction to the dilaton. This seems to contradict the bosonic solution by suggesting that the expansion for the central charge c_3 terminates at one loop. This contradiction is resolved by noting that for an $N = 4$ worldsheet supersymmetric solution[13] the bosonic contribution to the central charge is given by

$$c_3 = \frac{3k'}{k'+2}\,, \tag{19}$$

where $k' = k - 2$. This reduces to

$$c_3 = 3 - \frac{6}{k} = 3 - \frac{6\alpha'}{Q}, \tag{20}$$

which indeed terminates at one loop order. The exactness of the splitting then requires that c_1 not get any corrections from $(\partial\Phi)^2$ so that $c_1 + c_3 = 4$ is exact for the tree-level value of the dilaton[13].

In a similar manner, we obtain an exact heterotic multi-monopole solution[14]. We do so by singling out a direction in the transverse space (say x_4). Since the derivation is essentially identical to that for the heterotic instanton, we simply write down the solution

$$
\begin{aligned}
g_{\mu\nu} &= e^{2\phi}\delta_{\mu\nu}, \qquad g_{ab} = \eta_{ab}, \\
H_{\mu\nu\lambda} &= \pm\epsilon_{\mu\nu\lambda\sigma}\partial^\sigma\phi, \\
e^{2\phi} &= e^{2\phi_0} f, \\
A_\mu &= i\overline{\Sigma}_{\mu\nu}\partial_\nu \ln f,
\end{aligned} \tag{21}
$$

where in this case $f = 1 + \sum_{i=1}^N \frac{m_i}{|\vec{x}-\vec{a}_i|}$ and $\vec{x}$ and $\vec{a}_i$ are vectors in the space (123). If we set $\Phi \equiv A_4$, then the gauge and scalar fields in the subspace (0123) may be simply written in terms of the dilaton as

$$
\begin{aligned}
\Phi^a &= \mp\frac{2}{g}\delta^{ia}\partial_i\phi, \\
A_k^a &= -\frac{2}{g}\epsilon^{akj}\partial_j\phi.
\end{aligned} \tag{22}
$$

The above solution represents an exact multimonopole solution of heterotic string theory[14], and is stable as a result of its saturation of a Bogomol'nyi bound between ADM mass and charge[15]. The saturation of the Bogomol'nyi bound in turn owes to the existence of partially unbroken spacetime supersymmetries.

In the string effective action, the term $-\alpha' F^2$ diverges near each source. However, this divergence is precisely cancelled by the term $\alpha' R^2(\Omega_\pm)$ in the $O(\alpha')$ action. This result follows from the exactness condition $A_\mu = \Omega_{\pm\mu}$ which leads to $dH = 0$ and the vanishing of all higher order corrections in α'. Another way of seeing this is to consider the higher order corrections to the bosonic action[12]. All such terms contain the tensor T_{MNPQ}, a generalized curvature incorporating both $R(\Omega_\pm)$ and F. The ansatz is constructed precisely so that this tensor vanishes identically[7,14]. The action thus reduces to its finite lowest order form and can be calculated directly for a multi-source solution from the expressions for the massless fields in the gravity sector.

The divergences in the gravitational sector in heterotic string theory thus serve to cancel the divergences stemming from the field theory solution. This solution thus provides an interesting example of how this type of cancellation can occur in string theory, and supports the promise of string theory as a finite theory of quantum gravity. Another point of interest is that the string solution represents a supersymmetric multimonopole solution coupled to gravity, whose zero-force condition in the gravity sector (cancellation of the attractive gravitational force and repulsive antisymmetric field force) arises as a direct result of the zero-force condition in the gauge sector (cancellation of gauge and Higgs forces of exchange) once the gauge connection and generalized connection are identified. The fulfillment of the exactness condition depends crucially on the existence of unbroken spacetime supersymmetries.

3. Dynamics of String Solitons

In earlier work[16], Dabholkar *et al.* presented a low-energy analysis of macroscopic superstrings and discovered several interesting analogies between macroscopic superstrings and solitons in supersymmetric field theories. The main result of this work centers on the existence of exact multi-string solutions of the low-energy supergravity super-Yang-Mills equations of motion. In addition, Dabholkar *et al.* find a Bogomolnyi bound for the energy per unit length which is saturated by these solutions, just as the Bogomolnyi bound is saturated by magnetic monopole solutions in ordinary Yang-Mills field theory. The solution may be outlined as follows. The action for the massless spacetime fields (graviton, axion and dilaton) in t he presence of a source string can be written as

$$S = \frac{1}{2\kappa^2}\int d^D x\sqrt{g}\left(R - \frac{1}{2}(\partial\phi)^2 - \frac{1}{12}e^{-2\alpha\phi}H^2\right) + S_\sigma, \tag{23}$$

with the source terms contained in the sigma model action S_σ given by

$$S_\sigma = -\frac{\mu}{2}\int d^2\sigma(\sqrt{\gamma}\gamma^{mn}\partial_m X^\mu \partial_n X^\nu g_{\mu\nu}e^{\alpha\phi} + \epsilon^{mn}\partial_m X^\mu\partial_n X^\nu B_{\mu\nu}), \tag{24}$$

with $\alpha = \sqrt{2/(D-2)}$ and γ_{mn} a worldsheet metric to be determined. The sigma model action S_σ describes the coupling of the string to the metric, antisymmetric tensor field and dilaton. The first part of the action S above represents the effective action for the massless fields in the spacetime frame and whose equations of motion are equivalent to conformal invariance of the underlying sigma model. The combined action thus generates the equations of motion satisfied by the massless fields in the presence of a macroscopic string source. The static solution to the equations of motion is given by

$$\begin{aligned} &ds^2 = e^A[-dt^2 + (dx^1)^2] + e^B d\vec{x}\cdot d\vec{x} \\ &A = \frac{D-4}{D-2}E(r) \qquad B = -\frac{2}{D-2}E(r) \\ &\phi = \alpha E(r) \qquad\qquad B_{01} = -e^{E(r)}, \end{aligned} \tag{25}$$

where x^1 is the direction along the string, $r = \sqrt{\vec{x}\cdot\vec{x}}$ and

$$e^{-E(r)} = \begin{cases} 1 + \frac{M}{r^{D-4}} & D > 4 \\ 1 - 8G\mu\ln(r) & D = 4 \end{cases}$$

for a single static string source. The solution can be generalized to an arbitrary number of static string sources by linear superposition of solutions of the $(D-2)$-dimensional Laplace's equation. The existence of this multi-soliton solution depends on the the zero-force condition which arises from the cancellation of long-range forces of exchange of the massless fields of the string (the graviton, axion and dilaton). This is a perfect analog to the zero-force condition of Manton for magnetic monopoles, which requires that the attractive scalar exchange force precisely cancel the repulsive vector exchange force when the Bogomolnyi bound is attained. Dabholkar *et al.* show that a similar Bogomolnyi bound is satisfied by their string soliton solutions, further strengthening the analogy with the monopoles.

In contrast to BPS monopoles, the string solitons also obey a zero *dynamical* force condition[2]. While the static force vanishes as a result of the cancellation of long-range forces of exchange, the force between two moving solitons is in general nonvanishing and depends on the velocities of the solitons. The most complete answer would be given by a full time-dependent solution of the equations of motion of the above action for the case of an arbitrary number of sources moving with arbitrary transverse velocities. These equations, however, are much more difficult to solve for moving sources than for a static configuration. Even a two-soliton solution is in general quite intractable for this class of actions. In what follows we briefly summarize some attempts to approximate the dynamical interaction of the solitons strings[2].

We first examine the scattering of these solitons using the above test-string approach. This entails solving the constraint equation for the worldsheet metric obtained by varying the worldsheet Lagrangian $\mathcal{L}$. The resultant solution for the worldsheet metric along with the static solution for the spacetime metric, antisymmetric tensor field and dilaton from the static ansatz for a single source string are then substituted into the Lagrangian, whose equations yield the dynamics of the test string in the source string background. We obtain a kinetic Lagrangian of the form $(\mu/2)\dot{X}^2 + O(\dot{X}^4)$. The absence of the potential term confirms the zero static force condition while the flatness of the leading order kinetic term suggests a zero dynamical force condition, i.e. trivial scattering, in the low velocity limit.

We now address the scattering problem from a string-theoretic point of view. The winding configuration described by $X(\sigma,\tau)$ describes a soliton string state. It is therefore a natural choice for us to compare the dynamics of these states with the above string solitons in order to determine whether we can identify these solitons with infinitely long fundamental strings. Accordingly, we study the scattering of $n=1$ winding states in the limit of large winding radius R. We find that the Veneziano amplitude $A_4 \to 0$ as $R \to \infty$ except at scattering angle $\theta = 0, \pi$. This result also indicates trivial scattering in this limit, providing evidence for the identification of the string solitons with infinitely long macroscopic fundamental strings.

Finally, we turn to soliton-soliton scattering. In the low-velocity limit, multi-soliton solutions trace out geodesics in the static solution manifold, with distance defined by the Manton metric on moduli space manifold. In the absence of a full time-dependent solution to the equations of motion, these geodesics represent a good approximation to the low-energy dynamics of the solitons. For BPS monopoles, the Manton procedure was implemented by Atiyah and Hitchin[17]. Computing the Manton metric on moduli space for the scattering of the soliton string solutions in $D=4$ (we expect that the same result will hold for arbitrary $D \geq 4$) we find that the metric is flat to lowest nontrivial order in the string tension. This result implies trivial scattering of the string solitons and is consistent with the above two calculations, and thus provides even more compelling evidence for the identification of the string soliton with the underlying fundamental string.

For the instanton and monopole fivebranes[18], both the test-fivebrane limit and the metric on moduli space also yield a zero dynamical force condition. Since a fundamental theory of fivebranes has not yet been constructed, there is no corresponding Veneziano amplitude computation with which to compare. For the instantons, it is sufficient to demonstrate Ricci flatness of the Manton metric to obtain trivial scattering while for the monopoles a flat metric can be explicitly computed. The zero dynamical force can be seen as a direct consequence of the exactness condition of equating the gauge and spin connections and is again a consequence of worldsheet supersymmetry[15].

4. Future Directions

While all the solutions we discussed are classical, one can still conceive of situations in which quantum corrections to the instantons, for example, drop out in nonperturbative computations (such as for vacuum tunnelling). To this end, a vertex operator representation of the instantons would be highly desirable. The most interesting feature of the heterotic monopole solution is the cancellation between gauge and gravitational singularities. If this is an intrinsically "stringy" feature, then it presumably occurs in a larger context within string theory, in which case the full quantum string loop extension of this solution promises to shed light on the nature of string theory as a finite theory of quantum gravity. If this cancellation is of a more accidental nature, then it would pay to concentrate more on the corresponding low-energy field theory, whose quantization is presumably far simpler. In either case, it would be interesting to see whether the singularity cancellation occurs in a quantized solution, or in the context of blackhole type solutions. We have compelling dynamical evidence for the identification of the string solitons with macroscopic fundamental strings, but an exact heterotic solution seems most natural in the context of the conjectured dual fundamental theory of fivebranes. While the construction of such a theory remains elusive, there is so far solid evidence to support the duality conjecture.

References

1. R. R. Khuri, *Phys. Rev.* **D46** (1992) 4526.
2. R. R. Khuri, *Geodesic Scattering of Solitonic Strings*, Texas A&M preprint CTP/TAMU-79/92; *Classical Dynamics of Macroscopic Strings*, Texas A&M preprint CTP/TAMU-80/92 (to appear in Nucl. Phys. B).
3. M. J. Duff, *Class. Quan. Grav.* **5** (1988) 189; M. J. Duff and J. X. Lu, *Nucl. Phys.* **B354** (1991) 129.
4. M. J. Duff, R. R. Khuri and J. X. Lu, *Nucl. Phys.* **B377** (1992) 281.
5. C. Montonen and D. Olive, *Phys. Lett.* **B72** (1977) 117; H. Osborn, *Phys. Lett.* **B83** (1979) 321.
6. G. 't Hooft, *Nucl. Phys.* **B79** (1974) 276; *Phys. Rev. Lett.* **37** (1976) 8; F. Wilczek, in *Quark confinement and field theory*, ed. D. Stump and D. Weingarten, (John Wiley and Sons, New York,1977); E. Corrigan and D. B. Fairlie, *Phys. Lett.* **B67** (1977) 69; R. Jackiw, C. Nohl and C. Rebbi, *Phys. Rev.* **D15** (1977) 1642; *Particles and Fields*, ed. David Boal and

A. N. Kamal, (Plenum Publishing Co., New York,1978) p.199.

7. R. R. Khuri, *Phys. Lett.* **B259** (1991) 261; *Proceedings of XXth International Conference on Differential Geometric Methods in Theoretical Physics*, ed. S. Catto and A. Rocha (World Scientific, Jan. 1992) p.1074.
8. I. Antoniadis, C. Bachas, J. Ellis and D. V. Nanopoulos, *Phys. Lett.* **B211** (1988) 393; *Nucl. Phys.* **B328** (1989) 117.
9. A. Strominger, *Nucl. Phys.* **B343** (1990) 167.
10. M. J. Duff and J. X. Lu, *Nucl. Phys.* **B354** (1991) 141.
11. J. M. Charap and M. J. Duff, *Phys. Lett.* **B69** (1977) 445.
12. E. A. Bergshoeff and M. de Roo, *Nucl. Phys.* **B328** (1989) 439; *Phys. Lett.* **B218** (1989) 210.
13. C. G. Callan, J. A. Harvey and A. Strominger, *Nucl. Phys.* **B359** (1991) 611.
14. R. R. Khuri, *Phys. Lett.* **B294** (1992) 325; *Nucl. Phys.* **B387** (1992) 315.
15. R. R. Khuri, *A Comment on the Stability of String Monopoles*, Texas A&M preprint CTP/TAMU-81/92.
16. A. Dabholkar and J. A. Harvey, *Phys. Rev. Lett.* **63** (1989) 719; A. Dabholkar, G. Gibbons, J. A. Harvey and F. Ruiz Ruiz, *Nucl. Phys.* **B340** (1990) 33.
17. N. S. Manton, *Phys. Lett.* **B110** (1982) 54; M. F. Atiyah and N. J. Hitchin, *Phys. Lett.* **A107** (1985) 21; *The Geometry and Dynamics of Magnetic Monopoles*, (Princeton University Press,1988).
18. R. R. Khuri, *Nucl. Phys.* **B376** (1992) 350; C. G. Callan and R. R. Khuri, *Phys. Lett.* **B261** (1991) 363; R. R. Khuri, *Phys. Lett.* **B294** (1992) 331; A. G. Felce and T. M. Samols, *Low-Energy Dynamics of String Solitons* ITP Santa Barbara preprint NSF-ITP-92-155.

SuperCosmology

SUPERSYMMETRY AND COSMOLOGY

David N. Schramm
The University of Chicago
5640 S. Ellis Avenue, Chicago, IL 60637
and
NASA/Fermilab Astrophysics Center
Fermi National Accelerator Laboratory
Box 500, Batavia, IL 60510-0500

ABSTRACT

Cosmology has provided some of the few "experimental" constraints on supersymmetry. The constraints come from (1) inflation, (2) big bang nucleosynthesis, and (3) dark matter. These constraints are briefly reviewed, and it is argued that supersymmetry does yield interesting and viable inflation models and viable dark matter candidates. The impact of the COBE results on these constraints is also discussed.

1. Introduction

The impact of cosmology on supersymmetry (SUSY) has grown over the last decade to become an exciting and active area. Rather than attempt to review the entire field, this paper will focus on three active areas at the interface. These areas provide interesting constraints on SUSY and in some cases even prefer SUSY solutions.

The three topics that are chosen are:
(1) Inflation,
(2) Big Bang Nucleosynthesis (BBN),
(3) Cosmic Structure Formation and Dark Matter.

These three topics are intimately and symbiotically interrelated and all three have been affected profoundly by recent experiments and astrophysical observations. In particular, we will discuss the impact of the COBE anisotropy measurements on inflation and on dark matter and structure formation. We will note the current potential problem of the age-expansion relations for inflation. And we will note the LEP neutrino counting results which give even greater confidence in the standard BBN model which has now become one of the three principle pillars of the Big Bang itself. Of course, one critical prediction of BBN is that the baryon density is low so that the critical density universe required by inflation mandates non-baryonic dark matter. One of the best motivated of the dark matter candidates is, of course, the lightest supersymmetric particle (LSP). This paper will not give the specific SUSY parameters that survive but will instead discuss the cosmological constraints in general broad brush terms.

2. Inflation

It is well known[1,2] that inflation[3,4,5] predicts a flat universe and produces gaussian density fluctuations with a power spectrum

$$(\frac{\delta\rho}{\rho})_k^2 \sim k^n \text{ with } n \sim 1$$

where k is the wave number of the fluctuation ($k \equiv 2\pi/L$ where L is the length scale). Ellis et al.[6] showed that the most reasonable particle models which gave a realistic amplitude to the fluctuations were SUSY models. Standard inflation yields the flat, $n = 1$, Harrison-Zeldovich spectrum with equal power on all scales. However, variants such as "natural inflation"[7] can yield a "tilted" spectrum with n slightly less than unity.

The recent COBE results[8,9] are certainly consistent with a gaussian spectrum and thus yield

$$n = 1.2^{+0.5}_{-0.6}$$

which is also consistent with a flat spectrum (or a slightly tilted one). Furthermore, since COBE observes fluctuations only on large angular scales, $\theta \gtrsim 7°$, which are outside the causal horizon ($\theta \sim 2°$) at the time of radiation decoupling, the observed fluctuations do support the need for inflation or something like it. In fact, minimal fluctuations, which exist only on the scales where galaxies and structures are seen, would naively add together in an incoherent manner on larger scales and would thus yield a power spectrum of $\sim k^4$. Therefore, COBE is telling us that the primordial spectrum is not just a superposition of the fluctuations that specifically made galaxies, clusters, etc., but that some larger scale primordial fluctuations did indeed exist. Whether or not some additional small scale (non-gaussian?) seeds also existed remains to be seen, but there is no question that COBE is consistent with inflation.

This consistency is a necessary but not sufficient statement. The idea of a flat $n = 1$ spectrum existed prior to the inflation idea. Obtaining such a spectrum then may not be unique to inflation (see, for example, the proposal of Nanopoulos).[10] Another necessary but not sufficient prediction of inflation is a flat universe, which, in its most natural mode, is just saying $\Omega = 1$. This prediction has gained support recently from the large scale velocity flow data[11,12] from the IRAS survey and the Great Attractor and potent work which seems to require $\Omega \sim 1$ and appears inconsistent with $\Omega \lesssim 0.3$. A further hint in this direction comes from the recent angular size versus redshift work of Kellerman[13] (see Figure 1) which is best fit by a flat universe, although evolutionary effects could alter such a conclusion.[14] The one potential cloud on the inflationary horizon is the value of the Hubble constant. Recall that the age of the universe is $t = \frac{2}{3H_0}$ for an $\Omega = 1$, matter-dominated universe. But globular cluster ages are best fit by $t_{GC} = 15 \pm 3$ Gyr.[15,16] Although recently Dearborn and Schramm[17] have argured that mass loss to fit a subset of the Pop II Li observations would create a downward shift of ~ 2 Gyr, nonetheless t_{GC} can still be fit only by an $\Omega = 1$ universe if $H_0 \lesssim 60$ km/sec/Mpc. (There is also a firm lower bound on t from nucleochronology,[15] $t > 10$ Gyr.) Thus, if astronomers determine a large H_0, then there is a problem with $\Omega = 1$. The simplest loophole of adding a cosmological constant to keep the universe flat may still be allowed but only if the large scale velocity flow data and the angular size data are shown to be wrong. The cosmological constant also has the repugnanace that its invocation today requires tuning at the level of $\sim 10^{-121}$.

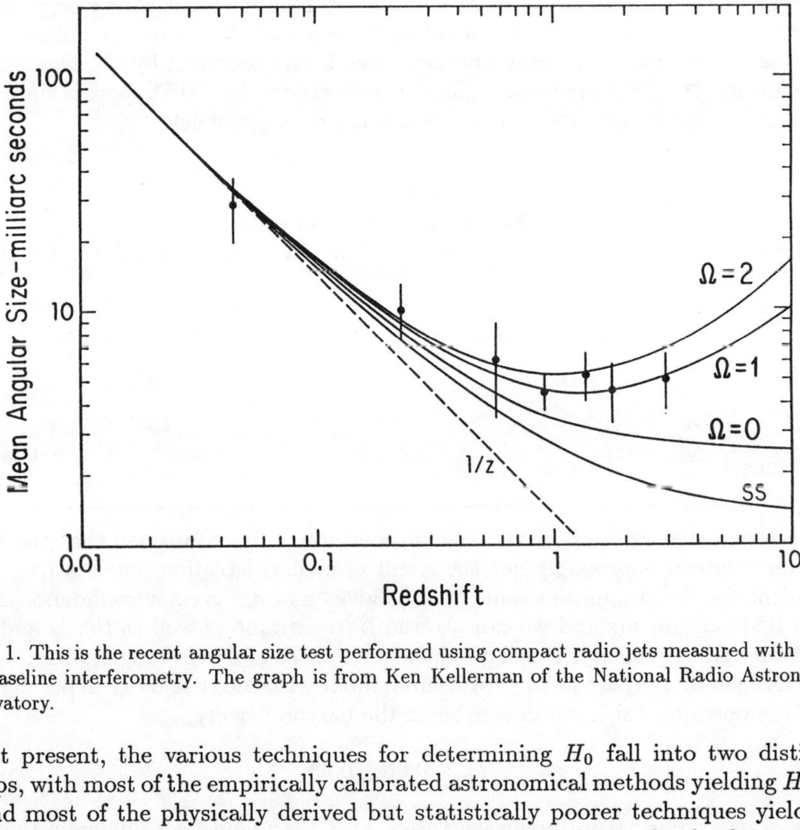

Figure 1. This is the recent angular size test performed using compact radio jets measured with very long baseline interferometry. The graph is from Ken Kellerman of the National Radio Astronomy Observatory.

At present, the various techniques for determining H_0 fall into two distinct clumps, with most of the empirically calibrated astronomical methods yielding $H_0 \sim 85$ and most of the physically derived but statistically poorer techniques yielding $H_0 \sim 50$. Obviously, inflation favors the $H_0 \sim 50$ camp, but only the future will truly decide.

3. Big Bang Nucleosynthesis

The third pillar of the Big Bang (after the Hubble expansion and the microwave background) comes from the light element abundance measurements and, most recently, the measurement at LEP of the number of families of neutrinos. In particular, the Big Bang predicts[18] that when the universe was at a temperature of $\sim$ 10 billion degrees and was about one second old, it should have started nuclear processes that would eventually yield certain well-specified abundances for the light isotopes (see Table 1). The abundances of these light elements have all been accurately determined to be in impressive agreement with the Big Bang predictions to the accuracy of the measurements. Of special interest is that even the one part in 10 billion for lithium works. Furthermore, the Big Bang predicted that the abundances would fit well only if there were no more than three families of neutrinos, and that was exactly what was observed at LEP. Thus, as is frequently emphasized, the Big Bang has made a variety of detailed predictions based on the nature of the universe at $\sim$ 1 sec, and the predictions have been confirmed by observation and experiment. The BBN neutrino arguments are relevant for SUSY models since they prohibit additional light ($\lesssim$ 1 MeV) weakly interacting particles.

TABLE 1
Light Element Abundances

Element (isotope)	Predicted Primordial Abundance	When Observed
^{1}H	~76% by mass	1960's
^{4}He	~24% by mass	1960's
^{2}H/H	2×10^{-5}	1970's
^{7}Li/H	10^{-10}	1980's
Neutrinos	$N_\nu = 3$	1990

A recent detection of ^{6}Li in a metal-poor star[19] has confirmed that the Spite Lithium Plateau is probably not the result of stellar depletion but is a true measurement for the primordial value.[20] This gives us even greater confidence in the basic BBN arguments and we can use the Li constraint as well as the D and ^{3}He constraint on the density. The light elements with abundances ranging from $\sim$ 76% for hydrogen to one part in 10^{10} for lithium all fit with the cosmological predictions, with the one adjustable parameter being the baryon density,

$$\Omega_B \sim 0.05 \pm 0.03.$$

Thus, if one prefers a universe with $\Omega_{TOTAL} = 1$, one must also demand that the bulk of the matter in the universe (greater than 90%) be something other than baryons. A favorite non-baryonic candidate is the LSP.

Recent attempts to find alternatives to this conclusion by introducing an inhomogeneous baryon distribution at the nucleosynthesis epoch have ended up reaching essentially the same constraint on Ω_B as in the standard homogeneous model.[21]

The baryon density, Ω_B, deduced from the abundances of light elements is within the range of dynamical estimates of the mean mass density in and around the bright parts of galaxies, including the dark massive halos associated with galaxies[22] (see Figure 2). The IRAS/GA point mentioned in the previous section shows that we are beginning to have some experimental/observational evidence that Ω_{TOTAL} exceeds Ω_B and that some non-baryonic dark matter is truly needed.

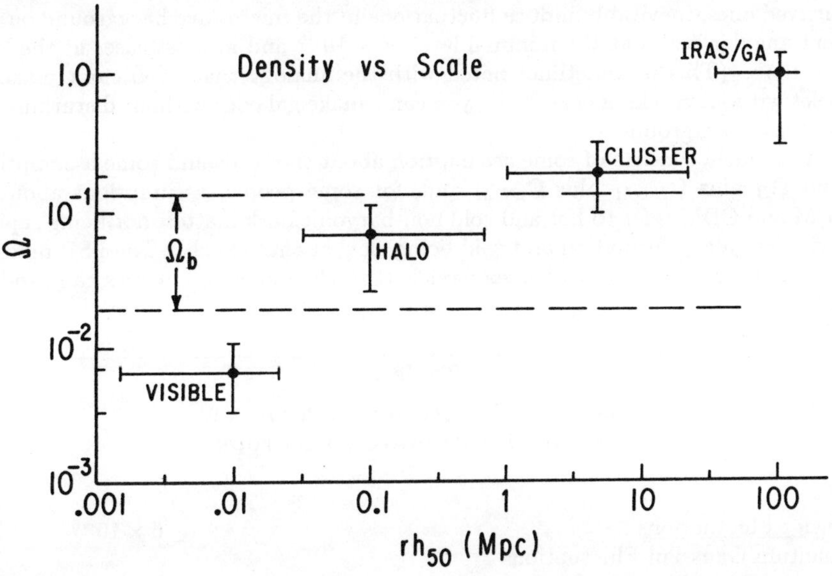

Figure 2. The density paramater Ω versus the scale on which it is determined. Note that the baryon density is consistent with galactic halo densities but exceeds visible matter densities. Note also that only on the largest scales is there any hint that Ω exceeds Ω_B.

An interesting additional bit of information on the distribution and relative proportions of matter in the universe has come from the ROSAT x-ray satellite. In particular, Mushotzky[23] has shown that rich clusters of galaxies have a mass fraction, $\sim 0.3h_{50}^{-3/2}$, of hot x-ray gas. Since $\Omega_{cluster} \sim 0.2$, this yields $\Omega_{baryon} \sim 0.06$ for these clusters in reasonable agreement with BBN (but only for $h_{50} = H_0/50 \sim 1$). However, such a high ratio of baryons to non-baryons in rich clusters and yet an overall low cosmic average would imply that even on the scale of clusters, baryons are preferentially being selected. Such selection is not present in most models for cluster formation and may turn out to be an important discriminator of structure formation models.

4. Cosmic Structure Formation

To make the observed objects in the universe such as galaxies, clusters, stars, planets and people requires something beyond just some combination of baryons and hot and/or cold exotic dark matter. There also had to be some "seeds" to get the matter, both baryonic and exotic, to begin clumping in some fashion. Having additional exotic matter does help accelerate the clumping process to form observable lumps, but all models require some sort of "seed" or density fluctuation to get from the smooth, early universe to the lumpy universe in which we live today. All seeds that are capable of producing the observed objects, even the most carefully contrived ones, inevitably induce fluctuations in the microwave background on different angular scales at the minimal level of $\sim 10^{-6}$ and in most cases at the level of $\sim 10^{-5}$. (This is sometimes noted with the analogy that "you can't make an omelet without cracking eggs." i.e., you can't make galaxies without disturbing the microwave background.)

A "model" consists of some assumption about the seeds and some assumptions about Ω_B plus Ω_{HDM} plus Ω_{CDM} and, for some people, assumptions about Λ. (HDM and CDM refer to hot and cold non-baryonic dark matter, hot being rapidly moving at galaxy formation and cold being slow at that epoch.) The LSP tends to be a fine CDM candidate, whereas models that yield neutrino masses can produce HDM.

Table 2

Seeds for Structure Formation and Minimal Microwave Anisotropy

Seed	$\frac{\Delta T}{T}$
Density Fluctuations [Quantum Gaussian Fluctuations From the End of Inflation]	$\geq 6 \times 10^{-6}$
Topological Defects* [structures form directly in cosmic phase transitions]	
A. "Cosmic Strings" from early phase transitions before decoupling of the cosmic background	$\gtrsim 5 \times 10^{-6}$
B. Defects from "late-time transitions" after decoupling	$\gtrsim 1 \times 10^{-6}$

*For topological defects, the fluctuations are not "randomly" distributed, so statistics on the average $\frac{\Delta T}{T}$ do not directly apply.

Table 2 summarizes the minimal predictions for some different possible "seeds." Note that these minimal predictions are actually below the level that the DMR

(Differential Microwave Radiometer) experiment[8,9] on COBE detected: $\frac{\Delta T}{T} \sim 10^{-5}$. This brings to mind two points. The first is that if COBE had failed to detect a signal, it would not yet have caused a serious problem since models existed which could generate structure with smaller values of $\frac{\Delta T}{T}$ than COBE was capable of measuring. The second is that we are very lucky that the DMR experiment was so well-designed that it could reach the level of sensitivity of 10^{-5}. At the time COBE was proposed 20 years ago, theories focused on anisotropy values near 10^{-3}. Fortunately, COBE was not designed to check just the in-vogue theories of the times, but went as far as it could go technically.

The general cause of the microwave fluctuations in apparent temperature from density fluctuations is merely gravity. If the density is higher than the average, then the gravitational field is higher and hence the background radiation climbing out of that higher potential well will be redshifted. (Note that regions of high density which eventually form structure would have produced temperature fluctuations to the low side, whereas low density fluctuations would yield "hot spots.") Different structure formation models have different distributions of the gravitational seeds or density fluctuations and hence predict different patterns and amplitudes for the temperature variations.

An additional difference between models is the relative size of fluctuations on different angular scales. For example, the density fluctuation model with CDM and $n = 1$ tends to predict a distribution something like Figure 3.

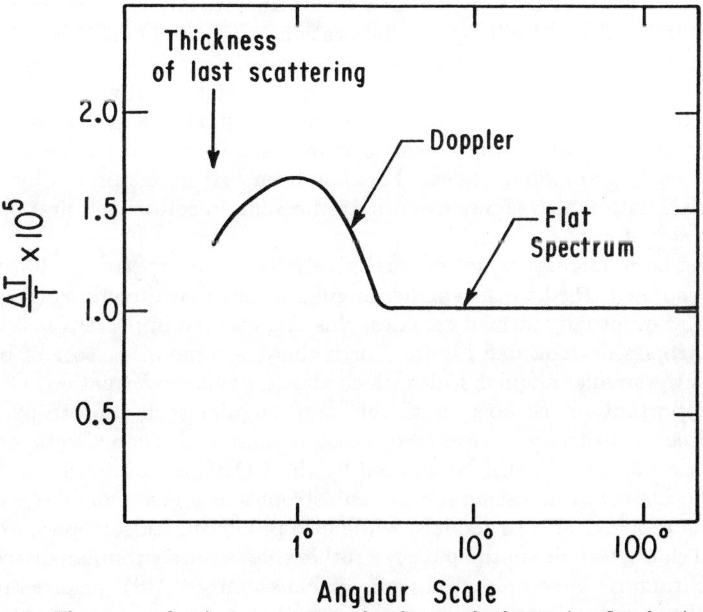

Figure 3. The expected anisotropy pattern for the standard gaussian flat density fluctuation model.

Note in Figure 3 that the flat spectrum does directly yield a "flat" prediction for large angles, but as the angular separation gets down to $\sim 2°$, the expected level of fluctations increases. As mentioned earlier, this is the angular size today of material that could be "causally connected" at the time the background radiation last interacted with matter. Since causally connected material could be influenced by the other material in that volume, significant motion and structure formation cannot occur on distance scales beyond the causally connected scale. Matter within the causally connected region is moving in a significant way and hence can create additional disturbances of the background radiation via the Doppler effect. On very small scales, the fluctuations in the radiation will get smeared out to uniformity due to the fact that the time, temperature and density of the last scattering of the photons with the matter is not a single, unique point but is spread out in these parameters. This causes the expected microwave anisotropy to be less on scales smaller than about 8 minutes of arc.

As mentioned earlier, the COBE measurement is consistent with a $n \sim 1$ spectrum with

$$\frac{\Delta T}{T} \sim 10^{-5}$$

for angular scales from 7° to 180°.

Of course, no scientific observation is believable until it is checked. Fortunately, the large scale COBE anisotropy has now been dramatically checked by a balloon experiment designed and built by a collaboration led by Stephan Meyer[24] of MIT (and Chicago). Their balloon mission also measured the anisotropy at large angular scales, and they found an anisotropy that both correlated with the directions of the COBE data and also had about the same size fluctuations. This dramatic confirmation means that we now seem to know that there really were primordial fluctuations on large angular scales. This has been further confirmed by the 2nd year of COBE data which shows essentially the same result as the first year, but with better statistics.[25]

With the large angular scales now relatively fixed (to experimental accuracy), attention has now turned to the smaller angular scales that directly correspond to the scales that eventually formed galaxies, clusters, etc. An important test is to see if the anisotropies do look like Figure 3 or if there is some other sort of behavior occurring at the smaller angular scales which affects structure formation. Of course, it is also important to see how "flat" the large angular scale anisotropy is since slightly "tilted" anisotropies which favor large or small scales can also be produced in some models and could still be allowed by the COBE measurements. One also would like to know how gaussian are the anisotropies at a given angular scale. For example, the topological seed models, while also predicting anisotropies, expect to have special characteristic spatial patterns rather than a purely random distribution. Luo and Schramm[26] have presented tests for gaussianity. SUSY inspired inflation models tend to produce gaussian fluctuations with $n \lesssim 1$.

Three groups recently announced somewhat ambiguous results about the 1-to-2 degree scale anisotropies. One of the groups,[27] led by Lubin of UC Santa Cruz, reported results from their previous year's South Pole observations. The second group, known as the MAX collaboration,[28] reported on their balloon-borne measurements. The third group is the Center for Astronomical Research at Antarctica which carried out two anisotropy experiments, one at small scales ($\lesssim 0.5$ deg.), $\frac{\Delta T}{T} < 10^{-5}$, and the other at 2.5 deg. $\frac{\Delta T}{T} \sim 3 \times 10^{-5}$. While the anisotropies observed are at levels not too different from that predicted by Figure 3, the relative levels reported were not uniform in all directions, hinting either to non-gaussianity or to source contamindation, but the error bars are also still quite large when fit to COBE's large angle data. Hopefully, with more South Pole and balloon data, the important 1-to-2 degree scales will be unambiguously mapped out in the near future. All microwave data on these scales can be directly compared with how galaxies cluster on the equivalent scales.

Several major surveys of the 3-dimensional positions of galaxies have mapped out some interesting nearby structure. For example, the Center for Astrophysics[29] survey has found a "Great Wall" of galaxies stretching for about 150 Mpc, and the APM survey[30] has also found evidence for very large structures of galaxies.

One general feature of all of these galaxy surveys is that they seem to find more really big structures than might have been estimated if galaxies were randomly distributed and if the mass of the systems really traced the emitted light. However, many have argued that it is possible that light is not a good tracer of the underlying mass distribution. Such a shift between light and mass distributions (biasing) makes it difficult to compare unambiguously galaxy distributions with theories. Furthermore, different types of galaxies (or clusters) may have different biasing factors and, at present, the statistics on large separations are poor since the numbers of objects studied is limiting. This latter point should be alleviated with the Sloan Digital Sky Survey (a Chicago-Princeton-Fermilab-Johns Hopkins-Japanese collaboration) when the 3-D positions for about a million galaxies will be determined (as contrasted with current surveys that see at most about ten thousand).

One way of comparing the microwave anisotropy data with the galaxy data is to look at the relative power implied on different separation scales. Because of the horizon cut-off, it is expected with flat gaussian density seeds that the power will peak on scales of about 1 or 2 degrees (see Figure 4) which today would correspond to distance scales of a few hundred Mpc. The dotted line in Figure 4 shows this behavior. The absolute amplitude of the initial fluctuations is very uncertain so the dotted curve is free to be moved up or down. For Figure 4, it has been fixed by the additional requirement that it fit the large scale data of COBE. The curves for the different galaxy surveys may be moved up or down, but not left or right via biasing and/or selection effects. Note that if the dotted curve is made to fit the small scale end of the galaxy data, then the large scale end doesn't fit. Note also that if we use the COBE fit at large scales and fit the IRAS or APM data at a couple of hundred Mpc, then the small scale IRAS and APM data falls below the dotted curve. Thus, no simple constant biasing can enable the dotted curve to work. Furthermore, there are hints from the radio galaxy[31] and cluster data[32] that there may be even more power near 100 to 200 Mpc, although the uncertainties are large and biasing may move the points coherently up and down.

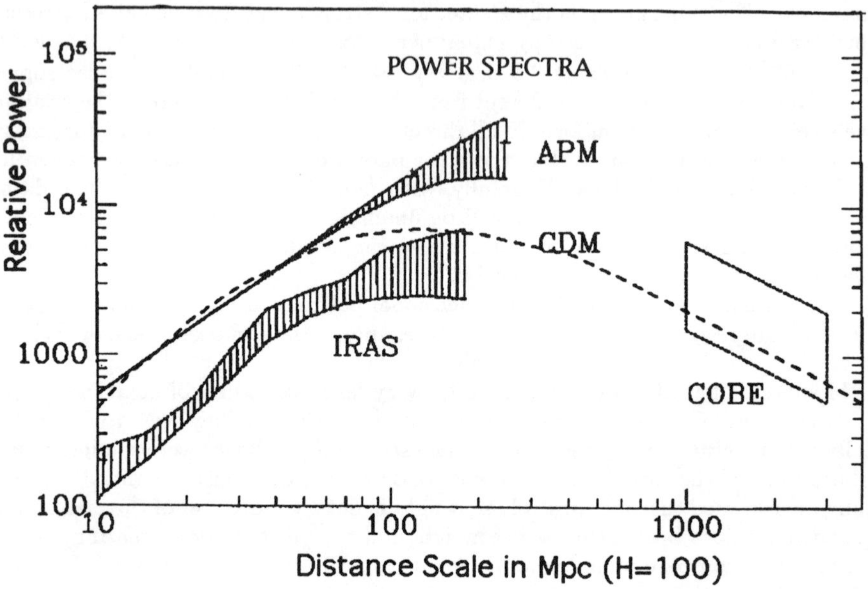

Figure 4. The relative power of producing structures on different scales. The dotted curve is for a traditional CDM model with gaussian density seeds. The galaxy data from different samples can be moved up and down via biasing and/or selection effects but the shapes are fixed. This figure was prepared by Jean Quashnock of the University of Chicago.

This conflict with the simple "CDM model" of an $n = 1$, gaussian density seed spectrum with CDM and constant biasing has led to several alternatives, all designed to fit the COBE data on large scales and the shape of the galaxy data on smaller scales. The possible models are shown in Table 3. As seen from the table, the surviving models each involve some assumptions beyond or instead of the traditional CDM constant biased model. For example, one can retain the gaussian density fluctuation picture by either adding a variable scale-dependent biasing (that goes negative at small scales!), or by adding some HDM to smooth out the small scale fluctuations and thereby decrease the small scale power, or by tilting the initial spectrum enough to fit the small scales but not violate the COBE error box. Another alternative is to go to topological defects. These can be either cosmic strings if the defects form in the early universe or almost any defect if the defects form after the microwave radiation decouples, but in this latter case, the COBE spectrum still requires some gaussian primordial fluctuations. The advantage of the defects plus HDM is that they tend to give a larger bump in the power spectrum near the epoch of galaxy formation which leads to an easier fit to the rich cluster and radio galaxy data (although this is the most uncertain part of the structure data and many models choose to ignore it at present).

Table 3

Models After COBE
With Critical Total Density and ~ 5% Baryons

Model Name	Non-Baryonic Dark Matter	Seed Type	Biasing
Modified CDM	CDM	flat-gaussian density	variable, going negative at small scales
Mixed (MDM)	CDM plus HDM	flat-gaussian density	negligible, except for cluster data
Tilted	CDM	slightly tilted spectrum increasing towards large scale	negligible, except for cluster data
Cosmic Strings	HDM	cosmic strings	negligible, may even fit cluster data
Late-Time Phase Transition	HDM	topological defects plus background flat-gaussian density	negligible

Note: models with cosmological constant or pure baryons are excluded here for reasons mentioned previously in text. Also, it is not obvious how many of the above models will fit the cluster x-ray gas analysis of Mushotzky.[23]

A resolution of which of the surviving models is correct (if any) can be accomplished with the next round of experiments and observations. Table 4 lists some key new pieces of information we should have by the end of the decade. Obviously, many of the solutions involve SUSY to one degree or another, whether for a solution to the inflation problems or to serve as dark matter or even to generate seeds at the end of inflation.

Table 4
Projects to Answer the Unsolved Questions

Topic	Projects
DARK MATTER	
• Baryonic Dark Matter	Gravitational Micro-lensing
• Cold Dark Matter	
-WIMPS	Underground direct searches
(SUSY)	Accelerator searches
-AXIONS	Direct Searches with Cavity
• Hot Dark Matter	Accelerator Neutrino Mixing
DISTRIBUTION OF GALAXIES	Dedicated Redshift Surveys
GALAXIES AT FORMATION	HST, Keck, AXAF, ROSAT, etc.
MICROWAVE ANISOTROPIES	South Pole, Balloons Future space mapping mission (away from Earth background)

This exciting and rapidly moving field may even come to some resolution of the problem of structure formation in the universe in the not-too-distant future. This resoution will come by an interdisciplinary effort of particle physicsts, nuclear physicists, radio, IR, optical, UV and x-ray astronomers and theorists. It is exciting to be doing cosmology in this "Golden Age" where progress really occurs.

5. Acknowledgements

Useful discussions with Ken Kellerman, Jean Quashnock, Don Lamb, Rich Mushofzky, Stephan Meyer, Michael Turner, Xiaochun Luo, Brian Fields, George Smoot, Keith Olive, Jim Truran and Lawrence Krauss are gratefully acknowledged. This work was supported in part by the NSF, DOE and NASA at the University of Chicago and by the DOE and by NASA through grant NAGW 2381 at Fermilab.

6. References

1. A. Linde, *Particle Physics and Inflationary Cosmology* (Harwood Academic Pub., Chur, Switzerland, 1990).
2. E. W. Kolb and M. S. Turner, *The Early Universe* (Addison-Wesley, Redwood City, CA, 1990).

3. A. Guth, *Phys. Rev* **D23** (1981) 347.
4. A. Albrecht and P. J. Steinhardt, *Phys. Rev. Lett.* **48** (1982) 1220.
5. A. Linde, *Phys. Lett.* **B108** (1982) 389.
6. J. Ellis, D. Nanopoulos, K. Olive and K. Tamvakis, *Phys. Lett.* **B118** (1982) 335.
7. K. Freese, J. A. Frieman, and A. V. Olinto, *Phys. Rev. Lett.* **65** (1990) 3233.
8. E. Wright et al., *ApJ Lett.* **396** (1992) L13.
9. G. Smoot et al., *ApJ Lett.* **396** (1992) L1.
10. D. Nanopoulos, private communication (1993).
11. A. Dekel, E. W. Bertschinger, et al., *ApJ* **412** (1993) 1.
12. F. Bouchet, M. Strauss, M. Davis K. Fisher, A. Yahil and J. Huchra, *ApJ* in press **417** (Nov. 1, 1993).
13. K. Kellerman, *Nature* **361** (1993) 134.
14. L. Krauss and D. N. Schramm, D.N., *ApJ Lett.* **405** (1993) L43.
15. D. Schramm, in *Astrophysical Ages and Dating Methods*, eds. E. Vangioni-Flam et al. (Editions Frontières, Gif-sur-Yvette, France, 1990) 365-383.
16. C. Delyannis et al., Yale University preprint (1992).
17. D. Dearborn and D. N. Schramm, Fermilab preprint (1993).
18. T. Walker, G. Steigman, D. N. Schramm, K. Olive and H.-S. Kang, *ApJ* **376** (1991) 51.
19. V. Smith, D. Lambert and R. Nissen, R., *ApJ* **408** (1993) 262.
20. K. Olive and D. N. Schramm, *Nature* **360** (1992) 434.
21. H. Kurki-Suonio, R. Matzner, K. Olive and D. N. Schramm, *ApJ* **353** (1990) 406.
22. J. R. Gott, J. Gunn, D. N. Schramm and B. Tinsley, *ApJ* **194** (1974) 543.
23. R. Mushotzky, in *Texas/PASCOS December 92: Relativistic Astrophysics and Particle Cosmology*, eds. C. Akerlof and M. Srednicki, *Annals of the N.Y. Academy of Sciences* **Vol. 688** (1993).
24. S. Meyer, in *Texas/PASCOS December 92: Relativistic Astrophysics and Particle Cosmology*, eds. C. Akerlof and M. Srednicki, *Annals of the N.Y. Academy of Sciences* **Vol. 688** (1993).
25. E. Wright, in *Texas/PASCOS December 92: Relativistic Astrophysics and Particle Cosmology*, eds. C. Akerlof and M. Srednicki, *Annals of the N.Y. Academy of Sciences* **Vol. 688** (1993).
26. X. Luo and D. N. Schramm, *ApJ* **408** (1993) 33.
27. P. Lubin et al., in *Texas/PASCOS December 92: Relativistic Astrophysics and Particle Cosmology*, eds. C. Akerlof and M. Srednicki, *Annals of the N.Y. Academy of Sciences* **Vol. 688** (1993).
28. MAX collaboration, in *Texas/PASCOS December 92: Relativistic Astrophysics and Particle Cosmology*, eds. C. Akerlof and M. Srednicki, *Annals of the N.Y. Academy of Sciences* **Vol. 688** (1993).
29. M. S. Vogeley, C. Park, M. Geller and J. Huchra, *ApJ Lett.* **391** (1992) L5.
30. J. A. Peacock, G. Efstathiou et al., *MNRAS* **253** (1991) 1.
31. J. A. Peacock and D. Nicholson, *MNRAS* **253** (1991) 307.

32. J. Bahcall and R. Soniera, *ApJ* **270** (1983) 20.

ASPECTS OF NEUTRALINO DARK MATTER

KAJIA YUAN
Center for Theoretical Physics, Department of Physics, Texas A&M University
College Station, TX 77843-4242, USA
Astroparticle Physics Group, Houston Advanced Research Center (HARC)
The Woodlands, TX 77381, USA

ABSTRACT

The possible solution of dark matter problem with neutralinos of supersymmetric models within the supergravity framework is reviewed. A novel correlation between the neutralino relic abundance Ω_χ and the soft supersymmetry breaking patterns is demonstrated. It is explained that, this generic result together with the proton-decay constraint could significantly reduces the allowed parameter space of the minimal $SU(5)$ supergravity model, and therefore makes this model more easily testable. The prospect of obtaining further cosmological constraints from underground experiments for the minimal $SU(5)$ supergravity model is also briefly discussed.

1. Introduction

Low-energy supersymmetry is now widely believed to be one of the most appealing ideas likely responsible for new physics beyond the Standard Model [1], as such, it has a very good chance to reveal itself at next generation of pp or e^+e^- colliders, one way or another [2]. If we indeed live in a "SuperWorld", whatever that might be, then low-energy supersymmetry must also have cosmological consequences. This motivated the study of "SuperCosmology", of which many exciting topics have been covered in previous talk by Prof. Schramm [3]. In this talk, I will confine myself to the proposal [5, 6] of solving the dark matter problem with neutralinos of supersymmetric models, as one typical example which ties up low-energy supersymmetry with cosmology.

Let me begin with some remarks about the so-called dark matter problem. First of all, it is a problem caused by the huge discrepancy between the amount of directly observed visible matter and the total amount of matter in the Universe indirectly inferred based on general relativity (mostly Newtonian dynamics) [4]. One should keep an open mind that, if at large scale the law of gravity is in fact somehow modified, the dark matter problem itself could cease to exist [7]. I will not entertain such a possibility here, but instead just follow the majority point of view. In terms of the present-day cosmological density parameter $\Omega \equiv \rho_0/\rho_{\rm crit}$, the visible matter of the Universe only amounts to $\Omega_{\rm vis} \lesssim 0.01$. Big-bang nucleosynthesis indicates that the fraction of critical density contributed by ordinary baryons is $0.011 \lesssim \Omega_B \lesssim 0.12$. However, current theoretical prejudice prefers a flat universe today, in particular, inflation predicts $\Omega = 1$ to a very high precision. Furthermore, without dark matter as an essential ingredient, the mechanism of structure formation in the Universe, based on the gravitational growth of primeval density inhomogeneities (Jeans instability), does not seem to work for galaxy formation. There are also observational evidences for dark matter coming from various measurements of the present-day mass density of the Universe, performed at different distance scales, among them the flat rotation curves of spiral galaxies appears to be most compelling. Based on these arguments, one concludes that dark matter must exist and be mostly in non-baryonic form.

Accepting the dark matter problem as a fact, now the question is: what is dark matter made of? So far, we do not have a definite answer, but we do have many speculations (maybe too many!). Instead of going through the list of all proposed dark matter candidates, which can be found elsewhere [4], I will only consider one type of hypothetical particle dark matter which is closely associated with low-energy supersymmetry, *i.e.*, the neutralino dark matter [5, 6]. It should be noted that, from particle physics point of view, massive neutrinos with mass around 30 eV and axions are also particularly attractive dark matter candidates in their own right.

We know that low-energy supersymmetry predicts many new particles yet to be discovered, it would be extremely fortunate if some of these particles in fact constitute the dark matter out there in the Universe. To be a possible dark matter candidate, a supersymmetric particle has to satisfy at least three requirements: (a) to be stable or at least have a life-time comparable with the age of the Universe so that it can be around today as a cosmological relic; (b) to have only gravitational and weak interactions; (c) to have right properties such that its relic abundance comes close to the critical value $\Omega = 1$. In most supersymmetric models, to eliminate the disastrous explicit baryon and lepton number violating interaction terms which lead to proton decay at unacceptable level, a discrete symmetry known as R-parity is often invoked, such that all ordinary particles are R-even, but all superparticles are R-odd. Therefore, if R-parity is conserved, superparticles always couple to the ordinary particles in pair, and as a result the lightest superparticle (LSP) is stable. The LSP is also likely to be colorless and electrically neutral, since otherwise it would have formed heavy anomalous isotope with ordinary matter which would have been observed already [8, 6]. These considerations tell us that, with unbroken R-parity, the LSP in fact could make a viable dark matter candidate, if requirement (c) is also satisfied. Unfortunately, no general remarks can be made about this requirement for LSP, mainly because we do not really know what is the LSP. The lightest neutralino χ, a mixture of neutral gauginos and higgsinos, which I will simply call neutralino in the rest of this talk, currently stands as a front runner for the LSP. The basic argument in favor of this theoretical prejudice is essentially that, so far there have been no cosmological, experimental or theoretical reasons against it. In any case, as a working hypothesis, I will assume that the lightest nuetralino is indeed the LSP. As we will see later, in the supergravity models, since the complete particle mass spectrum of the model can be specified rather economically in terms of only a few parameters, it becomes possible to consistently check the validity of this very assumption.

In the following, I will discuss some recent work regarding the prospects for neutralino dark matter within the supergravity framework, paying special attention to the correlation between the neutralino relic abundance Ω_χ and the supersymmetry breaking patterns. Taking the minimal $SU(5)$ supergravity model as an explicit example, in which the physics at grand unification scale come into play, I will demonstrate the implications of this novel correlation, and in passing also address the technical issue of appropriately treating the thermal average cross section. In addition, assuming that neutralinos constitute the dark matter in the Galactic halo, and using the upward-going muon events in underground detectors as a possible signal from such neutralinos captured in the Earth and the Sun, I will also briefly discuss the prospect of further exploring the minimal $SU(5)$ supergravity model.

2. Neutralino Relic Abundance

Now let me examine the requirement (c) for neutralinos in more detail. Will neutralinos provide just right amount of mass to close the Universe? From earlier work [5, 6, 16, 17], we have learned that the answer to this question is YES and NO. Yes, because it is possible to find regions in the parameter space of the supersymmetric models where neutralino relic abundance is just what we would like it to be ($\Omega_\chi \sim 1.0$); No, because it is equally true that there are regions where either $\Omega_\chi \ll 1.0$ or $\Omega_\chi \gtrsim 1.0$. Needless to say, regions with $\Omega_\chi \sim 1.0$ are cosmologically favored. The regions of $\Omega_\chi \gtrsim 1.0$ should be excluded on cosmological grounds, since where the Universe is uncomfortably younger than about 10 billion years, and this argument alone often leads to interesting constraints on the models under consideration. On the other hand, if $\Omega_\chi \ll 1.0$, the best one can learn is that now neutralinos can not be the sole source of dark matter. Despite this, I want to stress that models with too small values of Ω_χ are perfectly healthy, while on the contrary models which predict $\Omega_\chi \gtrsim 1.0$ are definitely in trouble. Clearly, one would like to be able to identify these three distinct cases as precisely as possible.

2.1 Calculation Procedures

The basic physics involved in calculating relic abundance for any massive stable particles was well understood some time ago [9], which can be applied to the neutralino case. At very early time the Universe was radiation dominated, all particles would be in thermal equilibrium, neutralinos annihilate into other particle species, and vice versa. As the temperature drops down below the neutralino mass m_χ, the annihilation process becomes dominant and the neutralino number begins to decrease due to Boltzmann suppression, until the interactions between neutralinos "freeze out", which happens when the annihilation is no longer able to keep pace with the expansion of the Universe. After "freeze out", the number of neutralinos essentially remains constant and the number density only reduces as a consequence of the cosmological expansion. All of these are neatly embodied into the Boltzmann equation [9]

$$\frac{dn}{dt} = -3Hn - \langle \sigma v_{M\phi l} \rangle (n^2 - n_0^2), \tag{1}$$

where n is the actual number density of the neutralinos, n_0 is the density they would have in thermal equilibrium at temperature T, $H = (dR/dt)/R$ is the Hubble expansion parameter, and $\langle \sigma v_{M\phi l} \rangle$ is the thermal-averaged product of annihilation cross section and the Møller velocity of the annihilating neutralinos in the cosmic comoving frame [14].

In practice, it is convenient to replace time t in Eq. (1) by the photon temperature T, this is because the present-day cosmic background radiation (CBR) temperature can be measured rather accurately, while determining the age of the Universe is a quite different story (see e.g. Ref. [10]). Using the conservation of entropy, one can recast Eq. (1) into an convenient form involving T. Although there exist slightly different approaches in the literature, I will closely follow Ref. [11] here. The "new" Boltzmann equation now reads

$$\frac{dq}{dx} = \lambda(x)(q^2 - q_0^2)(x), \tag{2}$$

with

$$\lambda(x) = (\frac{4}{45}\pi^3 G_N)^{-1/2} \frac{m_\chi}{\sqrt{g(T)}} [h(T) + \frac{1}{3} m_\chi x h'(T)] \langle \sigma v_{M\phi l} \rangle, \tag{3}$$

where $q \equiv n/(T^3 h(T))$, $q_0 \equiv n_0/(T^3 h(T))$, $x \equiv T/m_\chi$, and $g(T)$, $h(T)$ ($h'(T) = dh/dT$) are the effective degrees of freedom associated with energy and entropy density respectively [11]. Finally, the relic abundance is given by

$$\Omega_\chi h_0^2 = 1.555 \times 10^8 (m_\chi/GeV) h(0) q(0), \tag{4}$$

where $h_0 = H_0/(100 \text{kmsec}^{-1}\text{Mpc}^{-1})$ parameterize our ignorance of the present-day value of the Hubble parameter ($0.5 \lesssim h_0 \lesssim 1.0$); $h(0), q(0)$ are the present-day values of $h(T)$, $q(T)$ respectively.

From Eqs. (2)-(4) it is clear that the calculation of relic abundance typically involves three procedures: (I) Computing $h(T), g(T)$, in particular, $h(0)$; (II) Evaluating $\langle \sigma v_{M\phi l} \rangle$; (III) Solving Eq. (2) for $q(x)$ and find $q(0)$. Again, one can often find different ways of implementing these procedures in literature. For a detailed discussion of our approach to (I) and (III), see the Appendix of Ref. [12]. It is worthy of mentioning that, in solving Eq. (2), a WKB approximation [13] was used in Ref. [12], which matches the solution $q(x)$ with $q_0(x)$ and provides the initial condition for Eq. (2) at a point x_0 where the WKB solution fails. Although, it is the same in spirit to determine x_0 here as to determine the so-called freeze-out temperature in the standard Lee-Weinberg method [5, 6], however, one advantage of this new approach is that the accuracy of the solution can be easily controlled as one desires.

The most accurate treatment so far available for the thermal average factor $\langle \sigma v_{M\phi l} \rangle$ was given in Ref. [14]. In this remarkable paper, the authors were able to reduce the multiple integrals associated with the thermal average into a single integral, which can be rewritten in the following convenient Lorentz invariant from [41]

$$\langle\sigma v_{M\phi l}\rangle = \frac{1}{4m_\chi^5 x K_2^2(x^{-1})}\int_{4m_\chi^2}^{\infty} ds(s-4m_\chi^2)^{1/2}K_1(\sqrt{s}/xm_\chi)\,w(s), \quad (5)$$

in terms of the Lorentz invariant function [11]

$$w(s) = \frac{1}{4}\int dLPS\,|\mathcal{A}(\chi\chi \to all)|^2. \quad (6)$$

In Eq. (5), $K_i, (i = 1,2)$ are the modified Bessel functions of order i. Eq. (2) is only sensitive to the value of $\lambda(x)$ for $x \lesssim 0.1$ which roughly corresponds to the freeze-out temperature, and the center-of-mass energy $\sqrt{s} \geq 2m_\chi$, as a result the argument of K_1 in Eq. (5) $\sqrt{s}/xm_\chi \gtrsim 20$ so that K_1 dies away quickly with increasing $\sqrt{s}$, owing to the asymptotic behavior of the Bessel function ($K_1(y) \sim \sqrt{\pi/2y}e^{-y}, y \gg 1$). Therefore, replacing $w(s)$ in (5) by its series expansion around $\sqrt{s} = 2m_\chi$ and the Bessel functions by their asymptotic expansions for large argument, the resulting series of integrals can be readily carried out analytically leading to [11]

$$\begin{aligned}\langle\sigma v_{M\phi l}\rangle &= \frac{1}{m_\chi^2}[w - \frac{3}{2}(2w - w')x + \frac{3}{8}(16w - 8w' + 5w'')x^2 + \mathcal{O}(x^3)]_{s=4m_\chi^2} \\ &\equiv a + bx + cx^2 + \mathcal{O}(x^3), \end{aligned} \quad (7)$$

where primes denote derivatives with respect to $s/4m_\chi^2$ rather than s itself.

Contrary to Eq. (5) which inevitably requires numerical integration, the expansion such as Eq. (7) or its non-relativistic counterpart is simple to use and gives overall fairly good results. In fact, most of the relic abundance calculations were carried out using such thermal average expansion [5, 6, 16, 17]. Recently, however, the degree of accuracy of such expansions has been reexamined, and now it is clear that such treatment actually fails badly near the s-channel resonances and/or new-channel thresholds, basically because the expansion of $w(s)$ (or equally, cross section for this matter) at $\sqrt{s} = 2m_\chi$ becomes inappropriate in those cases [14, 15]. In addition, in Ref. [41] it is shown with explicit examples that, when close to the s-channel resonances the first-order expansion $a + bx$ is much better behaved than the second-order expansion $a + bx + cx^2$, in fact latter renders the thermal average factor negative right above the poles which doesn't make sense. Of course, in this case, one should abandon the expansions and use the exact integral expression (5) to get reliable results. Later in subsection 2.4, I will discuss a practical situation where a proper treatment of the thermal average factor is needed and present the results calculated with Eq. (5) and with first-order expansion $a + bx$.

2.2 Organizing Model Parameters: Supergravity Models

So far, I have kept my discussion fairly model-independent, obviously, to go further we need the information of specific supersymmetric model. Primarily, the details of the particle physics model enter the function $w(s)$, which is related to the annihilation cross section of neutralino pair into all kinematically accessible final states. What makes the analysis of these cross section fairly complicated is mainly the fact that supersymmetric models often contain many free parameters which to certain degree are rather arbitrary. For instance, in the minimal supersymmetric standard model (MSSM) [18], which happens to be the playing ground of almost all previous studies on neutralino dark matter [5, 6, 16, 17], even if the mixing between generations are neglected and only the Yukawa terms for the third generation are kept, there are still at least 21 free parameters, many of which describe the soft supersymmetry breaking at low energies. With so many parameters, any thorough analysis becomes hopeless. Nevertheless, to make life easy, in all these earlier works, several additional assumptions about these parameters are made, notably, the GUTs relation among different gaugino masses, and a common mass parameter for all sleptons and squarks. This situation is hardly satisfactory. We know that supersymmetry must be broken, however, just as we do not really understand the gauge symmetry breaking mechanism, our knowledge about supersymmetry breaking is even more limited. The phenomenology of low-energy supersymmetry should reflect how supersymmetry is broken, and therefore could in principle provides us with some information on the pattern of supersymmetry breaking. Since any such information is precious, it would thus be very important if one can also obtain constraints on the soft supersymmetry breaking terms from the cosmology of neutralino dark matter. Clearly, in the usual approach, particularly due to the ad-hoc assumption about the masses of sleptons and squarks, no "fine structure" of supersymmetry breaking is left to be found. On the other hand, as I mentioned earlier, to address such problem with all the soft-breaking parameters being arbitrary simply is not possible. One way out is provided by $N = 1$ supergravity [19]. Although it was realized long ago that local supersymmetry, instead of global supersymmetry, should be the natural framework in which to construct realistic low-energy supersymmetric models, only very recently has the study of neutralino dark matter in supergravity models attracted enough attention [21, 22, 12, 23, 24, 25, 26, 27, 28].

In supergravity models, in addition to the observable sector, which contains quarks, leptons, Higgs bosons, gauge bosons as well as their superpartners, and in case of a grand unified model also the extra particles, there is also a so-called hidden sector responsible for the spontaneous breaking of $N = 1$ local supersymmetry. At low energies, the breaking of supergravity taking place in the hidden sector transmits into the observable sector via gravitational interaction, and therefore leads to a globally supersymmetric effective theory with explicit soft-breaking terms. One of the virtue of the supergravity models is that the number of soft-breaking parameters are greatly reduced. In fact, the supersymmetry breaking pattern now can be specified mainly in terms of only three universal parameters at some unification scale M_U: (I) the scalar mass m_0; (II) the Majorana gaugino mass $m_{1/2}$; and (III) the trilinear A and bilinear B scalar couplings. Furthermore, in supergravity models the breaking of the electroweak gauge symmetry can be realized as a consequence of the supersymmetry breaking through radiative correction [20], with this radiative breaking mechanism implemented, other model parameters can be further reduced. In the model with minimal particle content same as that of MSSM, up to the sign of Higgs mixing parameter μ, one only needs two more parameters to describe the whole model, these can be chosen as the unknown top quark mass m_t and the ratio of Higgs vacuum expectation values $\tan\beta = v_2/v_1$, and finally all the couplings and superparticle masses are determined dynamically via the relevant renormalization group equations (RGEs) as functions of the above five parameters [29].

2.3 Supersymmetry Breaking and Dark Matter

I am now ready to describe the correlation between the neutralino relic abundance Ω_χ and the supersymmetry breaking patterns, first found in Ref. [23], as one of the most interesting features of the supergravity models we are considering.

The idea is very simple. Since now all the low-energy couplings and superparticle masses entering $w(s)$ depend explicitly upon soft-breaking parameters m_0, $m_{1/2}$ and A, the relic abundance Ω_χ itself becomes a function of these parameters as well. The most crucial factors here are the masses of the gauginos, squarks and sleptons at low energies, which are determined as functions of soft-breaking parameters through the renormalization equations. First, the gaugino masses ($M_i, i = 1, 2, 3$) are given by $M_i = (\alpha_i/\alpha_U)m_{1/2}$, where $\alpha_U = 0.0409$ is the gauge coupling at the unification scale $M_U \approx 10^{16}$ GeV (determined from following input low-energy values: $\alpha_3 = 0.111$, $\sin^2\theta = 0.233$ and $\alpha = 1/127.9$). Second, neglecting the Yukawa coupling contributions, the renormalization equations of the sfermion mass can be solved exactly giving [30]

$$\begin{aligned} m_{\tilde{f}}^2 &= m_f^2 + m_0^2 + m_{1/2}^2 c_{\tilde{f}} - M_W^2 \frac{\tan^2\beta - 1}{\tan^2\beta + 1}[(T_{3,f} - Q_f)\tan^2\theta_W + T_{3,f}] \\ &= m_f^2 + (1.22M_2)^2(c_{\tilde{f}} + \xi_0^2) + D_{\tilde{f}} \end{aligned} \quad (8)$$

where $\xi_0 \equiv m_0/m_{1/2}$, and the coefficients $c_{\tilde{f}}$ are determined to be: $c_{\tilde{e}_L,\tilde{\mu}_L,\tilde{\nu}} = 0.514$, $c_{\tilde{e}_R,\tilde{\mu}_R} = 0.150$, $c_{\tilde{u}_L,\tilde{c}_L} = c_{\tilde{d}_L,\tilde{s}_L} = 6.134$, $c_{\tilde{u}_R,\tilde{c}_R} = 5.720$, $c_{\tilde{d}_R,\tilde{s}_R} = 5.670$. The parameter A enters the off-diagonal terms of the third generation scalar masses. In Ref. [23] only the left-right mixing for stop quark masses is considered, and the diagonal contributions for all the third generation sfermions are approximated with the c's given above and Eq. (8). It is found that the major effect of A is to restrict the allowed parameter space, otherwise A does not change the neutralino abundance Ω_χ significantly. In what follows, I choose $A = m_0$.

In Figs. 1–3, the neutralino relic abundance Ω_χ is shown in the (μ, M_2) plane for $\tan\beta = 8$, $\mu > 0$ ($\mu < 0$ case is similar) *, and three representative values of ξ_0: (a) $\xi_0 = 10.0$, (b) $\xi_0 = 1.0$, (c) $\xi_0 = 0.1$. Here only the tree-level Higgs masses are used, and the lightest CP-even Higgs mass is taken to be $m_h = 45$ GeV, different choice of this parameter leads to similar results. I also take $h_0 = 0.5$, a favorite choice of cosmologists [10], and divide the allowed parameter space into three types of distinct regions: (1) regions represented by stars corresponds to $\Omega_\chi > 1.0$, which are excluded cosmologically; (2) regions represented by vertical crosses corresponds to $0.1 < \Omega_\chi < 1.0$, which are cosmologically favored; (3) regions represented by dots corresponds to $\Omega_\chi < 0.1$, where neutralinos can not even account for enough dark matter in Galactic halos.

*Here the sign convention of μ is opposite to that of Refs. [12, 23].

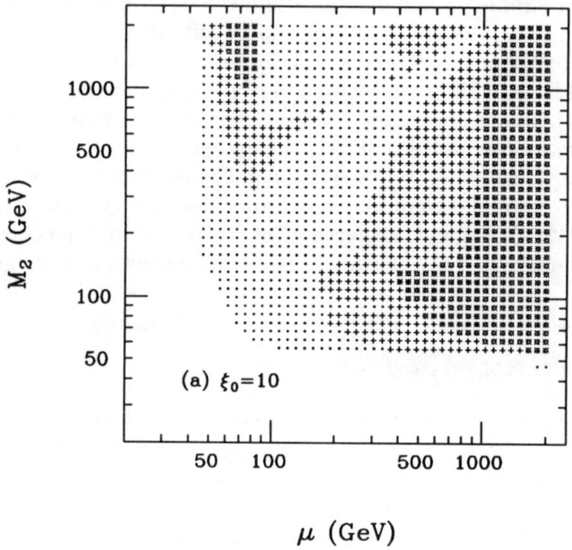

Figure 1: The neutralino relic abundance distribution for $\tan\beta = 8$, $h_0 = 0.5$, $\xi_0 = 10$. The meaning of the three different symbols: (1) stars ($\Omega_\chi > 1.0$); (2) crosses ($0.1 < \Omega_\chi < 1.0$); (3) dots ($\Omega_\chi < 0.1$).

In these figures, the shape of the allowed parameter space in each case was determined by several constraints. The direct experimental constraints used are: (1) The LEP lower bound on the chargino mass $m_{\chi^\pm} > 45$ GeV, and on the slepton mass $m_{\tilde{l}} > 43$ GeV; (2) The CDF lower bound on the gluino mass $m_{\tilde{g}} > 150$ GeV, which translates into $M_2 > 45$ GeV, and on the squark mass $m_{\tilde{q}} > 100$ GeV. In addition, the assumption about neutralino being the LSP imposes important consistency constraints on the allowed parameter space as well. For example, in Fig. 3 a portion of the upper right corner is excluded since there right-handed sleptons become lighter than the neutralino. Also, in all these figures, the allowed minimal value of M_2 is in fact bigger than that required by the gluino mass lower bound; and when ξ_0 decreases, this lower bound on M_2 increases. It is because that we do not want the sneutrinos to be lighter than the neutralino. Eq. (8) clearly shows that this constraint becomes stronger for small value of ξ_0. Finally, in the lower right corner of these figures, there is a triangle region where the left-right mixing term for stop quark masses could drive the $\tilde{t}_1$ mass below either the CDF bound or m_χ, and thus it should be excluded. This effect is only barely visible in Fig. 2, but it becomes more pronounced for small values of $\tan\beta$.

As in the case of MSSM [17], Figs. 1–3 once again exhibits the fact that the neutralinos relevant to cosmology should either contain a dominant higgsino component or a dominant bino component. In addition, two very interesting new features emerge in Figs. 1–3, due to the supergravity squark and slepton relation (8). First, in the pure higgsino region (upper left portion of the (μ, M_2) plane), we see that the relic abundance Ω_χ does not change with the different values of ξ_0. This is because there the contributions to the annihilation cross section due to the exchange of sfermion have already been considerably suppressed, so the variation of squark and slepton masses with supersymmetry breaking patterns essentially has no effect on Ω_χ for nearly pure higgsinos. Independent of how supersymmetry is broken, pure higgsino dark matter candidate will have a mass of roughly the order of M_W, but there the gluino would be heavier than about 1.5 TeV. However, if one insists that $m_{\tilde{g}} \lesssim 1$ TeV to insure the fine-tuning of the parameters occurs at two-orders-of-magnitude or less [29], this possibility of nearly pure higgsino dark matter would be eliminated. Second, in the pure bino region (lower right portion of the (μ, M_2) plane), however, Ω_χ changes with ξ_0 significantly, mainly because the sfermion exchange contribution is dominant in this region. From Eq. (8) we see that large value of ξ_0 implies all sfermions are heavy, so this contribution is suppressed, which leads to a large value of Ω_χ. When ξ_0 decreases, the annihilation due to sfermion exchange becomes increasingly efficient, and therefore Ω_χ reduces. Note that $m_{\tilde{f}}$ now vary throughout the (μ, M_2) plane due to their M_2 dependence, and in the case of stop quark $\tilde{t}_{1,2}$ their μ dependence as well.

The above discussion is mainly concerned with the impact of the supergravity relation Eq. (8) for the squark and slepton masses on the neutralino relic abundance Ω_χ. In this analysis, the radiative electroweak gauge symmetry breaking [20] and the radiative corrections to Higgs masses [31] were not considered. These two issues also have important consequences on the cosmology of neutralino dark matter

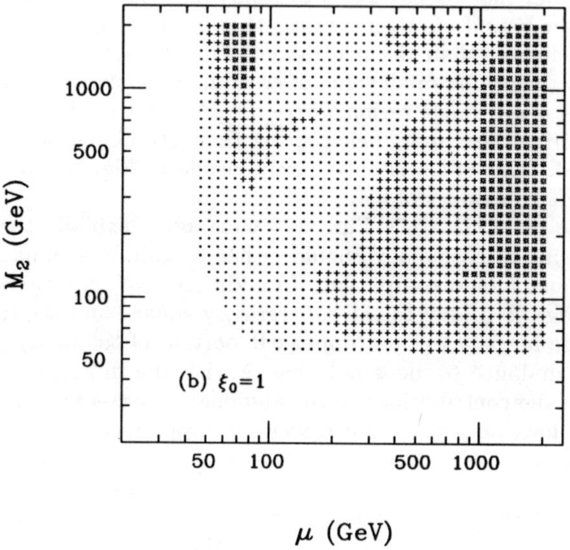

Figure 2: Same as Fig. 1, but with $\xi_0 = 1.0$.

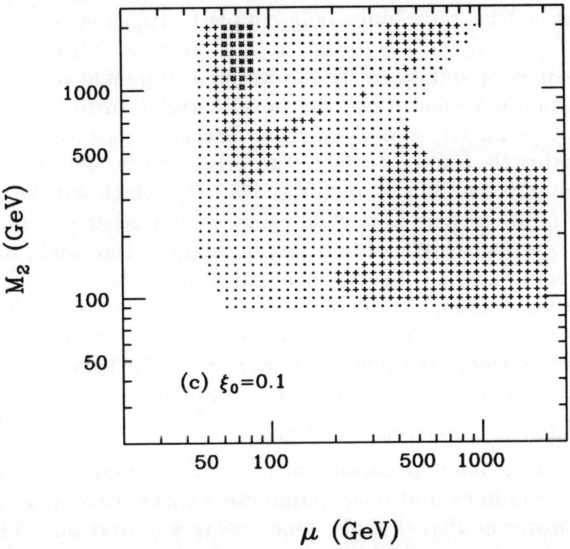

Figure 3: Same as Fig. 1, but with $\xi_0 = 0.1$.

[26, 27, 28]. The first new feature is that, after enforcing the requirement of radiative electroweak symmetry breaking, the allowed parameter space is considerably reduced [27]. In the usual (μ, M_2) plane, the two coordinates can no longer vary independently. For example, for $\xi_0 \lesssim 1.0$, this correlation between μ and M_2 could eliminate a rather large triangle region below the diagonal which are otherwise allowed, since $\mu \lesssim m_{1/2} = 1.22M_2$. In addition, since the one-loop radiative corrected Higgs masses vary continuously as one moves around the (μ, M_2) plane, and they are normally bigger than their tree-level values, the overall effect is a suppression of the relevant annihilation rate and hence an enhancement of the relic abundance Ω_χ relative to that shown in Figs.1–3, as long as the regions in comparison now are still allowed. Notice that pure binos only couple to Higgs very weakly, so such an enhancement of Ω_χ mainly occurs in the "mixed" regions. This makes the "mixed" neutralino a possible candidate to be the major component of the Galactic halo. Of course, the qualitative feature of the novel correlation between the neutralino relic abundance Ω_χ and the supersymmetry breaking patterns, as demonstrated in Figs.1–3, remains the same. In fact, the prospects for neutralino dark matter depend most strongly on the parameter ξ_0 [23, 27], which can be summarized as follows: (1) For $\xi_0 \sim 1.0$ there is a wide range of the other parameters such that $\Omega_\chi \sim 1.0$; (2) For $\xi_0 \ll 1.0$ the relic abundance normally is too small, but it is possible in this case that some "mixed" neutralinos may still be able to account for the dark matter in the Galactic halo; (3) For $\xi_0 \gg 1.0$ the relic abundance is almost always much too large in conflict with current cosmological observations, except the accidental circumstances where the relic abundance could be locally diluted due to the presence of resonances and thresholds in the annihilation cross section.

2.4. The Minimal SU(5) Supergravity Models

In the supergravity models, one of the crucial assumptions is that an unification of gauge couplings and mass parameters takes place at some high-energy scale M_U not far from the Planck scale. One simple way to realize this is to invoke a grand unification type of symmetry. ‡ Up to now, my discussion of the correlation between the neutralino relic abundance Ω_χ and the soft supersymmetry breaking patterns has not relied on the details of any specific grand unification models. Now, I would like to consider the implications of this generic result in the minimal $SU(5)$ supergravity model [32].

At low-energies, the minimal $SU(5)$ supergravity model consists of the normal light particles of the MSSM. Again, the masses and couplings of these light particles are completely specified in terms of a few parameters as described in subsection 2.2. The new feature of this model (as in all grand unified models) is the existence of the additional heavy degrees of freedom arisen around $M_U \sim 10^{16}$ GeV, which would lead to new phenomena such as proton decay [33]. In the minimal $SU(5)$ supergravity model, due to its supersymmetric nature, the usual proton decays through dimension-six operators mediated by exchange of either heavy gauge bosons or

‡However, in the context of superstring theories such unification arises naturally even in the absence of a grand unification group.

heavy triplet Higgs bosons, which plague the ordinary non-supersymmetric $SU(5)$ model [33], are strongly suppressed and place no danger as compared with the experimental limit. However, the new proton decays through the genuinely supersymmetric dimension-five operators [34] mediated by exchange of heavy higgsinos are not very strongly suppressed. Note, for the dimension-five-induced amplitude the suppression factor is $1/(M_U M_W)$, while for the normal dimension-six amplitude it is $1/M_U^2$. These dimension-five-induced processes could lead to rather dangerous proton decays, typically in the modes $p \to \bar{\nu}_{\mu,\tau} K^+$ [35].

Recently, in light of the current experimental limit $\tau_{p\to\bar{\nu}K^+} > 10^{32}$ yr, detailed analysis of the dimension-five-induced proton decay constraints have been carried out [36, 37, 38] in the minimal $SU(5)$ supergravity model. Assuming the exchanged triplet higgsino mass to be bounded above by $M_{\tilde{H}_3} < 3M_U$ [36], it is found that the proton decay constraint is rather restrictive. In particular, as for the soft supersymmetry breaking patterns, the proton deacy favors a large value of ξ_0 [36, 38]. Again, this result can be easily understood in view of Eq. (8). Since squarks and sleptons appear in the loops of the box diagrams responsible for the dimension-five-induced proton decays, large value of ξ_0 implies that such processes are suppressed. From the discussion in previous subsection, however, we see that the cosmology of neutralino dark matter, on the other hand, disfavors large value of ξ_0. Therefore, a delicate balance have to be attained in order to satisfy these two constraints simultaneously in the minimal $SU(5)$ supergravity model.

The conflict between these two types of constraints in the minimal $SU(5)$ supergravity model was first pointed out in Ref. [39], which has since spurred further investigations on this subject [38, 40, 41]. Considering the combined effect of these two constraints, it is found that the allowed parameter space of the minimal $SU(5)$ supergravity model is dramatically reduced (see below). However, there is still a region in the parameter space where both constraints are simultaneously satisfied. It turns out that, in this region the neutralino χ is actually very close to the lightest CP-even Higgs (h) and/or Z-boson resonances, *i.e.*, $m_\chi \approx \frac{1}{2} m_{h,Z}$. This fact has cast doubts [40] on the accuracy of the results obtained in Refs. [39, 38], since there the first-order expansion of form (7) has been used to approximate the thermal average factor, which in general is not a valid approach (see subsection 2.1). Two groups [40, 41] have recently reassessed this problem by treating the thermal average factor properly.

In Ref. [41], an extensive search of the parameter space of the minimal $SU(5)$ supergravity model has performed. In practice, a data set of the five-dimensional parameter space is generated first, which gives adequate radiative electroweak gauge symmetry breaking and satisfies all known current phenomenological constraints as described in Ref. [29]. Then, the proton decay constraint is used to further reduce the allowed points in the data set. In this step, the effects of two-loop gauge coupling unification and light supersymmetric thresholds have also been included [42]. Finally, for all the remaining points in parameter space, which have $\tan\beta = 1.5, 1.75, 2.0$, the neutralino relic abundance Ω_χ is computed in two different approaches: (a) the thermal average factor is treated accurately using Eq. (5); (b) the thermal average

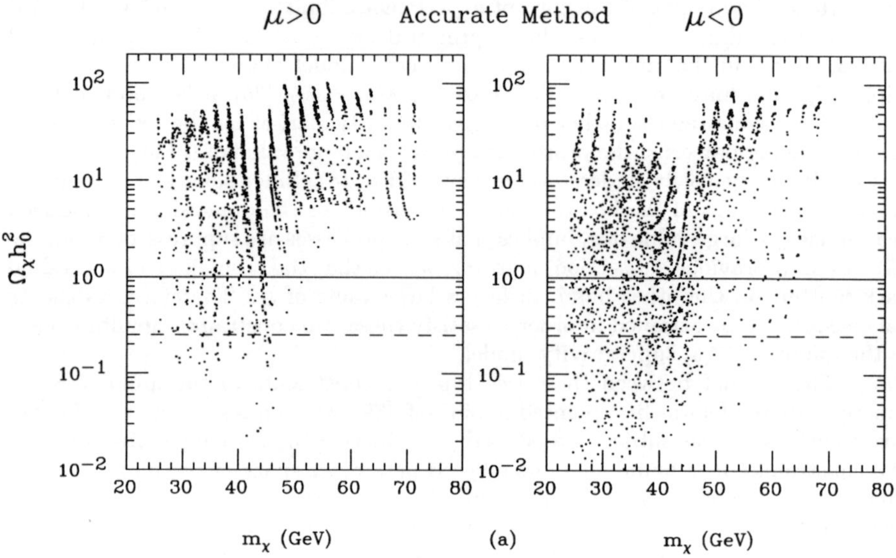

Figure 4: The neutralino relic abundance in the minimal $SU(5)$ supergravity model as a function of m_χ calculated using the exact thermal average procedure. The points above the solid (dashed) line are excluded if $h_0 = 1.0$ ($h_0 = 0.5$).

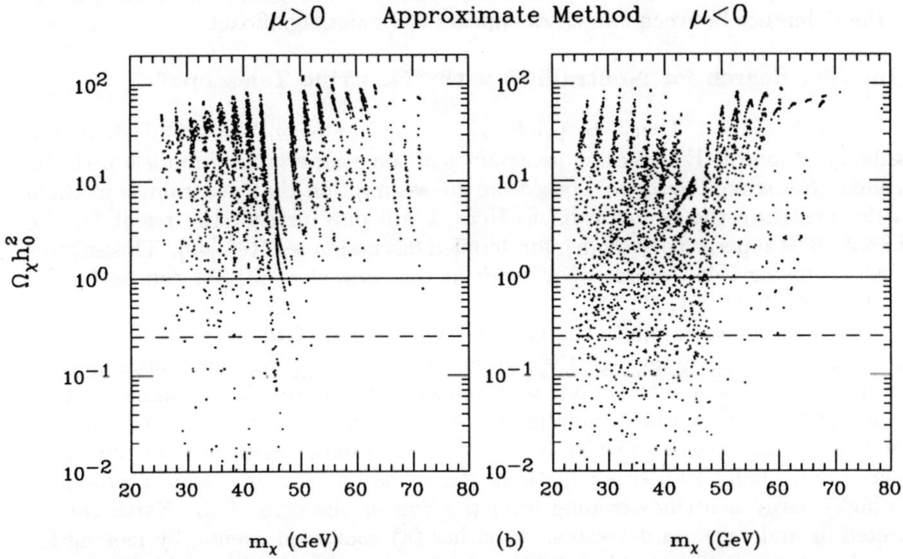

Figure 5: Same as Fig. 4, but the thermal average factor is approximated with the first-order expansion.

factor is treated approximately using first-order expansion of form Eq. (7). The results are shown in Fig. 4 and Fig. 5 respectively. These figures indicate that indeed the cosmological constraint is very powerful: all the points above the solid (dashed) line, allowed by other constraints, are now excluded if $h_0 = 1.0$ ($h_0 = 0.5$). This result does not change even if the exact thermal average procedure is followed. From these two figures, it is clear that some shifts of the points are noticeable, and the actual structure around the resonances in Fig. 4, in particular that close to the Z-pole ($m_\chi \sim \frac{1}{2}M_Z$), is broader, shallower and asymmetric relative to Fig. 5. Moreover, the cosmologically allowed points in Fig. 4 are increased relative to that in Fig. 5. However, the qualitative distributions of the points in Fig. 4 and Fig. 5 are similar, and the difference between two cases appears to be not significant.

3. Indirect Search for Neutralinos with "Neutrino Telescope"

In this section, I wish to briefly report on our recent work [43] about the possibility of indirectly detecting neutralinos of some supergravity models with the so-called "Neutrino Telescope", based on the assumption that neutralinos in these models constitute the Galactic halo. Here, I will only present the result for the minimal $SU(5)$ supergravity model, for detailed discussion see Ref. [43]. This subject has a rich history [46], more recent work in the context of MSSM can be find in Refs. [47, 48, 49, 50, 51, 52, 53].

If neutralinos populate the Galactic halo, some of them could be trapped [44, 45] when traveling through the Sun or Earth, after losing substantial amount of energy during their interactions with nuclei. The captured neutralinos sink to the core of the Sun or Earth, annihilate with one another therein, then produce a shower of ordinary particles which further decay or interact with solar or terrestrial material and finally lead to energetic neutrinos as one sort of ultimate products. The high-energy neutrinos coming from the Sun or the core of the Earth can be detected in underground detectors, via either (a) contained events, by looking for charged leptons within the detector (useful for detecting both ν_e and ν_μ); or (b) through-going events, by searching for upward-going muons which are the products of the interaction of neutrinos with rock below the detector (useful for detecting ν_μ). I will only consider the neutrino-induced upward-going muon events here, as it is most promising.

In the primary direct capture, a neutralino with mass m_χ passing through the Sun or Earth loses enough energy due to its elastic scattering with nuclei, so that its velocity falls below the escape velocity at one particular point inside the Sun or Earth. The capture rate can be written as

$$C = \left(\frac{2}{3\pi}\right)^{1/2} M_B \frac{\rho_\chi \bar{v}_\chi}{m_\chi} \sum_i \frac{f_i}{m_i} \sigma_i X_i, \tag{9}$$

where M_B is the mass of the Sun or Earth, ρ_χ and $\bar{v}_\chi$ are the local neutralino density and velocity in the halo respectively, σ_i is the elastic scattering cross section of the neutralino with the nucleus of element i, and X_i is a kinematic factor which can be

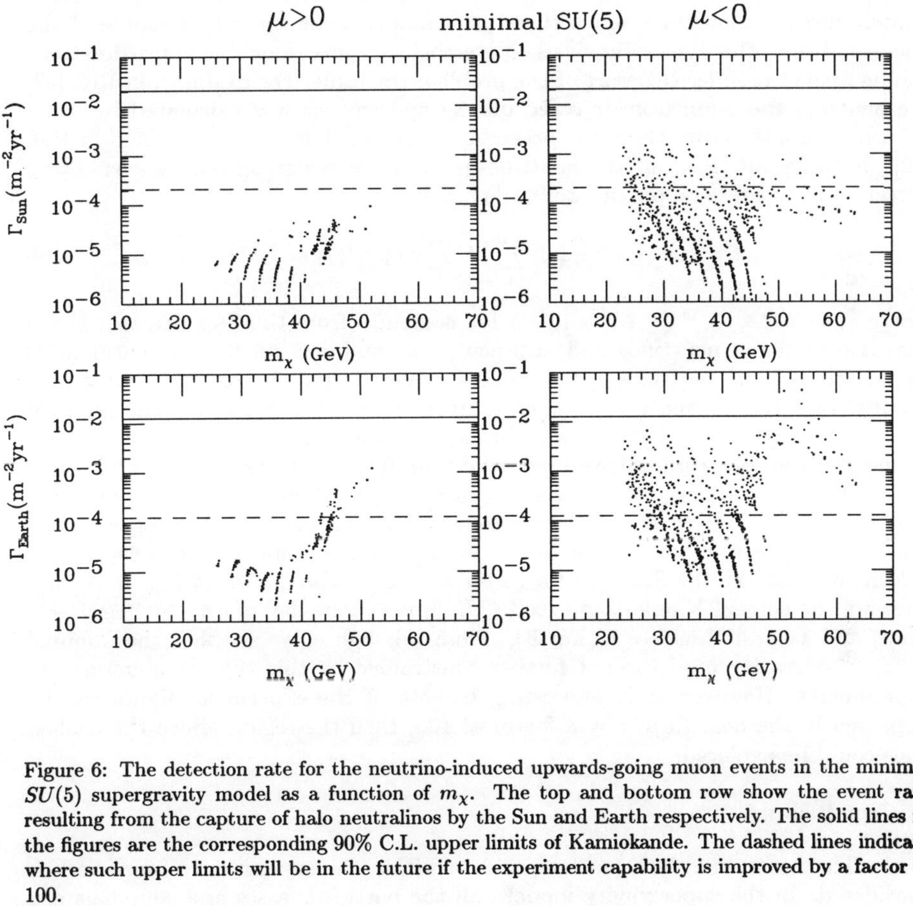

Figure 6: The detection rate for the neutrino-induced upwards-going muon events in the minimal $SU(5)$ supergravity model as a function of m_χ. The top and bottom row show the event rate resulting from the capture of halo neutralinos by the Sun and Earth respectively. The solid lines in the figures are the corresponding 90% C.L. upper limits of Kamiokande. The dashed lines indicate where such upper limits will be in the future if the experiment capability is improved by a factor of 100.

obtained from Eq. (A10) of the second paper in Ref. [45]. The annihilation process normally takes a time scale much shorter than the age of the Sun or Earth to reach equilibrium with the capture process, in this case the neutralino annihilation rate, entering the high-energy neutrino flux, is just half of that given in Eq. (9).

The determination of neutrino flux or final detection rate is normally complicated, basically because the differential energy spectra of neutrinos depend on various subsequent physical processes that take place, see *e.g* Ref. [50]. The rather limited current knowledge about these processes remains the major source of the uncertainties in this type of analysis. Nevertheless, some resonable approximations can be made to render this part of the problem tractable. For example, in Ref. [47] the neutrino spectrum from injected quarks and leptons was calculated by using the Lund Monte Carlo, the same procedure then was followed and refined in Ref. [50]. It turns out that, for the neutrino-induced upward-going muons events, the detection rate can be approximated as [47, 50]

$$\Gamma = \kappa_B(\frac{C}{\mathrm{sec}^{-1}})(\frac{m_\chi}{\mathrm{GeV}})^2 \sum_i a_i b_i \sum_F B_F \langle Nz^2\rangle_{Fi} \ \mathrm{m}^{-2}\mathrm{yr}^{-1}, \tag{10}$$

where $\kappa_B = 1.27 \times 10^{-29}$ ($= 7.11 \times 10^{-21}$) for neutrinos from the Sun (Earth); the i-sum is over muon neutrinos and anti-neutrinos, and $a_i = 6.8$ (3.1), $b_i = 0.51$ (0.67) for neutrinos (anti-neutrinos); the F-sum is over all annihilation final states that produce high-energy neutrinos ($\tau\bar{\tau}$, $c\bar{c}$, $b\bar{b}$, $t\bar{t}$, WW, ZZ, hA and HA in the minimal $SU(5)$ supergravity model.), each with branching ratio B_F, and $\langle Nz^2\rangle_{Fi}$ is the second moment of the spectrum of tpye-i neutrino from final state F scaled by m_χ^2.

In Fig. 6, the predicted detection rate of upward-going muon events in the minimal $SU(5)$ supergravity model prediction is shown as a function of m_χ. The top and bottom row correspond to the event rate resulting from the capture of halo neutralinos by the Sun and Earth respectively. Also shown in Fig. 6 as solid lines are the current Kamionkande 90% C.L. upper limits [54] of $6.6 \times 10^{-14}\mathrm{cm}^{-2}\mathrm{sec}^{-1}$ (Sun) and $4.0 \times 10^{-14}\mathrm{cm}^{-2}\mathrm{sec}^{-1}$ (Earth). From this figure, we see that the minimal $SU(5)$ supergravity model is not further constrained by this type of underground experiments. However, it is interesting to note, if the experiment limits can be improved in the near future by a factor of 100, then the region above the dashed lines could be explored.

4. Conclusions

The neutralino dark matter problem within the supergravity framework is considered. In the supergravity models, all the particle masses and couplings can be specified in terms of three universal soft supersymmetry breaking parameters m_0, $m_{1/2}$ and A at the unification scale M_U, along with low-energy parameters m_t and $\tan\beta$. Also, the electroweak gauge symmetry can be broken radiatively. As a result, the neutralino relic abundance becomes strongly depends upon the way supersymmetry is broken. If the soft supersymmetry breaking seed is primarily a universal scalar mass ($\xi_0 \gg 1.0$), then the relic abundance is almost always much too large, and the cosmological constraint is very strong. The exception occurs if resonances or thresholds are present. On the other hand, if the soft supersymmetry breaking seed is a dominant universal gaugino mass ($\xi_0 \ll 1.0$), the cosmological constraint is rather weak. In the case of $\xi_0 \sim 1.0$ there is normally a wide range of the other parameters where neutralinos can provide a closure relic density $\Omega_\chi \sim 1.0$. In the minimal $SU(5)$ supergravity model, since the dimension-five-induced proton decay also imposes very restrictive constraint which is somewhat in conflict with the cosmological constraint, the parameter space of this model is dramatically reduced, therefore, makes this model more easily testable. Assuming neutralinos constitute the dark matter in the Galactic halo, the upward-going muon events in underground detectors provide yet anther probe to explore supergravity models. However, in order to obtain further useful constraints for the minimal $SU(5)$ supergravity model, the current experimental limits would have to improve by a factor of about 100 in the future.

5. Acknowledgments

It's a pleasure to thank my collaborators R. Gandhi, S. Kelley, J.L. Lopez, D.V. Nanopoulos, H. Pois and A. Zichichi, for their contributions on the subjects covered this talk. This work has been supported by a World-Laboratory Fellowship.

6. References

1. See e.g., H.E. Haber, in these proceedings.
2. See e.g., V. Barger, in these proceedings.
3. D.N. Schramm, in these proceeding.
4. For reviews on dark matter and its detection, see e.g., V. Trimble, *Ann. Rev. Astron. Astrophys.* **25** (1987) 425; J.R. Primack, B. Sadoulet, and D. Seckel, *Ann. Rev. Nucl. Part. Sci.* **38** (1988) 751; D.O. Caldwell, *Mod. Phys. Lett. A* **5** (1990) 1543; P.F. Smith and J.D. Lewin, *Phys. Rep.* **187** (1990) 203.
5. S. Weinberg, *Phys. Rev. Lett.* **50** (1983) 387; H. Goldberg, *Phys. Rev. Lett.* **50** (1983) 1419; L.M. Krauss, *Nucl. Phys.* **B227** (1983) 556.
6. J. Ellis, J.S. Hagelin, D.V. Nanopoulos, K.A. Olive and M. Srednicki, *Nucl. Phys.* **B238** (1984) 453.
7. See e.g., M. Milgrom, *Comm. Astrophys.* **13** (1989) 215; R.H. Sanders, *Astron. Astrophys. Rev.* **2** (1990) 1.
8. S. Wolfram, *Phys. Lett.* **B82** (1979) 65; P.F. Smith and J.R.J. Bennett, *Nucl. Phys.* **B149** (1979) 525.
9. Ya.B. Zeldovich, *Adv. Astron. Astrophys.* **3** (1965) 241; B. Lee and S. Weinberg, *Phys. Rev. Lett.* **39** (1977) 165; M.I. Vysotsky, A.D. Dolgov, and Ya.B. Zeldovich, *JETP Lett* **26** (1977) 188.
10. M.S. Turner, *FERMILAB-Conf-92/313-A*.

11. M. Srednicki, R. Watkins and K.A. Olive, *Nucl. Phys.* **B310** (1988) 693.
12. J. L. Lopez, D. V. Nanopoulos, and K. Yuan, *Nucl. Phys.* **B370** (1992) 445.
13. J. Bernstein, L. Brown, and G. Feinberg, *Phys. Rev.* **D32** (1985) 3261.
14. P. Gondolo and G. Gelmini, *Nucl. Phys.* **B360** (1991) 145.
15. K. Griest and D. Seckel, *Phys. Rev.* **D43** (1991) 3191.
16. K. Griest, *Phys. Rev.* **D38** (1988) 2357; *Phys. Rev.* **D39** (1989) 3802 (E); K. Griest, M. Kamionkowski, and M.S. Turner, *Phys. Rev.* **D41** (1990) 3565.
17. K.A. Olive and M. Srednicki, *Phys. Lett.* **B230** (1989) 78; *Nucl. Phys.* **B355** (1991) 208; J. McDonald, K.A. Olive, and M. Srednicki, *Phys. Lett.* **B283** (1992) 80.
18. For a review, see e.g., H.E. Haber and G.L. Kane, *Phys. Rep.* **117*C*** (1985) 75.
19. For reviews, see e.g., H.P. Nilles, *Phys. Rep.* **110** (1984) 1; A.B. Lahanas and D.V. Nanopoulos, *Phys. Rep.* **145** (1987) 1.
20. For a review, see e.g., L.E. Ibáñez and G.G. Ross, *CERN-TH.6412/92.*
21. J. Ellis and F. Zwirner, *Nucl. Phys.* **B338** (1990) 317.
22. M.M. Nojiri, *Phys. Lett.* **B261** (1991) 76.
23. J. L. Lopez, K. Yuan, and D. V. Nanopoulos *Phys. Lett.* **B267** (1991) 219.
24. J. Ellis and L. Roszkowski, *Phys. Lett.* **B283** (1992) 252.
25. S. Kelley, J. L. Lopez, D. V. Nanopoulos, H. Pois, and K. Yuan, *Phys. Lett.* **B273** (1991) 423.
26. M. Dress and M.M. Nojiri, *Phys. Rev.* **D47** (1993) 376.
27. S. Kelley, J. L. Lopez, D. V. Nanopoulos, H. Pois, and K. Yuan, *Phys. Rev.* **D47** (1993) 2461.
28. R. Roberts and L. Roszkowski, RAL-93-003 and UM-TH-33/92.
29. See e.g., S. Kelley, J. L. Lopez, D. V. Nanopoulos, H. Pois, and K. Yuan, *Nucl. Phys.* **B398** (1993) 3.
30. L.E. Ibáñez and C.López, *Nucl. Phys.* **B233** (1984) 511; C. Kounnas, A. Lahanas, D.V. Nanopoulos, and M. Quiros, *Nucl. Phys.* **B236** (1984) 438.
31. Y. Okada, M. Yamaguchi, and T. Yanagida, *Prog. Theor. Phys.* **85** (1991) 1, *Phys. Lett.* **B262** (1991) 54; J. Ellis, G. Ridolfi, and F. Zwirner, *Phys. Lett.* **B257** (1991) 83; H.E. Haber and R. Hempfling, *Phys. Rev. Lett.* **66** (1991) 1815.
32. A. Chamseddine, R. Arnowitt, and P. Nath, *Phys. Rev. Lett.* **49** (1982) 970.
33. H. Georgi and S.L. Glashow, *Phys. Rev. Lett.* **32** (19438) 74.
34. S. Weinberg, *Phys. Rev.* **D26** (1982) 287; N. Sakai and T. Yanagida, *Nucl. Phys.* **B197** (1982) 533.
35. S. Dimopoulos, S. Raby, and F. Wilczek, *Phys. Lett.* **B112** (1982) 133; J. Ellis, D.V. Nanopoulos, and S. Rudaz, *Nucl. Phys.* **B202** (1982) 43.

36. R. Arnowitt and P. Nath, *Phys. Rev. Lett.* **69** (1992) 725; P. Nath and Arnowitt, *Phys. Lett.* **B287** (1992) 89, *Phys. Lett.* **B289** (1992) 368.
37. M. Matsumoto, J. Arafune, H. Tanaka, and K. Shiraishi, *Phys. Rev.* **D46** (1992) 3966.
38. J.L. Lopez, D.V. Nanopoulos, and H. Pois, *Phys. Rev.* **D47** (1993) 2468.
39. J.L. Lopez, D.V. Nanopoulos, and A. Zichichi, *Phys. Lett.* **B291** (1992) 255.
40. R. Arnowitt and P. Nath, *Phys. Lett.* **B299** (1993) 58 and Erratum; P. Nath and R. Arnowitt, *Phys. Rev. Lett.* **70** (1993) 3696.
41. J.L. Lopez, D. V. Nanopoulos, and K. Yuan, *Texas A&M preprint CTP-TAMU-14/93, to appear in Phys. Rev. D.*
42. J.L. Lopez, D.V. Nanopoulos, H. Pois, and A. Zichichi, *Phys. Lett.* **B299** (1993) 262.
43. R. Gandhi, J.L. Lopez, D.V. Nanopoulos, K. Yuan, and A. Zichichi, *Texas A&M preprint CTP-TAMU-48/93.*
44. W.H. Press and D.N. Spergel, *Astrophys. J.* **296** (1985) 679.
45. A. Gould, *Astrophys. J.* **321** (1987) 560, 571; **328** (1988) 919; **388** (1992) 338.
46. Some earlier papers are: J. Silk, K. Olive, and M. Srednicki, *Phys. Rev. Lett.* **55** (1985) 257; T. Gaisser, G. Steigman, and S. Tilav, *Phys. Rev.* **D34** (1986) 2206; J. Hagelin, K. Ng, and K.A. Olive, *Phys. Lett.* **B180** (1987) 375; M. Srednicki, K. Olive, and J. Silk, *Nucl. Phys.* **B279** (1987) 804; L. Krauss, M. Srednicki, and F. Wilczek, *Phys. Rev.* **D33** (1986) 2079; K. Freese, *Phys. Lett.* **B167** (1986) 295.
47. S. Ritz and D. Seckel, *Nucl. Phys.* **B304** (1988) 877.
48. G.F. Giudice and E. Roulet, *Nucl. Phys.* **B316** (1989) 429.
49. G. Gelmini, P. Gondolo, and E. Roulet, *Nucl. Phys.* **B351** (1991) 623.
50. M. Kamionkowski, *Phys. Rev.* **D44** (1991) 3021.
51. F. Halzen, M. Kamionkowski, and T. Stelzer, *Phys. Rev.* **D45** (1992) 4439.
52. M. Mori *et al.* (Kamiokande Collaboration), *Phys. Lett.* **B270** (1991) 89.
53. A. Bottino, V. de Alfaro, N. Fornengo, G. Mignola and S. Scopel, *Phys. Lett.* **B265** (1991) 57; *Astroparticle Phys.* **1** (1992) 61.
54. M. Mori *et al.* (Kamiokande Collaboration), *Phys. Lett.* **B289** (1992) 463.

Experimental Prospects

SEARCHES FOR SUPERSYMMETRY AT FERMILAB

James T. White
D-Zero Collaboration
Department of Physics, Texas A&M University
College Station, Texas 77843, USA

ABSTRACT

The Fermilab Tevatron, a 1.8 TeV E_{cm} $p\bar{p}$ collider, has the initial state diversity to produce a variety of the sparticles predicted in supersymmetric (SUSY) grand unification models. The collider is currently in operation with two large general purpose detectors (D-Zero and CDF) collecting data at unprecedented rates. An overview of the types of SUSY signatures possible at the Tevatron and review of some of the current work in progress is presented.

1. Introduction

The "Tevatron" at Femilab is a $p\bar{p}$ machine which is currently running at a center-of-mass energy of 1.8 TeV with an average luminosity of $\approx 5 \times 10^{30} cm^{-2}s^{-1}$. Both the CDF and D0 detectors are collecting data and have already accumulated over 10 pb^{-1} of data each. By the end of this running period (April 1992 - May 1993), CDF expects a total integrated luminosity of $\geq 20pb^{-1}$ and D0 expects $\geq 15pb^{-1}$. The current Fermilab plans are to deliver another 100 pb^{-1} from fall 1993 through summer 1994, a few hundred pb^{-1} around 1996 and about $1fb^{-1}$ by the end of the century.

Although the coming supercolliders are expected to settle the question of the existance of supersymmetry (SUSY) once and for all, given the above mentioned integrated luminosity and CM energy, the Fermilab Tevatron actually has a significant reach. In this report, the main SUSY analyses currently in progress are briefly described. Both the number and sophistication of analysis efforts will certainly increase in the future, however.

2. SUSY Production at the Tevatron

At a $p\bar{p}$ collider, it is possible to produce several members of the zoo of expected new sparticles. Some of the reactions that are potentially accessible at the Tevatron include:

- **Gluinos and Squarks:** $\tilde{g}\tilde{g}$, $\tilde{g}\tilde{q}$, $\tilde{q}\tilde{q}$
- **Associated Production:** $\tilde{g}\widetilde{Z}_1$, $\tilde{g}\widetilde{W}_1$, $\tilde{g}\widetilde{Z}_2$, $\tilde{q}\widetilde{Z}_1$, $\tilde{q}\widetilde{W}_1$, $\tilde{q}\widetilde{Z}_2$, ...

- **Charginos and Neutralinos:** $\widetilde{W}_1\widetilde{Z}_1$, $\widetilde{W}_1\widetilde{Z}_2$, $\widetilde{W}_1\widetilde{W}_1$, $\widetilde{Z}_2\widetilde{Z}_1$, $\widetilde{Z}_2\widetilde{Z}_2$

- **Sleptons:** $\tilde{l}\tilde{l}$, $\tilde{l}\tilde{\nu}$

This list is only partially complete. For example, it does not include interesting reactions involving SUSY Higgs particles and reactions in which "R" parity is not conserved. For R conserving models the lightest supersymmetric partner (LSP) is stable and weakly interacting, so signatures for these processes all include missing transverse momentum. Here, all discussions will be based on the assumption that the LSP is the lightest neutralino, $\widetilde{Z}_1$, and is stable. In general then, all signatures involve $m\ jets + n\ leptons + \not{p}_T$.

In supergravity models, the masses of the sparticles are not random, but instead, are strongly correlated and determined by relatively few parameters. This effect greatly reduces the number of- and constrains the nature of possible SUSY signatures. An extremely useful new tool that allows experimenters to address this class of models directly is the ISASUSY [1] program. With this tool, SUSY particles can be generated and decayed within the Minimal Supersymmetric Standard Model (MSSM), allowing optimization of cuts and determination of detector acceptance as a function of the most relavant variables.

An example of the mass relationships (from ISASUSY) in the MSSM for a selection of sparticles is shown in Fig. 1. Shown is the correlation between $\widetilde{W}_1$ - $\widetilde{Z}_1$, $\widetilde{W}_1$ - $\widetilde{Z}_2$, and $\tilde{g}$ -$\widetilde{W}_1$. For specific models the mass correlations can be even tighter for the parameter space allowed once experimental constraints are imposed. For example, Fig. 2 shows the allowed correlations between the gluino mass and the $\widetilde{W}_1$ and $\widetilde{Z}_1$ masses for minimal SU(5) supergravity [2]. Similarly, Fig. 3 shows the allowed mass relationships between the χ_1^+ ($\widetilde{W}_1$), χ_1^0 ($\widetilde{Z}_1$) and χ_2^0 ($\widetilde{Z}_2$) in the flipped SU(5) model [3].

For tests of these "minimal" models, such strong correlations translate into well defined spectra for leptons, jets and missing transverse momentum. One cannot arbitrarily assign the mass of the LSP to be zero, which, in turn, would greatly distort, for example, the $\not{p}_T$ spectrum relative to a "model allowed" set of parameters. This greatly simplifies experimental MSSM searches and allows for greater sensitivity.

3. Current Analyses

3.1 Gluinos and Squarks $\rightarrow n\ jets + \not{p}_T$

Gluinos and squarks are produced strongly in $p\bar{p}$ colliders. Historically, because of this, the standard $\not{p}_T$ SUSY search has received the greatest attention to date at the Tevatron. The idea is that for "light" gluinos and squarks, the dominant decays are $\tilde{g} \rightarrow q\bar{q}, \widetilde{Z}_1$, and $\tilde{q} \rightarrow q, \widetilde{Z}_1$. Hence, the most promising signature is n jets + $\not{p}_T$, where the missing pt is reasonably large and n is ≥ 2 for $\tilde{q}\bar{\tilde{q}}$, $\tilde{g}\tilde{q}$, $\tilde{g}\tilde{g}$. It was realized [4], however, that once the sparticles become heavy ($\geq\approx$ 150 GeV), cascade decays involving heavier neutralinos and charginos become important. A plot [5] of

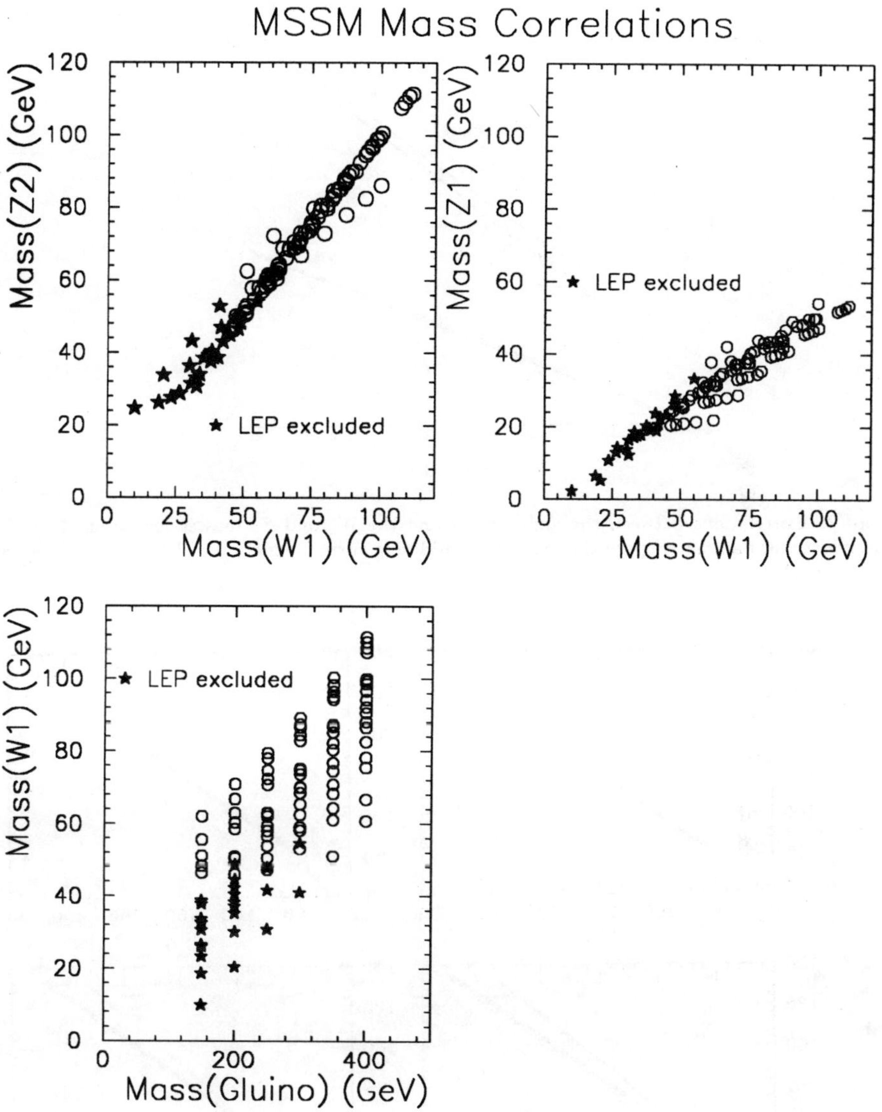

Figure 1: Correlations between a) $\widetilde{W}_1$ and $\widetilde{Z}_1$, b) $\widetilde{W}_1$ and $\widetilde{Z}_1$ and c) $\tilde{g}$ and $\widetilde{W}_1$ from the ISASUSY program. Values shown are for $-1000 \leq \mu \leq 1000$ in steps of 100, and $\tan\beta = 2$ and 10. "Star's" indicate points already excluded by experiment.

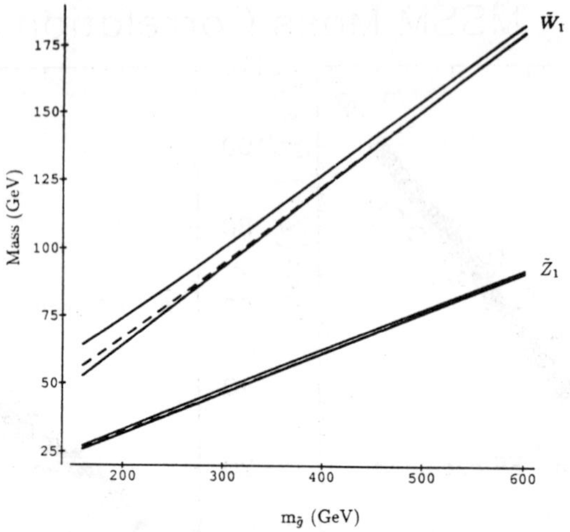

Figure 2: Correlations between the gluino mass and the $\widetilde{W}_1$ and $\widetilde{Z}_1$ masses for minimal SU(5) supergravity for parameter space allowed after considering current experimental constraints. (figure 3 from ref. 2)

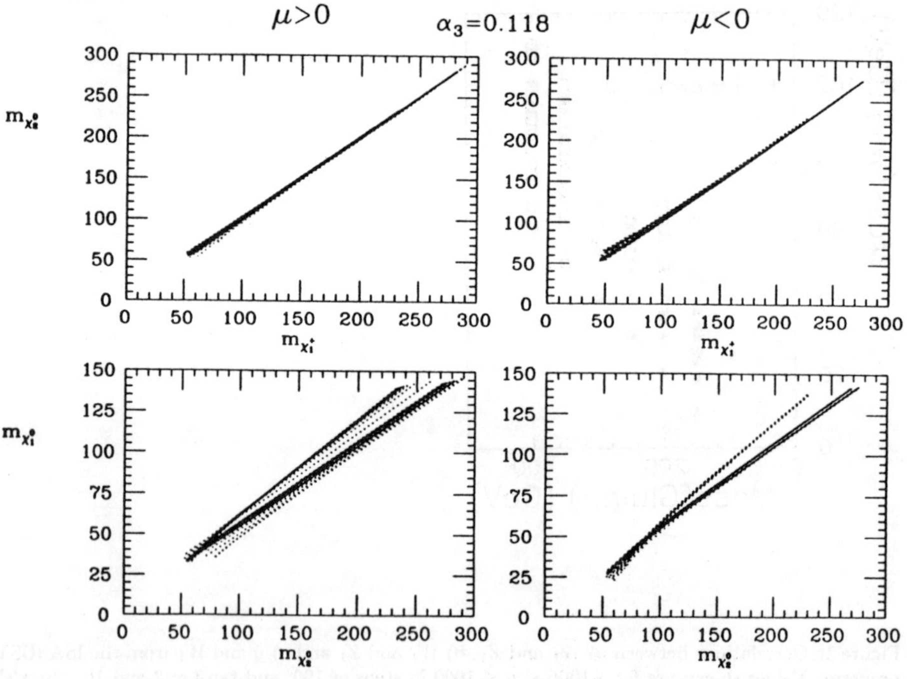

Figure 3: Allowed correlations between the χ_1^+ ($\widetilde{W}_1$), χ_1^0 ($\widetilde{Z}_1$) and χ_2^0 ($\widetilde{Z}_2$) in the the flipped SU(5) supergravity model. (figure 5 from ref. 3)

the cross sections as a function of gluino mass for n jets, n=1 thru 7, is shown in Fig. 4. Here the curves are the "accepted" cross sections after applying "CDF-like" cuts as explained in the reference. It is difficult to estimate the experimental reach of the Tevatron from these curves, since backgrounds from QCD and reactions such as $Z^0 \to \nu\nu + jets$ must be considered with great care. Also, the signal depends strongly on the masses of other sparticles including the squarks and other gauginos. Results from the current run are expected soon, but the most recent public result is the CDF limit shown in Fig. 5. Shown is the mass limit [6] in the $\tilde{g}\ \tilde{q}$ plane based on $\approx 4pb^{-1}$ data from the 1988-89 run. With the full data set from the current run, and with the similar analysis from DZero, one can expect a significant increase in excluded parameter space, if not discovery.

3.2 Gluinos and Squarks $\to leptons + jets + \not{p}_T$

Since the decay products of the squarks and gluinos can include charginos and neutralinos, signatures involving leptons are also expected. Figure 6 shows the branching fractions of the gluino [7] as a function of mass for a region of parameter space. For decays involving single leptons, large backgrounds are expected from sources such as $W \to l\nu + jets$. A result from a study [5] of dileptons + jets is shown in Fig. 7. For the SUSY parameter space chosen here, the rates are not large and the study showed that the same-sign dilepton signal seems most promising.

It is interesting to note, however, that there are regions of parameter space where the signal can be much larger. In fact, if nature were to so choose, the dilepton signal could be quite similar in both rate and appearance to that expected from $t\bar{t}$ decays!

This interesting scenerio can occur under the following conditions. First, if the gluino is not too heavy ($<\approx 300 \to 350$ GeV), then production of u and d squarks via a t-channel gluino exchange between the valence quarks of the $p\bar{p}$ pair can dominate over s-channel production. For reasonably light $\widetilde{W}_1, \widetilde{Z}_2$ the left and right u, d squarks can decay with $U_L, D_L \to \widetilde{Z}_1 + u, d$ ($\approx 5\%$), $\to \widetilde{Z}_2 + u, d$ ($\approx 30\%$), $\to \widetilde{W}_1 + u, d$ ($\approx 65\%$), and $U_R, D_R \to \widetilde{Z}_1 + u, d$ ($\approx 100\%$). Next, if the sleptons are light compared to the squarks, the $\widetilde{W}_1$ and $\widetilde{Z}_2$ can decay with large branching fractions to leptons. For a specific example, consider the case where $M_{squark} = 220$ GeV, $M_{gluino} = 230$ GeV, $M_{slepton} = 75$ GeV, $M_{W_1} = 70$ GeV. In this case, $\sigma * BF \to e + \mu + 2\ jets + \not{p}_T \approx 0.1 \to 0.2$ pb. This is comparable to the rate expected for a 140 GeV top quark. Figure 8 shows a comparison of the lepton and jet pt distributions and the missing pt distribtuion for the 220 GeV Squark vs the 140 GeV top quark. Notice that the lepton tends to have a larger pt for the top quark while the situation is reversed for the jets. This is because for the top, the leptons come from $W \to l\nu$, a two body decay, while for the squark, the lepton comes mostly from $\widetilde{W}_1 \to l\nu\widetilde{Z}_1$. The squark jets have higher pt because the squark is just heavier than the top quark, while the $\widetilde{W}_1$ has a mass similar to the W. It is also interesting that this "scenerio" happens to fit quite well into the parameter space allowed in the flipped SU(5) type models [3].

It is curious that both the first "top candidate" presented by CDF (at DPF'92), and the first presented by D-Zero (Aspen, Jan. '93) have a jet with

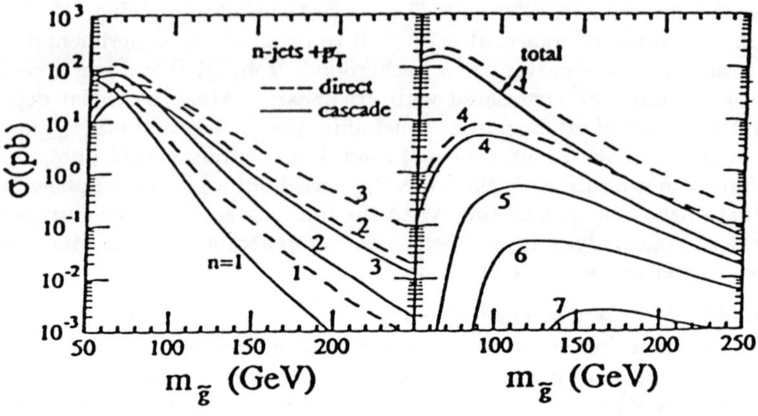

Figure 4: Cross sections for 0 leptons + n jets + $\not{p}_T$ from gluino pair production at the Tevatron with (solid) and without (dashed) cascade decays. (Figure 3 in ref.5)

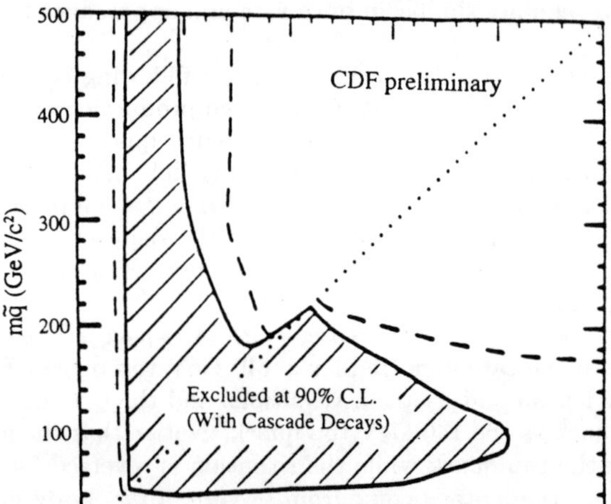

Figure 5: Excluded region in the squark-gluino mass plane including cascade decays for a selection of SUSY parameters.

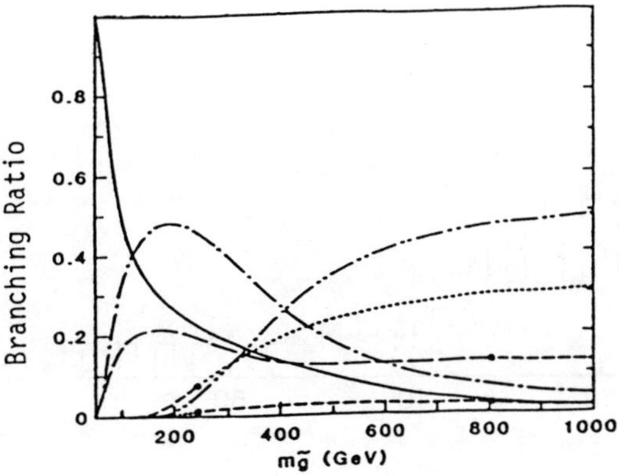

Figure 6: The branching fractions for the gluino as a function of gluino mass for the case $m_{\tilde{q}} = 2m_{\tilde{g}}$, $\tan\beta = 0.4$, and $2m_1 = M_w$.(figure 3 of ref. 7)

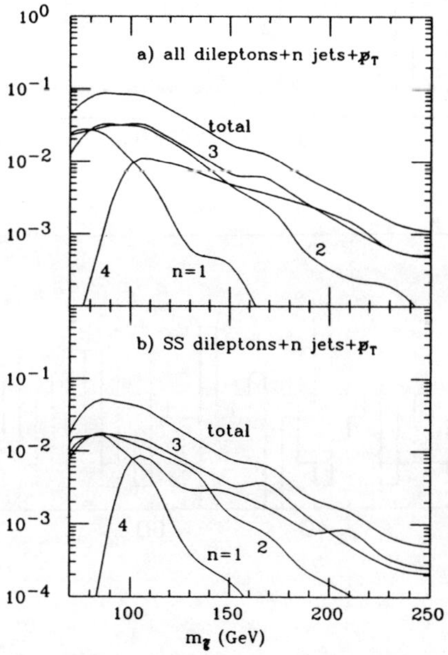

Figure 7: Dilepton plus n jets + $\not{p}_T$ cross sections from gluino pair production at the Tevatron as a function of gluino mass.(figure 9 of ref. 5)

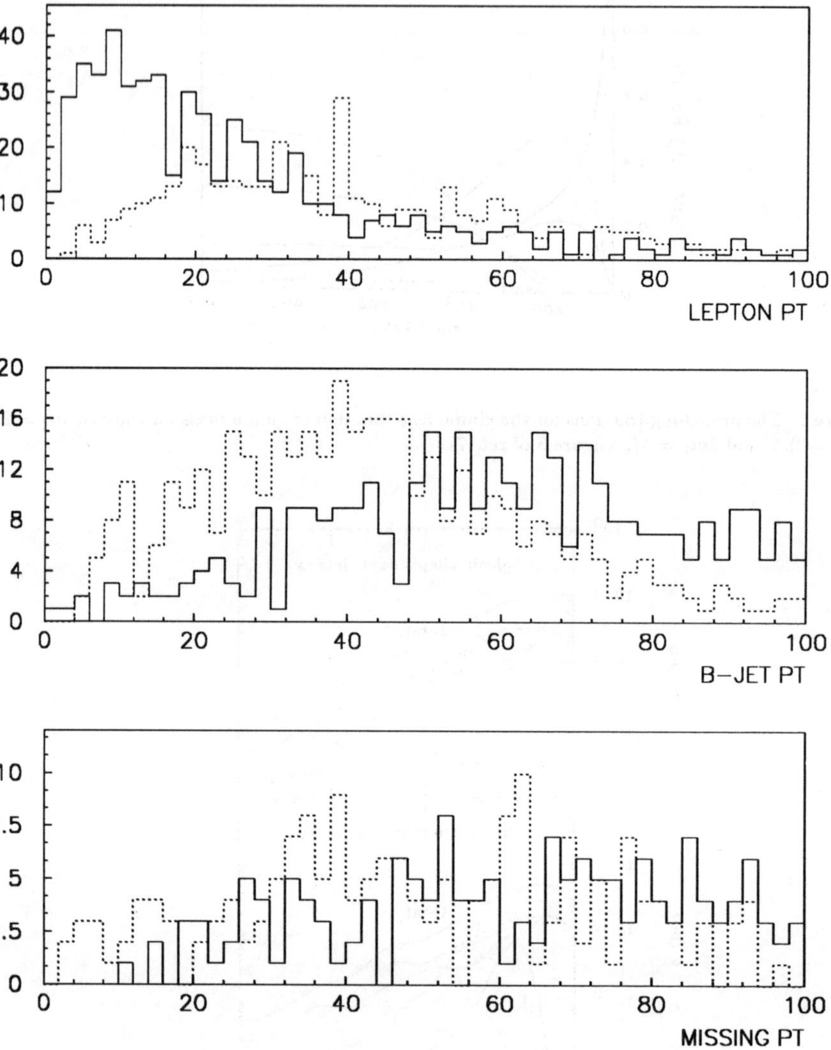

Figure 8: Comparison of pt spectra from 140 GeV $t\bar{t} \to e\mu + 2\ jets + \not{p}_T$ decays (dotted lines) and 220 GeV squarks (solid).

pt around 100 GeV and fit the expections from squark decays better than the top! These leptonic decay channels are now being fully investigated at the Tevatron and results should appear soon.

3.3 Associated Production

Reactions in which squarks or gluinos are produced in association with charginos and neutralinos would result in a large number of interesting signals involving the usual leptons + jets + missing pt. Again, standard model backgrounds are expected from the W + jets and Z + jets reactions for decays involving leptons, but many decays result in signals not too different from the squark-gluino combinations. Consequently, these reactions are currently being lumped in with the gluino-squark analyses.

3.4 Charginos and Neutralinos

The lightest chargino, $\widetilde{W}_1$, and second lightest neutralino, $\widetilde{Z}_2$, can be produced at the Tevatron, but the reactions involve "weak" couplings. Because of this, little attention was paid to these objects at the Tevatron until recently. They were searched for at CERN as decay products of the newly discovered W and Z bosons, and it was realized early [8] that one of the possible signals was a trilepton mode from the decays $\widetilde{W}_1 \to l\nu\widetilde{Z}_1$ plus $\widetilde{Z}_2 \to l\bar{l}\widetilde{Z}_1$. It was first pointed out by Arnowitt and Nath [9], however, that they can also be produced with reasonable efficiency at the Tevatron via offshell W's and Z's and through other mechanisms. An example [10] of the production cross sections for several chargino-neutralino combinations at the Tevatron is shown in Fig. 9 as a function of the parameter μ. The $\widetilde{W}_1, \widetilde{Z}_2$ rates are the largest. The $\widetilde{W}_1, \widetilde{W}_1 \to$ dileptons + missing pt + no jets reaction is also promising, but standard model backgrounds may be fairly large.

Based on the suggestions in reference 10, studies within D-Zero [11] considering more realistic trigger thresholds, analysis cuts and backgrounds confirmed that the suggested trilepton signal indeed provided a new window of opportunity for observing SUSY at the Tevatron. It turns out that, because of the fact that the masses of the the charginos and neutralinos are strongly correlated with, and lighter than, the gluino mass (in supergravity models), the experimental reach in SUSY parameter space is at least as great for chargino-neutralino searches as for squark-gluino searches. A burst of recent theoretical work [12, 10, 13] has ignited a great interest in this approach. It is now one of the major new analyses at the Tevatron!

The $\widetilde{W}_1, \widetilde{Z}_2$ total cross section (from ISASUSY) as a function of the mass of the $\widetilde{W}_1$ is show in Fig. 10 for a range of SUSY parameters, and can be seen to exceed one pb for masses up to $\approx$ 100 GeV. The branching fraction to trileptons, however, is a strong function of other SUSY parameters. The dominant decays of the $\widetilde{W}_1$ include the intermediate states $\widetilde{W}_1 \to W^*\widetilde{Z}_1$, $l\tilde{\nu}$, $\nu\tilde{l}$, and $q\tilde{\bar{q}}$, where $\tilde{\nu} \to \nu\widetilde{Z}_1$, $\tilde{l} \to l\widetilde{Z}_1$ and $\tilde{q} \to q\widetilde{Z}_1$. Similarly, $\widetilde{Z}_2 \to Z^*\widetilde{Z}_1$, $l\tilde{l}$, $\nu\tilde{\nu}$, and $q\tilde{\bar{q}}$. Because the intermediate states involve $\tilde{l}, \tilde{\nu}$ and $\tilde{q}$, the branching fractions depend strongly on the relative masses of these objects.

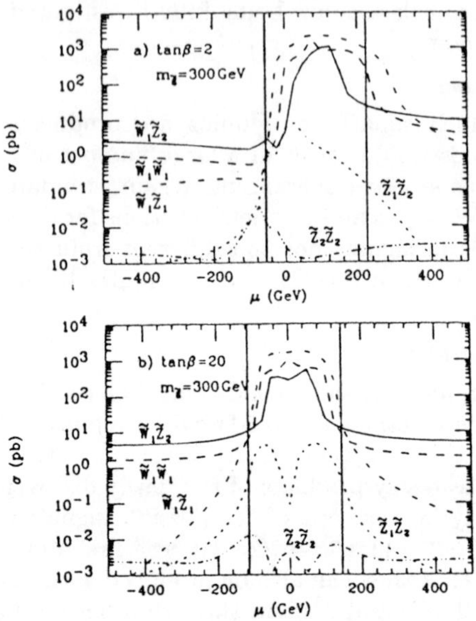

Figure 9: Production cross sections for chargino-neutralino combinations as a function of μ for the case tanβ = 2 and $M_{\tilde{g}}$ = 300 GeV (fig. 1 from ref. 10).

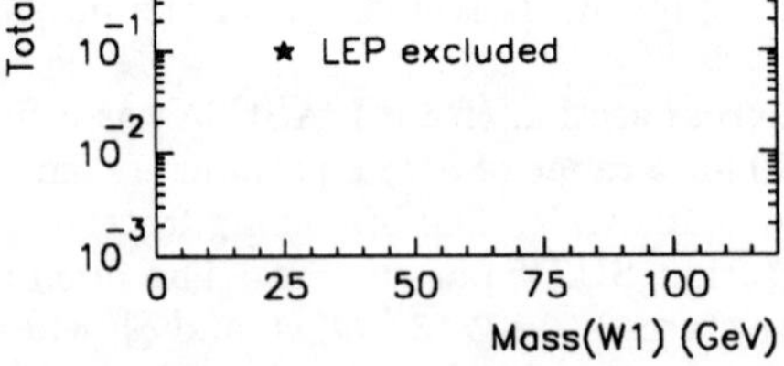

Figure 10: Production cross section for $\widetilde{W}_1\widetilde{Z}_2$ as a function of $M_{\widetilde{W}_1}$ for a variety of SUSY parameters.

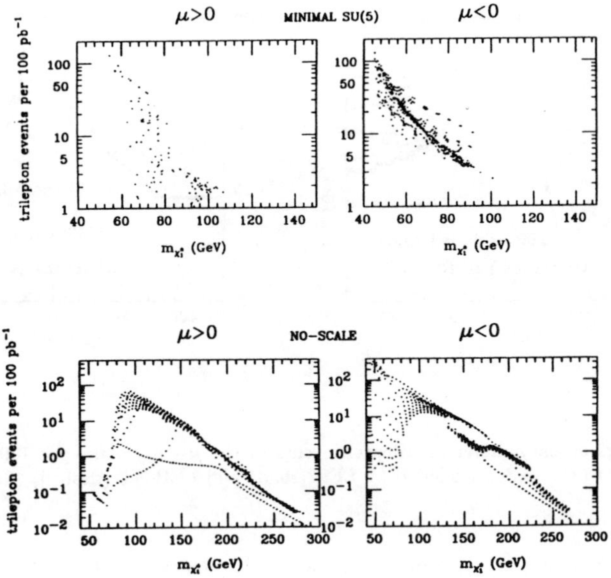

Figure 11: Plots of $\sigma \times BF \rightarrow$ trileptons in minimal and no-scale flipped SU(5) supergravity for phase space remaining after taking account of experimental limits (from ref. 13).

Grand unification models include two extremes. The minimal SU(5) model favors heavy sleptons and squarks [2], so the branching fractions are expected to be driven by the decay modes involving the virtual W and Z intermediate states, (leading to standard model brancing fractions). The flipped SU(5) models, however, favor light sleptons [3] relative to the squarks, so leptonic decays can dominate, approaching 100%. (The $\tilde{\nu}$, $\tilde{\nu}$ mode of the $\widetilde{Z}_2$ can render it invisible, however.) An example [13] of the $\sigma \times BF \rightarrow$ trileptons for allowed parameter space for the minimal SU(5) and the flipped SU(5) model is shown in Fig 11. With efficiencies for observing three clean, isolated leptons expected to vary from about 5% to over 20% from the low to high mass regions, it is clear that the Tevatron will have a significant reach if nature has chosen this model. Clearly, with 100 pb^{-1} accumulated data expected before the turn-on of LEP-200, and up 1 fb^{-1} before the supercollider era, the trilepton signal alone will allow the Tevatron to take a first look at a significant region of SUSY parameter space. For example, a plot giving an indication the reach of the Tevatron relative to LEP [10] is shown in Fig. 12.

An example of a trilepton candidate from D-Zero is shown in Fig 13. This "three electron" candidate is shown from an end view, and has three objects identified as possible electrons plus a fairly large missing pt. One must keep in mind that it is possible for jets and photons to fake electrons, so one cannot claim with

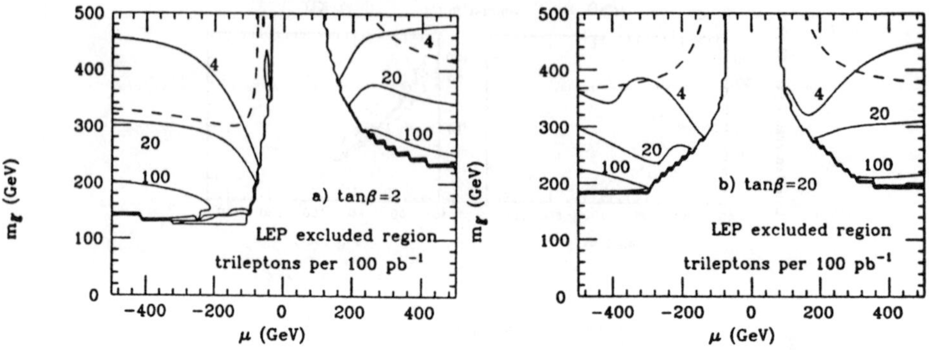

Figure 12: Contours of the number of trilepton events in the μ - $m_{\tilde{g}}$ plane per 100 pb^{-1} at the Tevatron compared to the region excluded by LEP (solid) and LEP-200 (dashed) (figure 4 of ref. 10).

100% certainty that the three objects are actually electrons. In fact, two the objects are high quality electron candidates, but the third is slightly questionable since hadronic activity is near by. Still, the event is interesting since the missing Pt is well measured, and is of the scale one would expect from a W event. A possible explanation from a known, but low rate background is W $\rightarrow e\nu$ + Drell-Yan $\rightarrow e\bar{e}$. No two electrons are near the Z mass, but CDF has a similar event in which two of the electrons do form a Z^0, and hence is thought to be a WZ event.

4. Sleptons

The final item on the list of "possibly observable" reactions listed above is the pair production of sleptons. A plot [14] of the production cross section of $\tilde{e}, \tilde{e}$ as a function of mass of the slepton is shown in Fig. 14. The signal from this reaction would be two leptons plus missing pt. Unfortunately, the cross section is quit small and would likely be hidden by standard model backgrounds such as $Z^0 \rightarrow \tau\bar{\tau}$ and WW production. One must conclude that direct observation of these particles is not possible at the Tevatron.

4. Summary

In conclusion, the two detectors at the Tevatron, CDF and D-Zero, are expected to record over 100 pb^{-1} combined by mid 1994. With such a data set, the Tevatron can explore a reasonably healthy chunk of SUSY parameter space via both gluino-squark decays and chargino-neutralino decays. The major portion of analysis

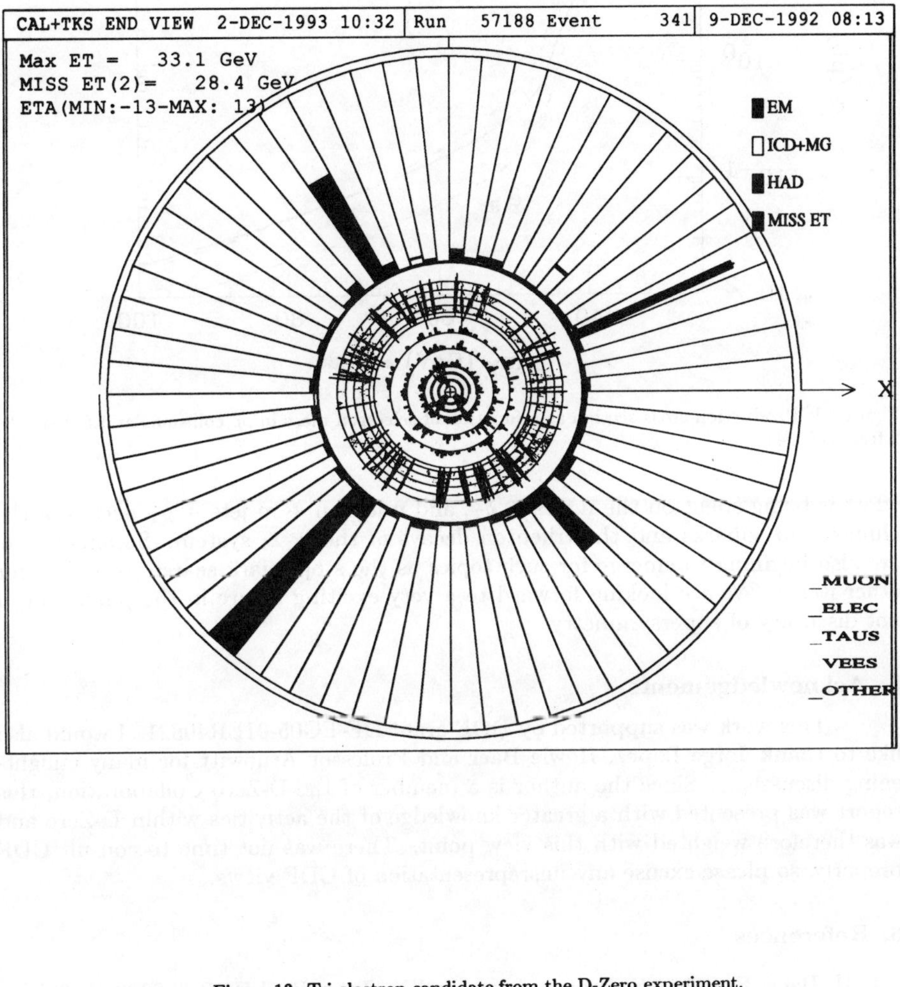

Figure 13: Tri-electron candidate from the D-Zero experiment.

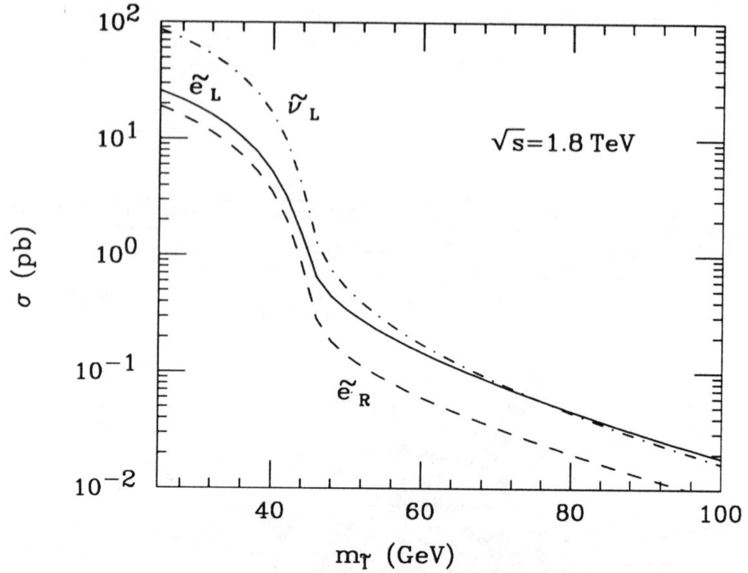

Figure 14: Production cross section as a function of selectron mass in $p\bar{p}$ collisions at 1.8 TeV (fig. 2 from ref. 14).

effort is being spent on the n jets + $\not{p}_T$, and n lepton + m jets + $\not{p}_T$ decays of the gluinos and squarks and the trilepton decays of the $\widetilde{W}_1\widetilde{Z}_2$ system. Serious efforts are also begining to emerge for such topics as the stop quark search as well as for other items. We are looking forward to a very exciting future in the pursuit of, if not discovery of supersymmetry.

5. Acknowledgements

This work was supported by DOE grant DE-FG05-91ER40621. I would also like to thank Jorge Lopez, Howie Baer and Professor Arnowitt for many enlightening discussions. Since the author is a member of the D-Zero Collaboration, this report was presented with a greater knowledge of the activities within D-Zero and was therefore weighted with this view point. There was not time to consult CDF properly, so please excuse any misrepresentation of CDF views.

6. References

1. H. Baer, F. Paige, S. Protopopescu and X. Tata, FSU-HEP-930329 (1993).
2. Arnowitt and Nath, Phys. Rev. Lett. **69**, 725 (1992).
3. J. Lopez, D. Nanopoulos and A. Zichichi, CTP-TAMU-68/92 (1992) (to appear in Phys. Rev. D).
4. H. Baer, X. Tata and J. Woodside, Phys. Rev. **D44**, 207 (1991).
5. H. Baer, X. Tata and J. Woodside, Phys. Rev. **D41**, 906 (1990).
6. CDF Collaboration (F. Abe *et al.*), Phys. Rev. Lett. **69**, 3439 (1992).

7. See, for example, Lectures by X. Tata, Mt. Sorak Symposium on the Standard Model and Beyond, Mt. Sorak, (1990).
8. J. Ellis, J. Hagelin, D. V. Nanopoulos, and M. Srednicki, Phys. Lett. B **127**, 233 (1983).
9. P. Nath and R. Arnowitt, Mod. Phys. Lett. **A2**, 331 (1987).
10. H. Baer and X. Tata, Phys. Rev. **D47**, 2739 (1993).
11. J.T. White, D-Zero Note 1205, McPhysics II, Sept. (1991).
12. R. Barbieri et. al., Nucl. Phys. **B367** 28 (1991).
13. J. Lopez, D. Nanopoulos, X. Wang and A. Zichichi, Phys. Rev. **D48**, 2062 (1993).
14. H. Baer, C. Chen, F. Paige, and X. Tata, FSU-HEP-931104.

Closing Remarks

THE INFN ELOISATRON PROJECT

A. Zichichi
CERN, Geneva, Switzerland

1. Why 200 TeV Now
2. The Steps towards ELN
3. The ELN Conceptual Design
4. Results on Energy and Luminosity Limits
5. Superconducting Dipole Magnet Studies
6. R&D for Detectors: the LAA Results
7. Physics Scenarios and MonteCarlo Simulation Studies
8. Future Plans for Theoretical Studies
9. Conclusions
10. Acknowledgements

INFN ELOISATRON PROJECT

A. ZICHICHI

1. Why 200 TeV Now

A possible title of this report could be "Why 200 TeV now". There are two possible scenarios; one I will define PESSIMISTIC, the other OPTIMISTIC.

1) Pessimistic: LEP, HERA, LHC, SCC will prove the validity of the Standard Model at the highest possible energy and accuracy. In this case Subnuclear Physics will not survive. Therefore a group is necessary to think about the highest possible energy and the utmost luminosity now.

2) Optimistic: LEP, HERA, LHC, SCC will provide great discoveries. In this case the energy jump would be even more compulsory. But the implementation of a (100+100) TeV collider needs at least ten years.

Conclusion: in any case - be the truth on the pessimistic or optimistic side - we need now to think seriously, - i.e. devoting a non-negligible fraction of our time - about the highest possible levels of energy and luminosity.

2. The Steps towards ELN

The Eloisatron Project (ELN) started in 1979. The basic steps are shown in Fig. 1. We have been able to build, in Italy the Superconducting Cyclotron and 50% of the HERA superconducting dipole magnets(*) for the 800 GeV proton ring.

An intensive R&D project to study new subnuclear technologies has been implemented at CERN.

These basic ELN steps have been the merging point for an effective collaboration between various Universities, Institutions and Industries, as reported in Fig. 2.

(*) The original proposal was 100%, but German industries did not allow DESY to accept 100%.

3. The ELN Conceptual Design

The conceptual design of ELN has been performed by K. Johnsen and collaborators. This study shows that there are no conceptual difficulties in the realisation of a (100+100) TeV collider able to work with a luminosity $L = (0.9 \div 1.8) \times 10^{33}\ cm^{-2}s^{-1}$. The basic and lattice parameters of the collider are shown in Fig. 3, and the collider layout in Fig. 4. The 100 TeV ring is working as a 10 TeV injector to the same ring. The circumference is 300 KM and the intersections are (3+3) on two opposite sides.

4. Results on Energy and Luminosity Limits

Our "machine" group has reached the following theoretical results:

- A collider designed for a luminosity L_1 cannot be operated at $100 \times L_1$ without spending a large fraction of the original cost of the main rings of the collider. Capability for high luminosity operation must be part of the initial design of a supercollider.
- To be specific:
  - The upgrade potential of SSC designed for $L=10^{33}\ cm^{-2}s^{-1}$ will be strongly limited by existing design choices.
  - The central beam chamber of ELN should be designed for $10^{35}\ cm^{-2}s^{-1}$.
- The 200 TeV (ELN collider) is practical with existing accelerator technology.
- A theoretical study of the maximum energy and luminosity levels attainable yields

$$E_{MAX} = (500 + 500)\ TeV$$

and

$$L_{MAX} = 10^{36}\ cm^{-2}s^{-1}$$

These ultimate ELN features are shown in Fig. 5 where all other colliders are included for a comparative analysis.

5. Superconducting Dipole Magnet Studies

The initial design studies of a dipole magnet for the ELN collider have been made. The design goal is to reach a central field of

13.5 Tesla at 4.2 K

Using Nb_3Sn superconducting material and cables, it is possible to reach 13.5 T with a quite good field uniformity:

$$(5.8\div13.5)\ \mathrm{T},\ \frac{\Delta B}{B_o} \leq 10^{-4}$$

$$(3\div5.8)\ \mathrm{T},\ \frac{\Delta B}{B_o} \leq (9\div1)\times10^{-4}$$

A dipole design is shown in Fig. 6 for Scheme 3 with

4-layer coils

By choosing an average current density of

250 A/mm^2 at 14.28 T

it is possible to get the needed Nb_3Sn cables with existing technique.

6. R&D for Detectors: the LAA Results

Let me first mention the LAA basic points:

1.	RADIATION HARDNESS	•••
2.	HERMETICITY	•••
3.	RATE CAPABILITY	•••
4.	MOMENTUM RESOLUTION	•
5.	ENERGY RESOLUTION	•
6.	TRACK & SPACE RESOLUTION	•
7.	TIME RESOLUTION	•
8.	PARTICLE IDENTIFICATION	•

Note that hermeticity is an essential feature as the discovery potential of a detector lies essentially in its capability to see "missing" momenta. A (4π) detector has never been built. The challenge to study how to build a detector without holes in the 4π solid angle coverage, able to handle high rates and be radiation hard, has been our main (R&D) task during the last four years.

A synthesis of the main results is shown in Fig. 7 where the crucial components needed for a (4π) detector working in the (20-200) TeV energy range are indicated, together with the number of important results obtained in each component. It is impossible to report telegraphically and in a non-boring way the work of 200 specialists during four years.

Let me just say that is thanks to this effort that many problems connected with the possibility of doing physics in the multi-TeV energy range, with high luminosity and rates have been solved. If it were not for this component of the ELN project, i.e. LAA, we would not be in a position to compete - here in Europe - with our American colleagues. In fact we are - in many areas of technological detector developments - ahead of them.

7. Physics Scenarios and MonteCarlo Simulation Studies

The bulk of soft processes (i.e. minimum bias events) in hadron-hadron collisions is interesting from different viewpoints:

- "leading" hadron and effective energy for multiparticle production;
- dynamics of strong interactions in the nonperturbative domain;
- overwhelming "background" in the search for new Physics phenomena.

The present theoretical situation is paradoxical.

Rare processes (high-p_T jets, W/Z, etc.) are by far easier to describe than low momentum transfer processes which do not have so far a correct treatment.

The description of soft processes is considerably complex since two colliding hadrons are:

- composite;
- extended.

PYTHIA (perturbative) and QGSM (non-perturbative) models have been compared in pp and $p\bar{p}$ inelastic collisions at :

$\sqrt{s}$ = 62 GeV [ISR]
$\sqrt{s}$ = 630 GeV [$Sp\bar{p}S$]
$\sqrt{s}$ = 1.8 TeV [Tevatron]
$\sqrt{s}$ = 16 TeV [LHC]
$\sqrt{s}$ = 40 TeV [SSC]
$\sqrt{s}$ = 200 TeV [ELN]

The comparison has been done in terms of inclusive distributions of multiplicity and other variables, such as x_F, rapidity, etc., for different secondary hadrons, in particular for light and heavy baryons.

We have focused our simulations on Heavy Higgs search at [LHC (16), SSC (40) and ELN (200)] TeV energies.

Heavy Higgs Search at $\sqrt{s} = 16, 40, 200$ TeV ($M_H > 2\,M_Z$)

The main production mechanisms investigated are:

$gg \to H^o$ $W^+W^- \to H^o$

The results are reported in Fig. 8. How a heavy Higgs can be observed in a real (LAA-like) set-up has been extensively studied. Let me show the results, in terms of signal-to-background, for the case of heavy Higgs production, $M_H = 1$ TeV ($m_t = 130$ GeV), at three total c.m. energies: $\sqrt{s} = (16, 40)$ TeV (Fig. 9) and $\sqrt{s} = 200$ TeV (Fig. 10).

If a Heavy Higgs is there, there is no question that ELN is the best collider.

8. Future Plans for Theoretical Studies

To think about problems arising in the multi-hundred TeV energy range is a crucial part of the ELN project. If there was no physics in the extreme energy range, there would be no point to promote ELN.

This, as you know, is not the case. In fact we have every reason - it has never been so - to believe that the higher the energy, the more we will learn.

The present plans are as follows:

- Phenomenologically, the signature and possible background for non-perturbative electroweak processes deserve further studies.
- The question whether B- and L-violation can be disentangled in such a multiparticle process is far from being solved.
- The possibility of large σ for non-pertubative electroweak B-and L-violating processes needs further work. The final goal is to compute the total cross-section for B-violation to exponential accuracy at arbitrary energies.
- Another interesting problem is to understand whether similar non-perturbative interactions also contribute significantly to total cross sections in QCD.

9. Conclusions

In order to be sure that Subnuclear Physics will stay in the forefront of scientific research, we need to study now the multi-hundred TeV physics in terms of theoretical goals, machine problems and new detector technologies. Fig. 11 shows the present status of the ELN working group structure.

Acknowledgements

To all my collaborators (Fig. 12a)b)c)), I would like to express my deep gratitude.

FIGURE CAPTIONS

Fig. 1.	Basic steps of ELN.
Fig. 2.	Universities, Institutions and Industries collaborating with ELN.
Fig. 3.	Basic and lattice parameters of ELN.
Fig. 4.	ELN Collider layout.
Fig. 5	The ultimate energy and luminosity of ELN, compared with all other colliders in the world.
Fig. 6.	An example of detailed dipole design.
Fig. 7.	R&D for new detector technologies (LAA). The numbers of main results (D,R,N, I) are shown for each component.
Fig. 8	Higgs production cross-section (pb) (versus Higgs mass) at $\sqrt{s}$ = (16, 40) TeV following two different production processes: (gg) and (WW) fusion.
Fig. 9	Heavy Higgs: signal/background at $\sqrt{s}$ = (16, 40) TeV.
Fig. 10	Heavy Higgs: signal/background at $\sqrt{s}$ = 200 TeV.
Fig. 11	Working group structure.
Fig. 12a)b)c)	List of collaborators.

Steps towards ELOISATRON

◊ *1979* - *CS Project (Superconducting Cyclotron) resumed.*

◊ *1981* - *Start of the CS construction (Ansaldo - LMI - Zanon).*
The HERA Project is presented to the INFN Council.

◊ *1982* - *First Meeting of the HERA designers with Ansaldo - LMI - Zanon.*

◊ *1985* - *The construction in Italy of the superconducting magnets (prototypes) for HERA begins.*

◊ *1986* - *R&D for Detectors - LAA Project is approved and financed.*

◊ *1987* - *R&D for LSCM (Long SuperConducting Magnet) prototypes - 10% ELOISATRON Model - starts at Ansaldo.*

◊ *1989* - *The construction of the HERA magnets is completed.*

◊ *1990* - *Magnet construction for the 10% ELOISATRON Model is expected to start.*

◊ ☐ - *The Full Scale ELOISATRON PROJECT:*
- *1st Conceptual Design.*
- *2nd R & D in fundamental technologies.*
- *3rd Construction of the Superconducting Long Magnets.*

Figure 1

The following Universities, Institutions and Industries are collaborating with the various components of the ELN Project:

Universities (20):

UNIVERSITY OF BOLOGNA, BOLOGNA, ITALY
UNIVERSITY OF CALABRIA, CALABRIA, ITALY
MIT, CAMBRIDGE, MASS., USA
UNIVERSITY OF COIMBRA, COIMBRA, PORTUGAL
UNIVERSITY OF FIRENZE, FIRENZE, ITALY
UNIVERSITY OF GLASGOW, GLASGOW, UNITED KINGDOM
UNIVERSITY OF KARLSRUHE, KARLSRUHE, FRG
EPFL, LAUSANNE, SWITZERLAND
COLUMBIA UNIVERSITY, NEW YORK, USA
CORNELL UNIVERSITY, NEW YORK, ITHACA, USA
UNIVERSITY OF PALERMO, PALERMO, ITALY
COLLEGE DE FRANCE, PARIS, FRANCE
UNIVERSITY OF PARIS LPNHE, PARIS, FRANCE
UNIVERSITY OF PERUGIA, PERUGIA, ITALY
UNIVERSITY OF RIGA, RIGA, LATVIA
UNIVERSITY OF CALIFORNIA, SAN DIEGO, USA
UNIVERSITY OF CALIFORNIA, SANTA CRUZ, USA
UNIVERSITY OF TORINO, TORINO, ITALY
UNIVERSITY OF TOULOUSE, TOULOUSE, FRANCE
UNIVERSITY OF UPPSALA, UPPSALA, SWEDEN

Institutions (19):

CCAST, BEIJING, PEOPLE'S REPUBLIC OF CHINA
CNR, ITALY
COPPE, RIO DE JANEIRO, BRAZIL
CRN, STRASBOURG, FRANCE
DESY, HAMBURG, FRG
IHEP , SERPHUKOV, PROTV., USSR
INFN, BOLOGNA, ITALY
INFN, FIRENZE, ITALY
INFN, TORINO, ITALY
INFN, LABORATORI NAZIONALI DI FRASCATI, FRASCATI, ITALY
IN2P3, FRANCE
KERN FORSCHUNGSZENTRUM, KARLSRUHE, FRG
LIP, LISBON, PORTUGAL
LLNL, LIVERMORE, USA
MAX PLANCK INSTITUT, HEIDELBERG, FRG
PAUL SCHERRER INSTITUTE, ZURICH, SWITZERLAND
RUTHERFORD APPLETON LABORATORY, CHILTON, UNITED KINGDOM
WEIZMANN INSTITUTE, REHOVOTH, ISRAEL
WORLD LABORATORY, LAUSANNE, SWITZERLAND

Industries (8):

ANSALDO, GENOA, ITALY
CENTER FOR INDUSTRIFORSK., OSLO, NORWAY
DIGITAL ITALIA, ROME, ITALY
DRAPER LABORATORIES, CAMBRIDGE, MASS., USA
EUROPA METALLI - LMI, FIRENZE, ITALY
IMEC/ASP, LEUVEN, BELGIUM
NMRC, CORK, IRELAND
SMART SILICON SYSTEM, LAUSANNE, SWITZERLAND

Figure 2

BASIC PARAMETERS

Energy per beam	100 TeV
Number of bunches	39600 per beam
β-value at interaction point	1.25 - 0.6 m
Normalized emittance	$0.75\pi \times 10^{-6}$ m
r.m.s. beam radius at interaction point	$1.25 - 0.9 \times 10^{-6}$ m
Circulating current	16.43 mA
Particles per bunch	2.56×10^{9}
Beam-beam tune shift per crossing (with 6 active crossings)	1.67×10^{-3}
Bunch spacing	25×10^{-9} s
Stored beam energy	1.623×10^{9} J
Luminosity ($cm^{-2}s^{-1}$)	$0.9 - 1.8 \times 10^{33}$
Energy loss per turn due to synchrotron radiation	23.34 MeV
Radiated power (per beam)	385 KW
Power per unit length of one beam	1.89 W/m
Transverse em. damping time	1.2 h

LATTICE PARAMETERS

Length of period	200 m
Phase advance per period	$\pi/3$
Betatron wavelength	1200 m
Bending angle per normal period	4.7 mrad
Number of quads per period	2
Effective length of each quad	13.6 m
Nº of quadrupoles (without insertion)	2664
Maximum dipole field	10 Tesla
Bending radius	33356 m
Number of dipoles per normal period	12
Effective dipole length	13.1 m
Nº of dipoles	15984

Figure 3

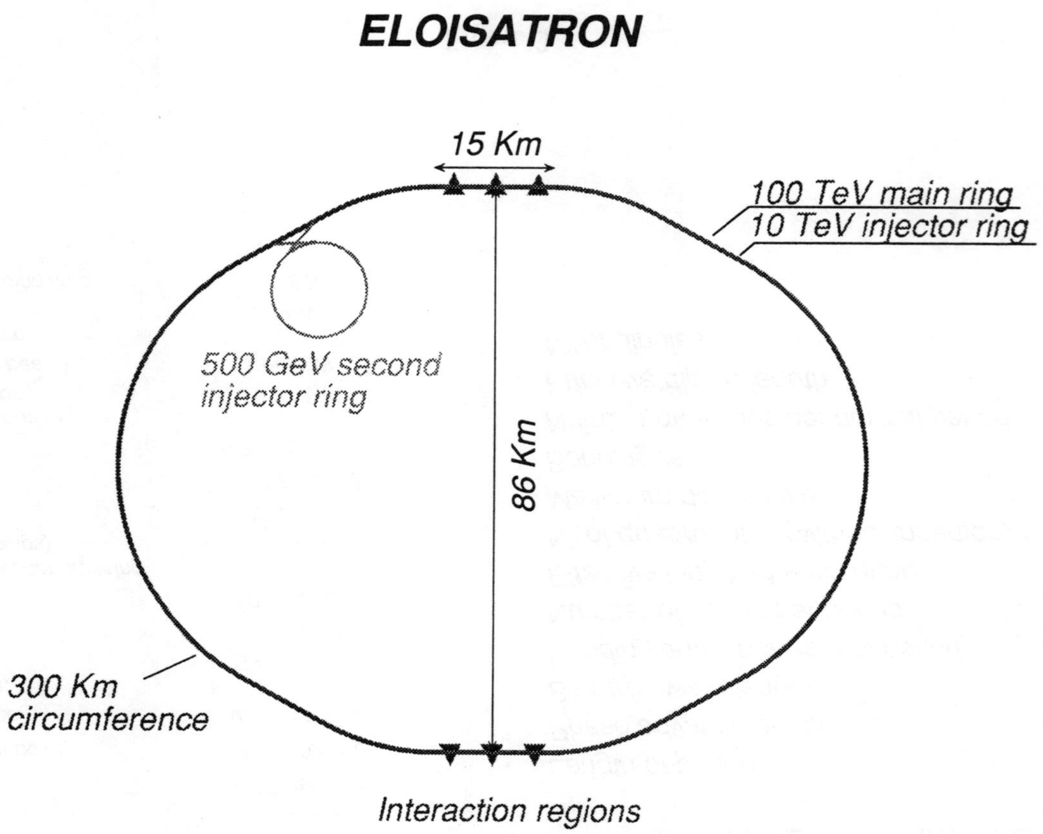
ELOISATRON
15 Km
100 TeV main ring
10 TeV injector ring
500 GeV second
injector ring
86 Km
300 Km
circumference
Interaction regions

Figure 4

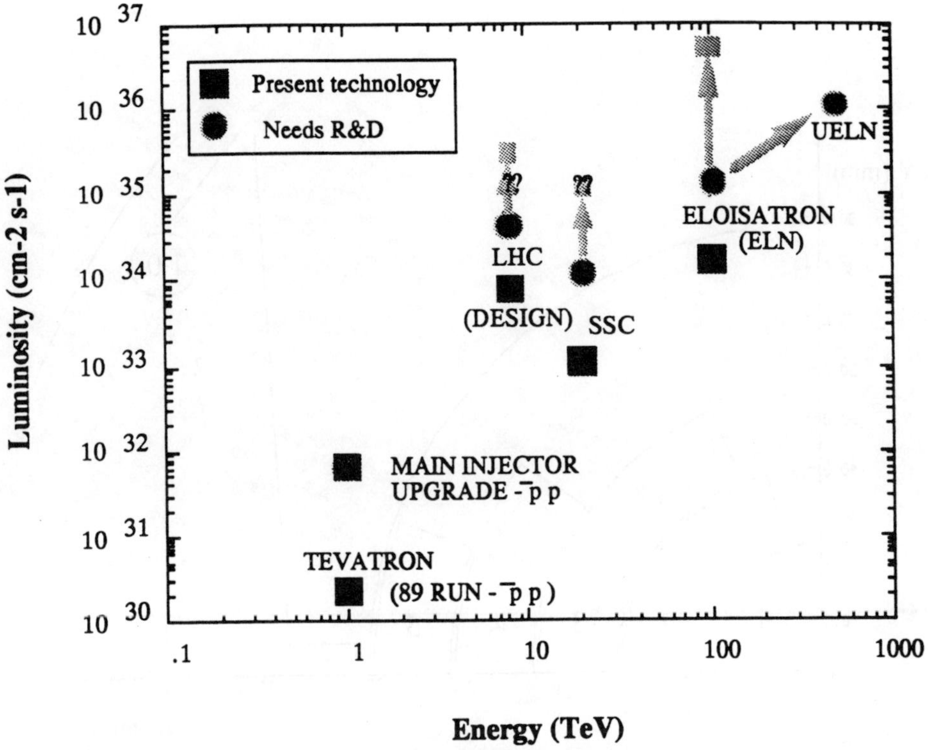
Present technology
Needs R&D
Luminosity (cm-2 s-1)
Energy (TeV)
10 37
10 36
10 35
10 34
10 33
10 32
10 31
10 30
.1
1
10
100
1000
??
??
LHC
(DESIGN)
SSC
ELOISATRON
(ELN)
UELN
MAIN INJECTOR
UPGRADE - $\bar{p}$ p
TEVATRON
(89 RUN - $\bar{p}$ p)

Figure 5

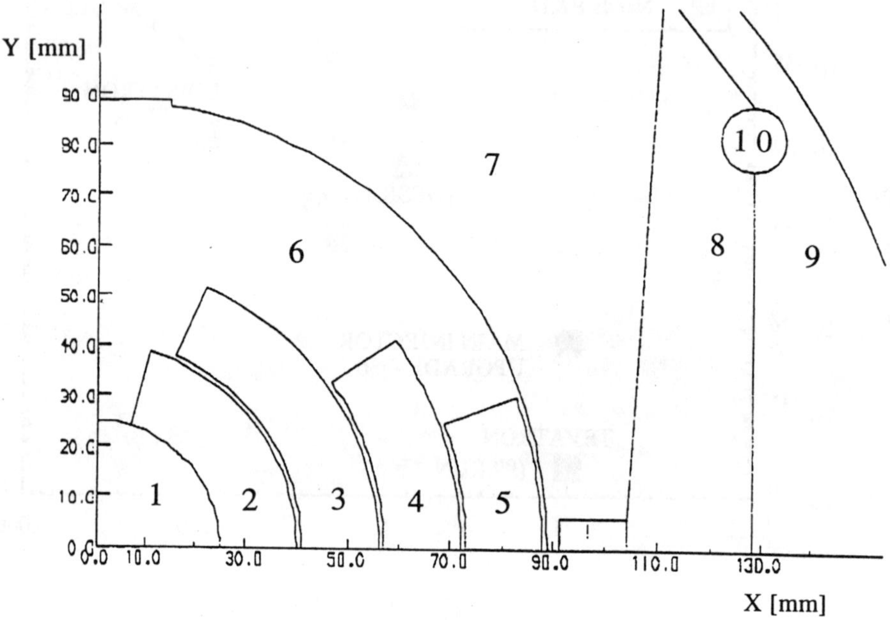

Fig. 6 One - quarter dipole cross - section. Scheme (3) for field uniformity calculation.

1)	Beam Pipe
2, 3, 4, 5)	(Nb_3Sn)-Cu Superconducting Cable
6)	Aluminium Collar
7, 8, 9)	Iron Lamination
10)	Helium Channel

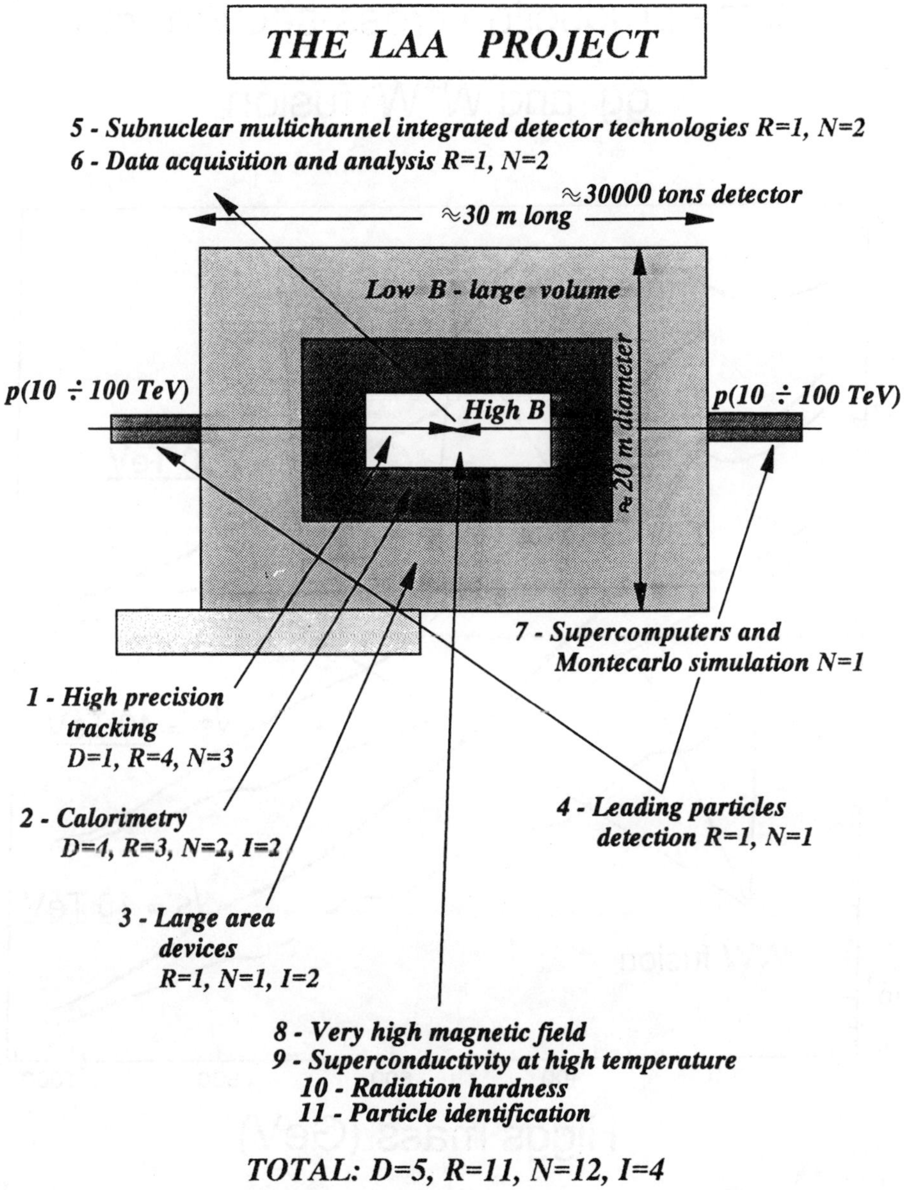
THE LAA PROJECT
5 - Subnuclear multichannel integrated detector technologies R=1, N=2
6 - Data acquisition and analysis R=1, N=2
≈30000 tons detector
≈30 m long
Low B - large volume
p(10 ÷ 100 TeV)
High B
≈20 m diameter
p(10 ÷ 100 TeV)
7 - Supercomputers and Montecarlo simulation N=1
1 - High precision tracking D=1, R=4, N=3
2 - Calorimetry D=4, R=3, N=2, I=2
3 - Large area devices R=1, N=1, I=2
4 - Leading particles detection R=1, N=1
8 - Very high magnetic field
9 - Superconductivity at high temperature
10 - Radiation hardness
11 - Particle identification
TOTAL: D=5, R=11, N=12, I=4

Figure 7

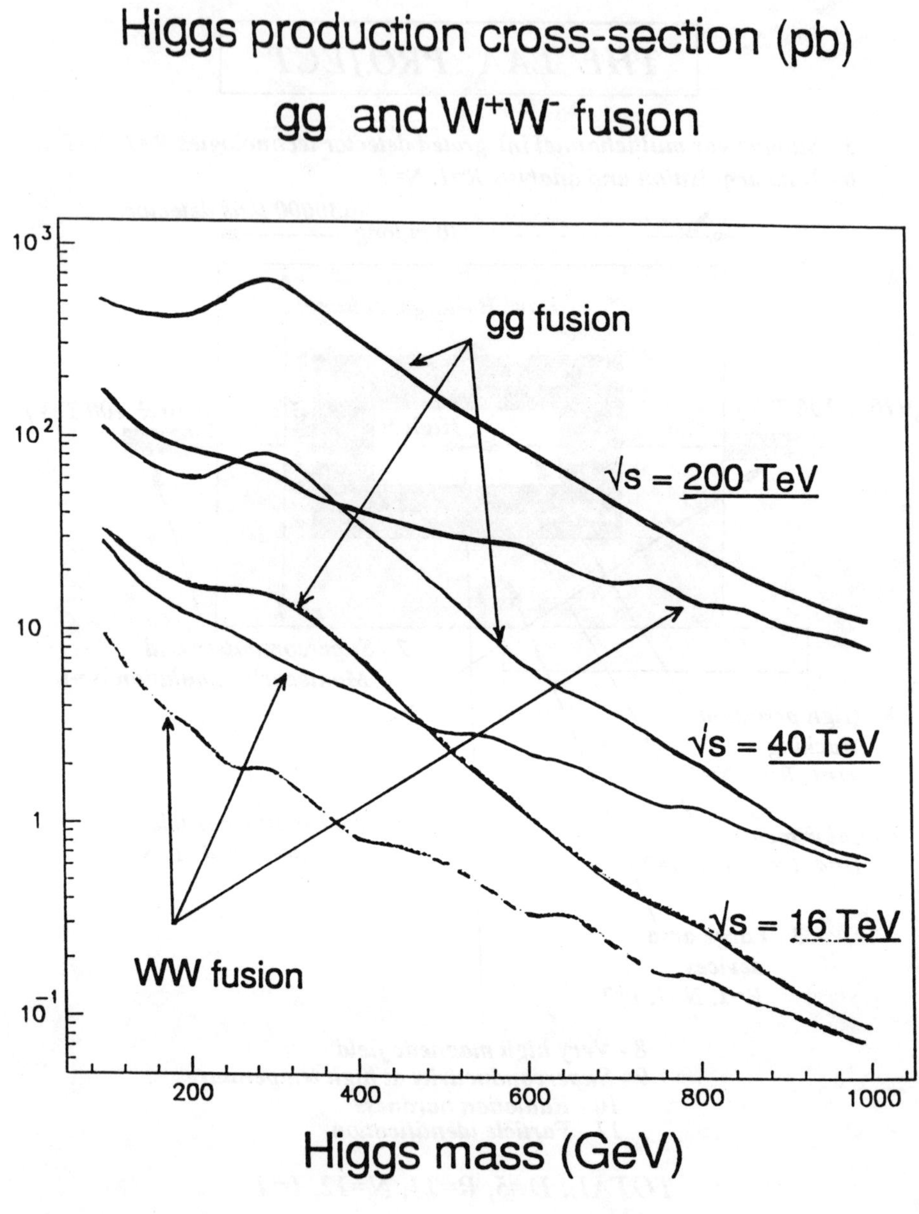
Higgs production cross-section (pb)
gg and W+W- fusion
gg fusion
√s = 200 TeV
√s = 40 TeV
√s = 16 TeV
WW fusion
10^3
10^2
10
1
10^-1
200
400
600
800
1000
Higgs mass (GeV)

Figure 8

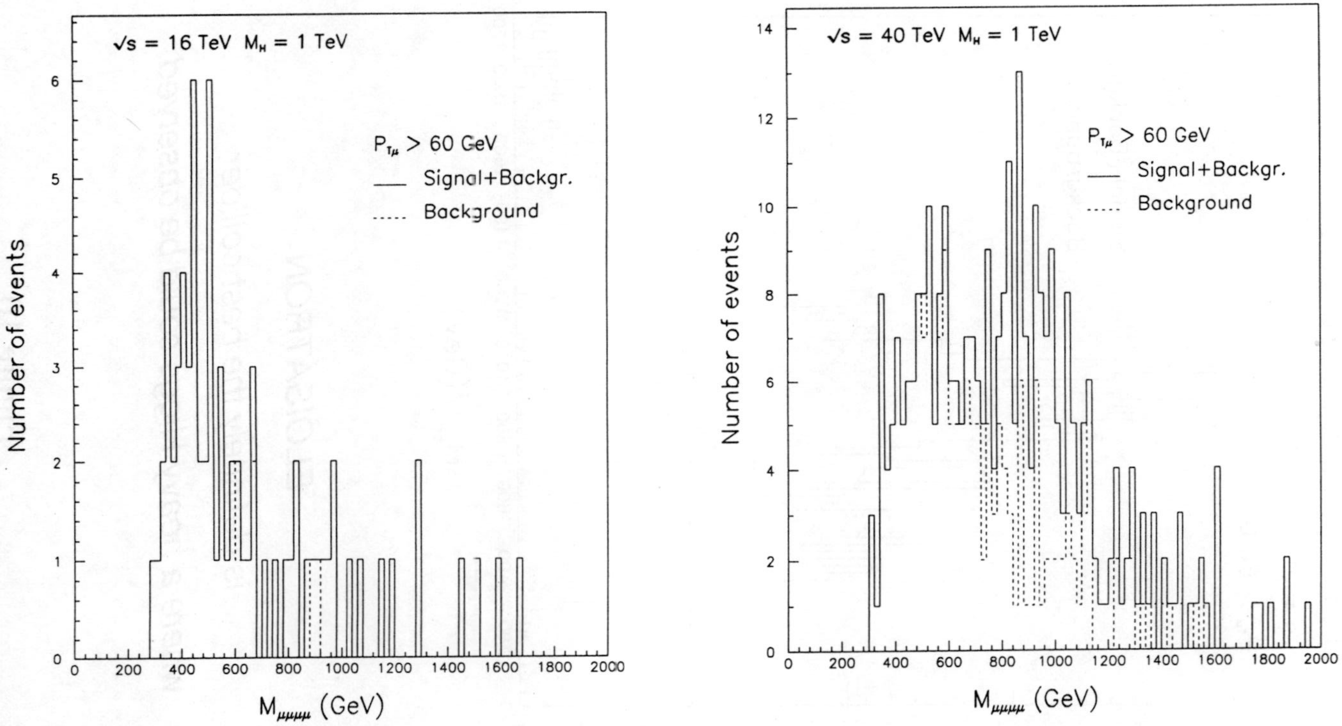
√s = 16 TeV M_H = 1 TeV
$P_{T\mu}$ > 60 GeV
Signal+Backgr.
Background
Number of events
$M_{\mu\mu\mu\mu}$ (GeV)
0
1
2
3
4
5
6
0
200
400
600
800
1000
1200
1400
1600
1800
2000
√s = 40 TeV M_H = 1 TeV
$P_{T\mu}$ > 60 GeV
Signal+Backgr.
Background
Number of events
$M_{\mu\mu\mu\mu}$ (GeV)
0
2
4
6
8
10
12
14
0
200
400
600
800
1000
1200
1400
1600
1800
2000

Figure 9

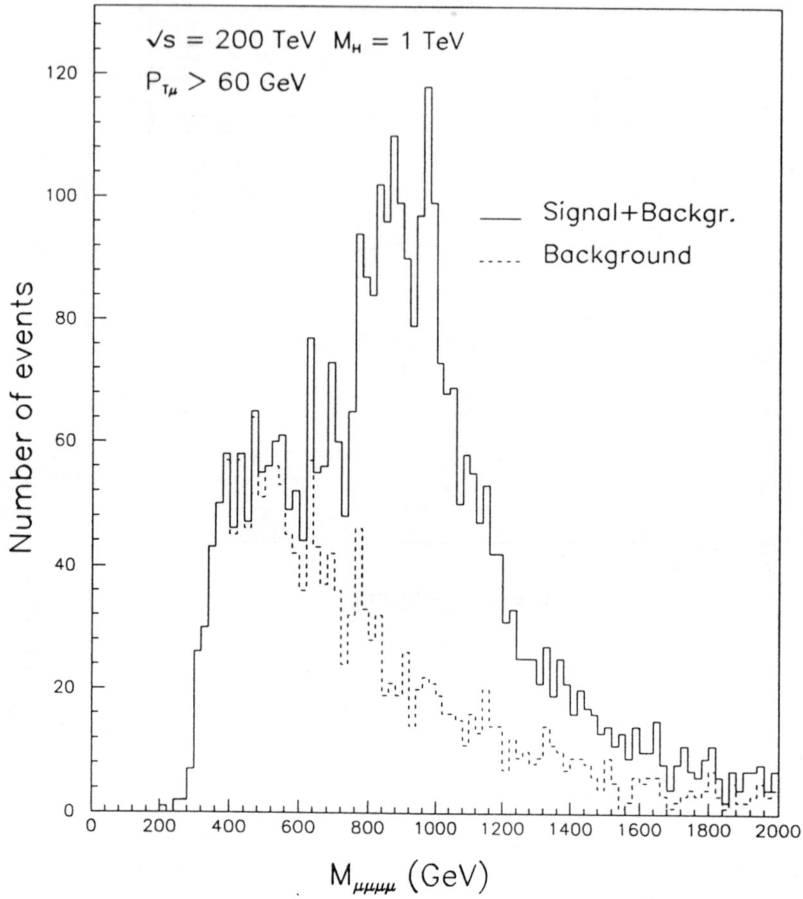

ELOISATRON

is definitely the best collider

where a heavy Higgs could be observed.

Figure 10

Working Groups Structure

1. MACHINES

- Theoretical (K. Johnsen).
- Studies at the highest levels of luminosity and energy (W. Barletta).

2. DETECTORS

- Research studies on Super Detectors capable of working at the highest energies and luminosities (H. Leutz).

3. PHYSICS

- MonteCarlo Simulations at 200 TeV and Supercomputers (L. Cifarelli, T.D. Lee).
- Theoretical studies (J. Ellis, V. Khoze, R. Peccei).
- World Data Base, Networking and Supercomputing (D. Söderman).

Figure 11

The following Scientists (Physicists, Engineers and Technicians) have participated in the work of the Project:

K.S. Adhao, Z. Aftab, K.B. Aglawe, C. Aglietta, P.B. Agrawal, Z. Ahsan, J. Alberty, M. Ali, B. Alpat, J. Alsford, C. Alvisi, E.D. Alyea, G. Ambrosi, Q. An, F. Anghinolfi, F. Anselmo, P. Antonioli, D. Antreasyan, G. Anzivino, A. Arefiev, M. Arneodo, R. Arnold, F. Arzarello, A.V. Asole, P. Aspell, B.H. Aurade, P.V.K.S. Baba, V.B. Badhe, G. Badino, S.D. Bagchi, M.K. Bagde, K.R. Bagree, D.R. Bahekar, F.N. Bai, J.Z. Bai, U.D.N. Bajpai, J.S. Bakare, C. Baldanza, Y. Ban, A.M. Band, G. Bandyopadhyay, S. Bandyopadhyay, H. Banerjee, S.K. Banerjee, A.N. Bannore, G. Barbagli, E. Barberio, G. Bari, T. Barillari, W. Barletta, M. Basile, M. Bassetti, A. Bassi, B. Basu, D. Basu, R. Battiston, C. Baudoin-Bijst, A. Bazzani, U. Becker, S.G. Bedi, S.Y. Bedore, L. Bellagamba, R.B. Belorkar, N.G. Belsare, M. Benot, P. Benvenuto, J. Berbiers, J. Berdugo, V.S. Berezinsky, L. Bergamasco, F. Bergsma, A. Bertin, R. Bertin, V.S. Bhagde, D.R. Bhakre, P.D. Bhalerao, K.S. Bhamra, N.C. Bhattacharya, T. Bhattacharya, S.N. Bhongle, S. Bianco, S.S. Bijwe, N. Bingefors, D. Bisello, A. Bizzetti, R.K. Bock, S.M. Borikar, M.S. Borkar, D. Boscherini, M. Bosteels, R. Bouclier, M. Bramhall, E. Braun, G. Bruni, P. Bruni, R. Bruzzese, F. Bugatti, J. Burger, D.K. Burghate, A. Bussolotti, V. Buzuloiu, N. Cabibbo, X.D. Cai, M. Campbell, Y. Cao, L. Caputi, G. Cara Romeo, R. Casaccia, C. Castagnoli, A. Castellina, A. Castelvetri, H. Castro, S. Ceresara, S.M. Chandekar, J.M. Chapuis, G. Charpak, U.K. Chaturvedi, B.F. Chen, C.Q. Chen, H.J. Chen, J.P. Chen, K. Chen, L.M. Chen, P.G. Chen, R. Chen, S. Chen, S.Y. Chen, W.Y. Chen, M. Cherian, E. Chesi, M. Chiarini, J.A. Chinellato, S.P. Chiney, J. Christiansen, E. Christofel, L. Cifarelli, F. Cindolo, G. Cini, F. Ciralli, M. Civita, A. Clare, Z. Coa, E. Colavita, G. Conforto, S. Cong, F. Coninckx, A. Contin, S. Cooper, M. Costa, I. Crotty, H.C. Cui, R.Y. Cui, S.Z. Cui, X.T. Cui, X.Y. Cui, G. D'Ali, C. D'Ambrosio, I. D'Antone, S. D'Auria, V.L. Dadykin, G.L. Dai, C.K. Damle, D.N. Darane, M. Dardo, T.K. Das, M. Dasgupta, B. Datta, B.K. Datta, R. Datta, U.R. Datta, S. Datta Majumdar, D. Davasumony, S. De Pasquale, V. de Sabbata, R. De Salvo, C. Del Papa, S. Deng, Y.P. Deng, S.L. Deoghare, V.S. Deogoankar, D.W. Deshkar, A.A. Deshmukh, B.T. Deshmukh, D.T. Deshmukh, K.C Deshmukh, K.G. Deshmukh, R.M. Deshmukh, V.R. Deshmukh, A.M. Deshpande, G.T. Deshpande, M.V. Deshpande, P.D. Deshpande, A. Desilva, P. Destruel, M. Deutsch, J. deWitt, J.A. Dhanaraj, S.B. Dhannasare, G.B. Dhanokar, R.V. Dharaskar, M.V. Dhargave, M.T. Dharmadhikari, T.N. Dhobe, O. Di Rosa, D. Dorfan, L.G. Dos Santos, M.T. Dova, W.F. Du, X.J. Du, E. Duchovni, J. Dupont, J. Dupraz, M. Dutta, J. Egger, T. Ekelöf, J. Ellis, R.B. Elgira, R.I. Enikeev, C.C. Enz, Y. Ermoline, F.L. Fabbri, J.P. Fabre, C. Fang, S.X. Fang, R.N. Faustov, P. Feraudet, S. Ferrara, L. Ferrario, S. Finelli, F. Fiori, E. Fishbach, D. Fong, P. Ford, F. Frasconi, M. French, M. Fuchs, M. Fukushima, W. Fulgione, K. Gabathuler, P.M. Gadkari, V.J. Gaikwad, Y. Galaktionov, S. Galassini, P. Galeotti, J. Galvez, C. S. Gao, W. Gao, W.X. Gao, M. Gasperini, J. Gaudaen, K.N. Gawande, S.G. Gawande, S.W. Gawande, M.H. Ghate, S.M. Ghatole, P. Ghia,, G.A. Ghike, S.G. Ghosh, S.N. Ghosh, G.T. Gillies, Y. Giomataris, J.P. Girod, P. Giusti, K. Goebel, V.G. Gokhale, V.M. Gokhale, R. Goldoni, N. Gopalkrishnan, Y Gorodkov, A. Gougas, M. Gourdin, F. Grianti, C. Grinnel, C. Gu, M.P. Gu, S.D. Gu, S.H. Gu, F.Q. Guang, M. Guanziroli, M. Guerzoni, W.X. Guo, Y. Guo, Z.Y. Guo, V.D. Gupta, H. Gústen, J.L. Guyonnet, T. Gys, E.S. Hafen, Q. Han, W. Hao, P. Haridas, A. Hasan, D. Hazifotiadu, M.J. He, E. Heijne, T. Henkes, G. Herten, W.B. Hirurkar, H. Hoorani, M. Hourican, G. Hu, J.L. Hu, J.W. Hu, L.D. Hu, K.X. Huang, L. Huang, N. Huang, Y.Z. Huang, P.P. Huddar, S.I. Hussain, G. Iacobucci, M. Ikram, M.F. Ingalgi, A.G. Ingle, B.G. Ingle, R.P. Ingole, N. Inove, M. Italiani, J. Jain, S.K. Jain, R.K. Jambhorka, P. Jarron,

Figure 12a

Y.W. Jia, C.H. Jiang, Y.L. Jiang, Q.S. Jin, Y.K Jin, A.S. Joalikar, J.P. Jobez, D.N. Joharapurkar, K. Johnsen, V.B. Johri, C. Joram, S.V. Joshi, D.K. Kadu, J.M. Kale, M.G. Kale, B. Kaleem Minhas, Y. Kamyshkov, T.M. Karadei, D.M. Kataria, M. Kaur, S.A.H. Kbatil, D.G. Kelkar, G.M. Kesharwani, G.S. Khadekar, M.T. Khadse, F.F. Khalchukov, G.A.N. Khan, M.Q. Khan, R.A. Khan, S. Khan, A.T. Khandare, M.L. Khandelwal, R.B. Kharade, T.M. Kharade, R.B. Kharat, U.U. Kharat, N.W. Khodragade, S.M. Khodragade, S. Khokhar, V. Khoze, T. Kiltamura, M.S. Kishirsagar, W. Kluge, E.V. Ko, M.S. Korde, P.V. Kortchaguin, V.B. Kortchaguin, A.P. Kowale, B.P. Kowale, W. Krisher, F. Krummenacher, B.L. Kuang, V.A. Kudryavtsev, A.B. Kulkarni, D.K. Kulkarni, R.S. Kulkarni, V. Kumar, G.N. Kumbhare, A. Kuzucu, G. La Commare, I. Laakso, J.C. Labbé, G.R. Labde, S.A. Ladhake, T. Lainis, G. Landi, P.F. Lang, R.B. Lanjewar, H. Larsen, K. Lau, G. Laurenti, T.D. Lee, A.R. Leo, M. Leo, M. Letheren, H. Leutz, G. Levi, L. Levinson, B.M. Li, C. Li, D.S. Li, D.Z. Li, F.B. Li, G.L. Li, J. Li, J.C. Li, J.G. Li, M. Li, Y.M. Li, Q. Lin, D.C. Ling, B. Lisowski, A. Litke, D.K. Liu, J. Liu, S.Y. Liu, T. Liu, X. Liu, Y. Liu, Y.C. Liu, C. Liverani, C. Ljuslin, G.C. Lokre, L. Lone, S.D. Long, C.G. Lu, J.H. Lu, L. Lu, W.D. Lu, G. Luches, D. Luckey, K. Luebelsmeyer, H.S. Lunge, Y.X. Luo, Z.H. Luo, J. Ma, J.M. Ma, Z. Ma, G. Maccarrone, J.K. Madhugiri, R.K.W. Mahalle, S.N. Mahapatra, M.A. Mahure, C.K. Majumdar, M. Malavasi, A.S. Malguin, R. Malik, B.N. Mandal, S.V. Manmode, J.L. Manputra, M. Mansoor Ali, H.S. Mao, A. Marchioro, A. Margotti, M. Marino, S. Marmi, T. Massam, C.G. Mathe, T. Matsuda, T. Matsuura, D. Mattern, B.R. Maurya, B. Mayes, P. Mazzanti, G. Meddeler, C.A. Mehare, K.H. Meier, V.N. Melnikov, S. Meneghini, R. Meng, N. Mengotti Silva, V.R. Metkar, J.L. Mi, R.S. Mi, Y. Mi, G. Mikenberg, G. Million, Y. Mir, I. Mirza, A. Misaki, K.D. Mishra, R.B. Mishra, R.S. Mishra, S.B. Mishra, V.G. Miskin, G.H. Mo, S.G. Modak, A.K. Modi, M. Mohan, G. Mohanty, S.C. Moholkar, G. Molinari, M.R. Mondardini, A. Montanari, B. Monteleone, C. Morello, K. Morey, G. Mörk, J. Moromisato, F. Motta, N.E. Moulai, R. Mount, A. Mukherjee, S. Mukherjee, K.V. Muley, V.D. Muley, A.M. Mundhada, B. Musso, S. Muthiah, A.B. Nadange, P.T. Naiekar, S.Y. Nandanwar, R. Nania, R.Y. Narkhede, S.G. Nasare, V. Nassisi, L. Natrajan, G. Navarra, C. Nemoz, K.A.N. Nerkar, S. Newett, H. Newman, M.A. Niaz, B.R. Nikhade, F.M. Nirwan, V. O'Shea, A. Oliva, A. Olsen, N. Ozdes, R.S. Pagrut, J.P. Pal, M. Pal, E. Pallante, F. Palmonari, H.B. Pan, L. Panaro, N.K. Pandc, S.N. Pandcy, T.N. Pandey, U.S. Pandcy, J.B. Pang, K.C. Pangarkar, G. Papini, S.C. Paranjape, Y.R. Paranjape, D.V. Parwate, G. Passardi, G. Passotti, C.B. Patil, M.B. Patil, P.D. Patil, S.P. Patil, B.A. Patki, A. Paul, K.V. Pawar, S.P. Pawar, R. Peccei, P.G. Pelfer, A.W. Pendharkar, M. Pereira, L. Periale, C. Peroni, E. Perotto, V. Peskov, P. Picchi, D. Piedigrossi, A. Pieretti, R. Pilastrini, P. Pinna, L. Pinsky, D. Pitzl, V. Pjidaev, I.A. Pless, V. Plyaskin, G. Pocci, R.B. Pode, H.M. Poharkar, O. Polgrossi, C.N. Potdar, N.B. Potdar, V.K. Pratapwar, G. Prisco, M. Pu, D. Puertolas, M. Puglisi, H.S. Pundkar, J. Pyrlik, S. Qian, Z.M. Qian, J.M. Qiao, J. Qin, H. Qiu, N. Qureshi, A. Racz, V.D. Raghatate, S.B. Raichur, V.B. Rajurkar, M. Rama, A.D. Rangari, S.P. Rant, V.A. Ratate, G.D. Rathod, A.D. Raut, D.R. Raut, W.B. Ren, Z.L. Ren, M. Ricci, G. Rinaldi, F. Rivera, H.A. Rizvi, C. Rizzuto, P. Rotelli, P. Roumeliotis, R. Roy, T.Z. Ruan, T. Ruf, M.V.G. Ryassny, O. Ryazhskaya, H. Rykaczewski, H. Rykaszewski, O. Saavedra, N. Sacchetti, H. Sadrozinski, E. Saletan, R. Salgne, D.L. Samudralwar, D. Sanders, A. Sandoval, L. Sandri, J.C. Santiard, S. Santini, G. Sartorelli, P. Sartori, S. Sarwar, N.N. Saste, R.V. Satpute, F. Sauli, W. Scandale, M. Schioppa, J. Schipper, H. Schönbacher, D. Scigocki, M. Scioni, R. Scrimaglio, J. Seguinot, R. Sehgal, A. Seiden, W. Seidl, J.M. Seixas, S.H. Selukar, D. Sen, M. Sen Gupta, N.D. Sen Gupta, H. Sens, S. Serra, G. Servizi, H.Y. Sha, D.I. Shahare, V. Sharan, A. Sharma, P. Sharp, W.R. Sheldon, V.P. Shelest, B.H. Shen, J.P. Shen, T.J. Shen, V. Shevchenko, Y.S. Shi, V.S. Shiramwar, S.P. Shrinkahnde, W.S. Shu, Y.D. Shu, J.B. Shukla, E. Shumilov, S. Siboni, A. Sigrist, G. Simonet, M.D. Singh, R.D. Singh, S.K. Singh, T. Singh, J. Singh Rana, K.P. Sinha, K. Smith, P.V. Soadekar, V.S. Soitkar, G. Soliani, Ke Songbai, D.N. Soni, M. Spadoni, E. Spencer,

Figure 12b

L. Sportelli, R.S.L. Srivastav, A. Staiano, M. Steuer, G. Sultanov, L. Sun, S.L. Sun, Y.L. Sun, S.N. Supe, G. Susinno, G.C. Susinno, J. Swain, S.P. Swami, A.A. Syed, V.A. Tabhane, S. Tailhardat, N. Takahashi, Y. Takano, V.K. Tale, V.P. Talochkin, R.K. Talwekar, T.M. Tamhane, H. Tanaka, B.P. Tang, C. Tang, H. Tang, J. Tang, S.M. Tang, M.S. Tapi, E. Tarkowski, M. Taufer, M. Tavlet, S. Tazzari, M.T. Teli, T.J. Telranhe, B.R. Tembhurne, K.J. Teng, A.E. Terraneo, S.C. Thaker, V.A. Thakhare, A.W. Thakre, R.V. Thakre, V.M. Thatte, F. Tian, W. Tian, S.D. Tikar, R. Timellini, S.C.C. Ting, J. Tischhauser, B.P. Tiwari, S.A. Tiwari, J. Tocqueville, G. Torelli, G.C. Trinchero, V.S. Tumram, G. Turchetti, A. Turtelli, D.U. Umak, I. Uman, V. Valencic, S. Valenti, P. Vallania, L. Vallini, G. Vanstraelen, G. Venturi, S. Vernetto, I. Vetliski, F. Vetrano, K. Vidya Gaikwad, P. Vikas, U. Vikas, F. Villa, A. Vishnoi, A. Vitale, M. Vivargent, E. von Goeler, L. Votano, T. Wada, M. Wadhwa, D.G. Wadke, S.D. Wadke, R.V. Waghamare, W.U. Waghmode, S.H. Walkey, A.B. Walkhade, B.S. Wang, D.C. Wang, F. Wang, H.F. Wang, H.J. Wang, J. Wang, L.L. Wang, L.Z. Wang, M. Wang, R. Wang, S. Wang, S.H. Wang, S.Q. Wang, T.J. Wang, X.P. Wang, Y.Y. Wang, Z. Wang, P.B. Wani, A.M. Wankhede, P.C. Wankhede, T.V. Warhekar, T.S. Wasnik, Cao Wei, K.Y. Wei, Z.Z. Wei, R. Weinstein, T. Wenaus, H. Wenninger, C. Werner, M. White, M. Widgoff, C. Williams, M. Willutzky, B.Z. Wu, C.C. Wu, M.A. Wu, S.X. Wu, W.M. Wu, W.T. Wu, Y.Z. Wu, Z.D. Wu, B. Wyslouch, B.X. Xi, J.W. Xi, L.G. Xiao, J.L. Xie, P.P. Xie, X.X. Xie, Y.Y. Xie, B. Xu, C. Xu, J.M. Xu, L. Xu, S.W Xu, X..K. Xu, Z.X Xu, J.X Xue, S.T. Xue, R. Yagannathan, V.F. Yakushev, I. Yamamoto, B.H. Yan, J. Yan, T.X. Yan, W.G. Yan, D.J. Yang, G. Yang, Y. Yang, X.G. Yao, C.H. Ye, Q.H. Ye, S.H. Yeh, A.K. Yelne, S.V. Yenkar, G. Yi, Z.S. Yin, Cai Yixing, J.M. You, K. You, T. Ypsilantis, Q.C. Yu, Q.F. Yu, Z.Q. Yu, Z.Z. Yu, W. Yue, N. Yunus, T.M. Zachariah, N. Zaganidis, A. Zallo, G. Zanetti, G. Zatsepin, M. Zeng, B.C. Zhang, C. Zhang, H.S. Zhang, Y. Zhang, Y.P. Zhang, Z.P. Zhang, B. Zhao, D. Zhao, F.G. Zhao, J.B. Zhao, J.J. Zhao, W.R. Zhao, Y.J. Zhao, G.Q. Zheng, L.S. Zheng, S.C. Zheng, Y.C. Zheng, Z.P. Zheng, Cai Zhiguo, S.C. Zhong, J.K. Zhou, S. Zhou, X. Zhou, X.G. Zhou, Y.H. Zhou, F.Q. Zhu, L.S. Zhu, R.Q. Zhu, R.Y. Zhu, S.G. Zhu, X. Zhu, Y.C. Zhu, B. Zhuan, A. Zichichi, A. Zucchini, M. Zuffa

Figure 12c

Participant List

Arnowitt, Richard
Texas A&M University
Department of Physics
College Station, TX 77843-4242
USA

Bahran, Moustafa
University of Oklahoma
Department of Physics and Astronomy
Norman, OK 73019
USA

Barger, Vernon
University of Wisconsin
1150 University Avenue
Madison, WI 53706
USA

Cobb, Bill
4 Goldthread Court
The Woodlands, TX 77381
USA

Dixon, John
Texas A&M University
Department of Physics
College Station, TX 77843-4242
USA

Erler, Jens
University of Pennsylvania
Department of Physics
209 S. 33rd Street
Philadelphia, PA 19104
USA

Gandhi, Raj
Texas A&M University
Physics Department
College Station, TX 77843-4242
USA

Giannakis, Iaonnis
Houston Advanced Research Center
4800 Research Forest Drive
The Woodlands, TX 77381
USA

Gui, Mei
Texas A&M University
Department of Physics
College Station, TX 77843
USA

Haber, Howard
University of California
Department of Physics
Santa Cruz, CA 95064
USA

Kagan, Alex
SLAC
P.O. Box 4349, Bin 81
Stanford, CA 94309
USA

Kaplunovsky, Vadim
University of Texas at Austin
Department of Physics
RLM 5.208
Austin, TX 78712
USA

Kelley, Stephen S.
Maharishi International University
1000 N. 4th Str., DB 1069
Fairfield, Iowa 52557
USA

Khuri, Ramzi
CERN-Theory Division
1211 Geneva 23
SWITZERLAND

King, Stephen F.
University of Southampton
Department of Physics
Southampton, England S09 5NH

Koures, Vasilios G.
University of Utah
Chemistry Department
Salt Lake City, UT 84112
USA

Leigh, Robert G.
Rutgers University
Department of Physics and Astronomy
Piscataway, NJ 08855-0849
USA

Leontaris, George
University of Ioannina
Department of Physics
P O Box 1186
Ioannina, GR-451-10, GREECE

Lopez, Jorge
Texas A&M University
Department of Physics
College Station, TX 77843-4242
USA

Participant List

Meshkov, Sydney
SSC Laboratory
MS 2007
2550 Beckleymeade Avenue
Dallas, TX 75237
USA

Nanopoulos, Dimitri
Houston Advanced Research Center, Texas
4800 Research Forest Drive
The Woodlands, TX 77381
USA

Norton, David
Houston Advanced Research Center
4800 Research Forest Drive
The Woodlands, TX 77381
USA

Park, Gye
Houston Advanced Research Center
4800 Research Forest Drive
The Woodlands, TX 77381
USA

Pois, Heath
University of California
Department of Physics
Davis, CA 95616
USA

Raby, Stuart
The Ohio State University
Department of Physics
174 W. 18th Avenue
Columbus, OH 43210
USA

Ramond, Pierre
University of Florida
Physics Department
Gainesville, FL 32611
USA

Rattazzi, Ricardo
Lawrence Berkeley Laboratory
1 Cyclotron Road
Berkeley, CA 34720
USA

Sarid, Uri
Lawrence Berkeley Laboratory
Theory Group 50A/3115
1 Cyclotron Road
Berkeley, CA 94720
USA

Schramm, David N.
University of Chicago
5640 S. Ellis Avenue-AAC 140
Chicago, IL 60637
USA

Seeba, Maureen
Idlewood Studio
27 Outervale Place
The Woodlands, TX 77381
USA

Stewart, Donna
Houston Advanced Research Center
4800 Research Forest Drive
The Woodlands, TX 77381
USA

Tye, Henry
Cornell University
Newman Laboratory
Ithaca, NY 14853-0269
USA

Wang, Meng Yuan
Harvard University
Department of Physics
Cambridge, MA 02138
USA

Wang, Xu
Texas A&M University
Physics Department
College Station, TX 77843-4242
USA

Webb, Robert
Texas A&M University
Department of Physics
College Station, TX 77843-4242
USA

White, James
Texas A&M University
Department of Physics
College Station, TX 77843-4242
USA

Wilson, Thomas L.
NASA/Johnson Space Center
Mail Code SN3
Houston, TX 77058
USA

Participant List

Yuan, Kajia
Houston Advanced Research Center
4800 Research Forest Drive
The Woodlands, TX 77381
USA

Zichichi, Antonino
CERN
1211 Geneva 23
Switzerland

